The Engineering of Digital Twins

John Fitzgerald • Cláudio Gomes
Peter Gorm Larsen
Editors

The Engineering of Digital Twins

Springer

Editors
John Fitzgerald
School of Computing
Newcastle University
Newcastle upon Tyne, UK

Cláudio Gomes
Department of Electrical
and Computer Engineering
Aarhus University
Aarhus, Denmark

Peter Gorm Larsen
Department of Electrical
and Computer Engineering
Aarhus University
Aarhus, Denmark

ISBN 978-3-031-66718-3 ISBN 978-3-031-66719-0 (eBook)
https://doi.org/10.1007/978-3-031-66719-0

This Springer imprint is published by the registered company Springer Nature Switzerland AG
The registered company address is: Gewerbestrasse 11, 6330 Cham, Switzerland

Preface

The concept of the Digital Twin (DT) is becoming ubiquitous. This powerful idea is that a virtual but faithful representation of a real-world system releases the potential to use computing power to marshal data, analyse events and explore future possibilities without disturbing "the real thing". Conferences, journals, magazine articles, books and news reports are focusing on this emerging technology. Nations and businesses are investing in DTs of the natural environment, of engineered systems and systems-of-systems on scales from the microscopic to the global.

Although the potential is evident, the methods and tools for developing DTs in a coherent, dependable way are in their infancy. To the business wishing to invest in this technology, it is not yet clear what the right questions are to ask, where the challenges lie, and what the processes might be for developing, deploying and using DTs successfully.

The story behind this book covers three decades of experience as practitioners and researchers in model-based design. Initially, we worked together on formal methods for software development. Significant advances came through the development of analytic tools that at least partially automated the analysis of software by exploiting formal semantics. Even here, a perpetual problem was that the software "as built" almost never remained the same as the software "as designed". Over time, models developed meticulously for a product would gather virtual dust while the software developed from them would be adapted to changing environments and modes of use so that the value of those models – and the expensive analyses done on them – diminished.

Once we stepped outside the purely cyber world to consider systems that combine computational and physical processes, we saw the need to combine quite heterogeneous models and to deal with the uncertainties of real-world environments. In engineering such *Cyber-Physical Systems (CPSs)*, the need seemed all the stronger to maintain the currency and value of complex design models through the life the system they purported to represent. This is where the potential of the DT comes to the fore.

For an organisation that can see, even in outline, some benefit in exploring DT technology, it is not clear where to start. How does one first think about the value of DTs, and what questions should one ask about how to create, maintain and operate them? Our goal in writing the book has been to go behind the glossy image of the DT to help readers – whether practitioners, students or researchers – think systematically about DT elements, architecture, construction and maintenance. In doing so, it is important to view DT engineering as a *systems engineering* discipline.

Our goal in creating this book was to bring together a sample of the methods, tools and experience that will contribute to the evolution of DT engineering. Much of the value in a DT derives from its integration of data and models from diverse sources and stakeholders. Consequently, our authors have contributed from the viewpoints of a wide range of formalisms and theories. Experience has been contributed by industry experts through a motivating roundtable discussion in Chap. 2 and by academics recounting case studies in developing DTs for CPSs that range from a tempeh incubator to a marine vessel, described in Chap. 12.

The breadth of the vision for DTs makes it difficult to offer a comprehensive account of DT engineering, and so we limit our scope to areas where DTs are being developed for products and systems that are themselves engineered. We do not focus on the undeniable potential of DTs in human or biological systems [1], natural phenomena such as weather and climate, or even social systems [3]. Within the scope of DTs for engineered CPSs, we focus on technical systems engineering processes, and not on business processes such as engaging organisations, negotiation, financial and economic aspects, or public engagement.

A plethora of disciplines and technologies come together in the engineering of DTs, including but not limited to: machine learning; co-simulation; noise filtering, time synchronisation, sensor fusion and state estimation; communication technologies; visualisations and game engines; run-time verification, fault detection and diagnosis. These are the subjects of existing specialist texts and we make no attempt at covering them in significant depth. Instead, this book will provide pointers into the specialist literature where appropriate. The emphasis of this book is marshalling these disciplines and technologies in a principled and effective way to deliver the benefits of the DT.

The book is structured in five parts:

Foundations: This part introduces the DT concept, the potential benefits of DTs seen from a business perspective, foundations for DT engineering, and considers how the engineering of DTs can be seen in a systems engineering context (Chap. 1 to 4). In order to demonstrate that DTs have potential in many other domains we include a dialogue with domain experts from consumer electronics, pumps, aerospace and the built environment.

Models and Data: This part introduces the range of models and data that form the core assets of DTs for CPSs. It covers ways in which models can be produced and calibrated, and considers how data is derived from a CPS and communicated to its DT. We briefly introduce first principles and data-driven models and give pointers to additional material, including guidance on how multiple disciplines and stakeholders can be involved (Chap. 5 to 7).

Services for Digital Twins: This part introduces some of the main services that a DT provide by building on the assets of models and data, including visualisation, fault detection and diagnosis and support for decision-making (Chap. 8 to 10).

Realising Digital Twins: This part covers the realisation of DTs, including a platform allowing engineers to construct DTs from reusable components. We report on case studies in food production, robotics and marine engineering using a systematic framework that aligns with the DT engineering concepts introduced in the earlier parts of the book (Chap. 11 and 12).

Advanced Topics in Digital Twins: The final part introduces advanced topics in delivering dependable DT-enabled systems. We focus on security and privacy, the capacity for autonomy, and a range of research topics as we look to the future of this emerging discipline (Chap. 13 to 15).

Each chapter has been written collaboratively by multiple authors, totalling contributions from 5 continents and 18 different countries. In order to make it easier to read chapters in different orders each chapter will introduce acronyms in the full on their first occurrence. However, a full list of acronyms is given for reference below. Each chapter includes its own list of references, and cross-references between the different chapters are made whenever these provide value.

Using the Book

The book is aimed at both engineering practitioners aiming to develop DTs and researchers in DT technology and design, including advanced (masters and doctoral) students. In keeping with the spirit of collaborative modelling, we assume that the reader has some experience in software development, control, systems engineering or modelling of dynamical systems, but we do not assume knowledge of more than one of these.

The book is also intended help those with non-computing backgrounds to develop an understanding of the digital technology that is becoming increasingly important in physical systems. We also aim to support educators offering an introduction to DTs as part of advanced courses in Computer Science/Engineering and Systems Engineering using Model Based Systems Engineering (MBSE). This support is offered through additional online materials available on the accompanying web site described below.

Throughout the text, acronyms are introduced in full at the first point of use in each chapter. Each chapter has its own list of references. There are many cross-references between chapters to allow the reader to jump between related topics. Material related to our running example is generally introduced as follows:

Incubator XXX

Example text

Practitioners with an interest in the techniques of DT engineering can follow the order presented in the first four parts. Most of the chapters in the technical flow of the book assume that preceding chapters have been read. The advanced topics in Part V are relatively independent of one another, and these topics are probably most relevant for researchers. Since we are trying to target a broad audience, the text includes boxes advising readers from different backgrounds on how they might approach potentially very unfamiliar material. These take the following forms:

A box with this symbol contains material aimed at readers with a background in "cyber" disciplines such as computing, control or software engineering.

If your background is in "physical" disciplines such as mechanical, civil or electrical engineering, physics or mechatronics, boxes with this symbol contain text aimed at your needs.

If you have a background in systems engineering, so that you are comfortable with approaches that integrate diverse engineering methods and tools, this box is for you.

Accompanying Web Site

DT engineering is a lively area of innovation and research, so we have tried to focus the main text on more enduring topics, with supporting materials on the more volatile areas on our web site. This includes our 'Twinpeh' running example and its DT [2], which readers will be able to download and run using a virtualised or self-built incubator box. The most common techniques described in the main text will be accessible via open source projects. We also offer further DT examples, exercises and solutions, as well as pointers to emerging standards, frameworks and platforms for practitioners. Classroom materials, exercises and solutions are available to bona fide teachers through the web site (`https://digital-twin-book.org/`).

Acknowledgments

We gladly acknowledge the sponsors of the research and innovation projects and the open source community that have enabled us to jointly establish the core of the technologies presented in this book and supported in the accompanying web site. We are particularly grateful to the Poul Due Jensen foundation, the European Commission, the Innovation Foundation Denmark and the European Space Agency (ESA), Aarhus and Newcastle Universities. Our special thanks to Nick Battle, Erik Kristiansen, Stefano Tonetta and Marcel Verhoef for reviewing drafts in various states of readiness. We would also like to thank our many teammates, collaborating companies and students who have been involved in the development and validation of the technologies presented here.

Newcastle upon Tyne, United Kingdom *John Fitzgerald*
Aarhus, Denmark *Cláudio Gomes*
Aarhus, Denmark *Peter Gorm Larsen*

References

1. Coveney, P., Highfield, R.: Virtual You: How Building Your Digital Twin Will Revolutionize Medicine and Change Your Life. Princeton University Press (2023)
2. Feng, H., Gomes, C., Thule, C., Lausdahl, K., Sandberg, M., Larsen, P.G.: The Incubator Case Study for Digital Twin Engineering (2021). URL `http://arxiv.org/abs/2102.10390`
3. Ketzler, B., Naserentin, V., Latino, F., Zangelidis, C., Thuvander, L., Logg, A.: Digital twins for cities: A state of the art review. Built Environment **46**(4), 547–573 (2020)

Contents

Part V Advanced Topics

List of Contributing Authors

Giuseppe Abbiati
Aarhus University, Aarhus, Denmark
e-mail: abbiati@cae.au.dk

Santiago Gil Arboleda
Aarhus University, Aarhus, Denmark
e-mail: sgil@ece.au.dk

Stylianos Basagiannis
Int. Hellenic Univ./ Collins Aerospace, Greece/Ireland
e-mail: stylianos.basagiannis@collins.com

Christian H. Bohlbro
Bohlbro.dk, Denmark
e-mail: hello@bohlbro.dk

Till Böttjer
Aarhus University, Aarhus, Denmark
e-mail: till.boettjer@ece.au.dk

Rob Charlton
Space Group, Newcastle upon Tyne, United Kingdom
e-mail: Rob.Charlton@spacegroup.co.uk

Henrik Ejersbo
Grundfos, Bjerringbro, Denmark
e-mail: hejersbo@grundfos.com

Lukas Esterle
Aarhus University, Aarhus, Denmark
e-mail: lukas.esterle@ece.au.dk

Hao Feng
Huawei, Shenzhen, People's Republic of China
e-mail: haof.au@outlook.com

John S Fitzgerald (*editor*)
Newcastle University, Newcastle upon Tyne, United Kingdom
e-mail: john.fitzgerald@newcastle.ac.uk

Mirgita Frasheri
Aarhus University, Aarhus, Denmark
e-mail: mirgita.frasheri@ece.au.dk

Cláudio Gomes (*editor*)
Aarhus University, Aarhus, Denmark
e-mail: claudio.gomes@ece.au.dk

Simon Thrane Hansen
Aarhus University, Aarhus, Denmark / University of Luxembourg, Luxembourg City, Luxembourg
e-mail: simon.hansen@uni.lu

Lars Ivar Hatledal
Norwegian University of Science and Technology, Ålesund, Norway
e-mail: laht@ntnu.no

Alexandros Iosifidis
Aarhus University, Aarhus, Denmark
e-mail: ai@ece.au.dk

Panagiotis Katsaros
Aristotle University of Thessaloniki, Thessaloniki, Greece
e-mail: katsaros@csd.auth.gr

Zahra Kazemi
Aarhus University, Aarhus, Denmark (now working at Vestas, Denmark)
e-mail: zahrakazemi1991@gmail.com

Klaus Kristensen
Bang & Olufsen, Struer, Denmark
e-mail: KRT@Bang-Olufsen.dk

Tomas Kulik
Sweet Geeks, Vejle, Denmark
e-mail: tku@sweetgeeks.dk

Peter Gorm Larsen (*editor*)
Aarhus University, Aarhus, Denmark
e-mail: pgl@ece.au.dk

Hugo Daniel Macedo
Aarhus University, Aarhus, Denmark
e-mail: hdm@ece.au.dk

Peter Høgh Mikkelsen
Aarhus University, Aarhus, Denmark
e-mail: phm@ece.au.dk

Bentley James Oakes
Polytechnique Montréal, Montréal, Canada
e-mail: bentley.oakes@polymtl.ca

Ken Pierce
Newcastle University, Newcastle upon Tyne, United Kingdom
e-mail: kenneth.pierce@newcastle.ac.uk

Daniel Enrique Lucani Rötter
Aarhus University, Aarhus, Denmark
e-mail: daniel.lucani@ece.au.dk

Michael Sandberg
Aarhus University, Aarhus, Denmark
e-mail: ms@mpe.au.dk

Prasad Talasila
Aarhus University, Aarhus, Denmark
e-mail: prasad.talasila@ece.au.dk

Valdemar Tang
Aarhus University, Aarhus, Denmark
e-mail: valdemar.tang@ece.au.dk

Daniella Tola
Aalborg University, Denmark
e-mail: dato@mp.aau.dk

Alessandro Ulisse
Collins Aerospace, Rome, Italy
e-mail: alessandro.ulisse@collins.com

Jim Woodcock
Aarhus University, Aarhus, Denmark / York University, York, United Kingdom
e-mail: jim.woodcock@york.ac.uk

Houxiang Zhang
Norwegian University of Science and Technology, Ålesund, Norway
e-mail: hozh@ntnu.no

Jonas Åkeson
Grundfos, Bjerringbro, Denmark
e-mail: jakeson@grundfos.com

Acronyms

AADL	Architecture Analysis and Design Language
AAS	Asset Administration Shell
ACL	Access Control List
ADC	Analog-to-Digital Converter
AI	Artificial Intelligence
AMQP	Advanced Message Queuing Protocol
ANN	Artificial Neural Network
API	Application Programming Interface
AR	Augmented Reality
ART	Applied Research & Technology
AV	Audio/Video
AWS	Amazon Web Services
B2B	Business-to-Business
B2C	Business-to-Consumer
BC	Boundary Condition
BDI	Belief-Desire-Intention
BIM	Building Information Model
BIP	Behaviour-Interaction-Priority
BMS	Building Management System
BRF	Belief-Revision-Function
CA	Collision Avoidance
CAD	Computer-Aided Design
CD	Collision Detection
CFD	Computational Fluid Dynamics
CI	Correct Information
CIA	Confidentiality, Integrity and Availability
CI/CD	Continuous Integration/Continuous Deployment
CPS	Cyber-Physical
CPU	Central Processing Unit

CSM	Computational Solid Mechanics
CSMA	Carrier Sense Multiple Access
CSV	Comma Separated Values
CUSUM	Cumulative Sum
DAM	Digital Asset Management
DB	Database
DDoS	Distributed Denial-of-Service
DevOps	Development & Operations
DI	Dependable Incubation
DoF	Degree of Freedom
DoS	Denial of Service
DLL	Dynamic Link Library
DM	Diversification of raw Materials
DoD	Department of Defence
DP	Dynamic Propusion
DR	Desktop Robotti
DRC	Dynamic Range Compression
DS	Digital Shadow
DT	Digital Twin
DT3	Digital Thread & Twins Technologies
DTA	Digital Twin Aggregate
DTCC	Digital Twin Cities Centre
DTDL	Digital Twin Definition Language
DTI	Digital Twin Instance
DTM	Digital Twin Monitor
DTP	Digital Twin Prototype
ED	Experimental Design
EDT	Edge Digital Twin
ESA	European Space Agency
Exec MS	Execution-manager Micro Service
FDI	False Data Injection
FDIIR	Fault Detection, Isolation, Identification, and Recovery
FDMA	Frequency Division Multiple Access
FEM	Finite Element Method
FI	Fault Injection
FMI	Functional Mockup Interface
FMU	Functional Mockup Unit
FPGA	Field Programmable Gate Array
FSM	Finite State Machine
FTP	File Transfer Protocol
FVM	Finite Volume Method
GA	Genetic Algorithm
GPS	Global Positioning System

HaaS	Hardware as a Service
HCI	Human Computer Interaction
HiL	Hardware-in-the-Loop
HMI	Human Machine Interface
HOCON	Human-Optimised Config Object Notation
HPC	High Performance Computing
HTTP	Hypertext Transfer Protocol
IC	Initial Condition
ICMP	Internet Control Message Protocol
IDE	Integrated Development Environment
IGDS	Interactive Graphic Design System
INCOSE	International Council on Systems Engineering
INTO-CPS	Integrated Tool Chain for Model-based Design of CPSs
IoT	Internet of Things
IP	Intellectual Property
IS	Ingredients Scalability
ISO	International Standards Organisation
IT	Information Technology
IVP	Initial Value Problem
JSON	JavaScript Object Notation
KF	Kalman Filter
KG	Knowledge Graph
LDAP	Lightweight Directory Access Protocol
LL	Lower Limit
LLM	Large Language Model
LSTM	Long Short-Term Memory
LTL	Linear Temporal Logic
LQG	Linear–Quadratic–Gaussian
MAC	Medium Access Control
MAPE-K	Monitor-Analyze-Plan-Execute over a shared Knowledge
MBSE	Model Based Systems Engineering
MCU	Micro-Controller Unit
MITL	Metric Interval Temporal Logic
MOR	Model Order Reduction
MPC	Model Predictive Control
MQTT	Message Queuing Telemetry Transport
MR	Mixed Reality
MRU	Motion Reference Unit
MS	Micro Service
MSE	Mean Square Error
NTP	Network Time Protocol
ODE	Ordinary Differential Equation
OMG	Object Management Group

OS	Operating System
OSA	Orchestration Synchronisation Administration
OSI	Open Systems Interconnection
OSP	Open Simulation Platform
PCA	Principal Component Analysis
PDE	Partial Differential Equation
PF	Particle Filter
PHM	Prognostic and Health Management
PI	Proportional-Integral
PLC	Programmable Logic Controller
PLM	Product Lifecycle Management
PPP	Point-to-Point
PSO	Particle Swarm Optimisation
PT	Physical Twin
PTM	Physical Twin Monitor
QoS	Quality of Service
RAM	Random Access Memory
RBAC	Role Based Access Control
RCC	Remote Control Centre
RDF	Resource Description Framework
ReLU	Rectified Linear Unit
REST	REpresentational State Transfer
RMQFMU	RabbitMQ Functional Mockup Unit
RNN	Recurrent Neural Network
ROC	Remote Operating Centre
RoI	Region of Interest
ROS	Robot Operating System
ROM	Reduced Order Model
RPC	Remote Procedure Call
RPi	Raspberry Pi
RPM	Revolutions Per Minute
RPN	Region Proposal Network
RUL	Remaining Useful Life
RV	Runtime Verification
R/V	Research Vessel
SaaS	Software as a Service
SASL	Simple Authentication and Security Layer
SCADA	Supervisory Control And Data Acquisition
SDE	Stochastic Differential Equation
SDG	Sustainable Development Goal
SE	Systems Engineering
SiL	Software-in-the Loop
SLA	Service Level Agreement

SME	Small Medium Enterprise
SMOL	Semantic Micro Object Language
SMTP	Simple Mail Transfer Protocol
SSD	Single Shot Detector
SSH	Secure SHell
SSL	Secure Sockets Layer
SPICE	Software Process Improvement and Capability Determination
SQL	Structured Query Language
SoS	System of Systems
SSH	Secure Shell
STRIDE	Spoofing, Repudiation, Information disclosure, Denial of service and Elevation of privilege
SUS	System Under Study
SVD	Single Value Decomposition
SVG	Scalable Vector Graphics
SysML	Systems Modelling Language
TC	Transformation Correctness
TCP	Transmission Control Protocol
TDMA	Time Division Multiple Access
TL	Temporal Logic
TLA	Temporal Logic of Actions
TLS	Transport Layer Security
TOSCA	Topology and Orchestration Specification for Cloud Applications
TSDB	Time-Series DataBase
UDP	User Datagram Protocol
UIO	Unknown Input Observer
UL	Upper Limit
UML	Unified Modelling Language
UniFMU	Universal Functional Mock-Up Unit
UQ	Uncertainty Quantification
UR	Universal Robots
URDF	Unified Robotics Description Format
USB	Universal Serial Bus
VDM	Vienna Development Method
VDM-RT	Vienna Development Method Real-Time
V&V	Verification and Validation
VR	Virtual Reality
VSC	Visual Studio Code
VV&A	Validation, Verification and Accreditation
WSM	Weighted Sum Model
YAML	Yet Another Markup Language
YOLO	You Only Look Once

Part I
Foundations

In this part of the book we aim to provide a broad view of the need for a disciplined approach to engineering Digital Twins (DTs) for cyber-physical systems. We introduce the DT concept in Chap. 1. We consider its potential benefits by "listening in" to a dialogue with experts from four industrial sectors in Chap. 2. We go on to provide conceptual foundations of DTs (and introduce a running example in tempeh fermentation!) in Chap. 3. Finally, in Chap. 4 we consider how DT engineering is seen as a form of systems engineering.

Chapter 1
Engineering Digital Twins for Cyber-Physical Systems

Peter Gorm Larsen, John Fitzgerald and Cláudio Gomes

Abstract In this chapter we introduce the concept of Digital Twin (DTs) for cyber-physical systems. We describe aspects of the technical engineering of DT-enabled systems, including their model-centric character, the need to maintain fidelity to the physical world, and the need to add value to the system being twinned. We examine the conceptual structure of a DT based on the assets it maintains and the services it offers, and consider the ways in which DTs may emerge through development and federation.

1.1 Introduction

DTs are a global phenomenon, the target of major investment and the subject of countless projects and publications. Part of their appeal is the simplicity of the concept, at least to those of us who have grown comfortable with digital, virtual worlds. The idea is that a real-world object has a virtual counterpart – a digital representation that we can explore, experience, analyse and change without risking damage to the "real thing". More than this, the virtual representation is kept up to date as its physical sibling changes through its life. This means we can be confident that the virtual world remains faithful enough to the physical to allow us to trust the conclusions of the analysis and experimentation done on the DT. Going one step further, we can envision the DT itself informing decisions that change the real physical world, if we trust it enough to let it do so.

The term "Digital Twin" was coined by Grieves and Vickers to describe a concept that arose in their work on managing product life cycles [13]. The idea has inspired visionary applications that range from better understanding and exploring human and other biological systems [7] to sustainable development [43]. DTs are increas-

J. Fitzgerald et al. (eds.), *The Engineering of Digital Twins*,
https://doi.org/10.1007/978-3-031-66719-0_1

ingly adopted in manufacturing [21], the automotive industry [30], energy [34, 29], agriculture [31] and even national infrastructure[1].

There is extensive, diverse and rapidly growing literature on DTs [17], alongside a burgeoning market for the technology. In 2018, DT technology was at the top of the Gartner Group's hype cycle[2] and in 2019 it was identified as one of the 10 most strategically important technologies[3], and in 2022 they predicted[4] that the DT market would reach $183 billion in revenue by 2031. Composite DTs were felt to present the largest opportunity. A case was made for developing libraries covering functionality and DT system templates for vertical markets in order to drive competitiveness. Another dimension that may make DT an interesting option is that "time to market" of products often is a critical point for many companies. DTs may enable companies to release a product before all services are available, if they have a possibility of subsequently providing the missing services via their DT. In the next generation of the product some of the services from the DT side may be moved over to the product as features. This kind of flexibility may be valuable in a competitive marked.

Realising the full potential of DTs is more of a challenge than it might at first seem. As we come to rely on decisions made with the aid of DTs, the need grows for twins that we can be confident will meet expectations. For the multidisciplinary teams involved in creating, maintaining and enhancing a DT, there are difficult questions to answer. What are the foundational elements of the DT? How do you scope and specify a DT so that it enhances and does not unduly compromise the thing that it is twinning? How do different engineering disciplines interact in creating a DT? How should the physical and virtual worlds be connected, and how do the imperfections and faults inherent in both worlds affect the DT's qualities? How can we use a DT to support decisions, and how do we maintain it through its life? The challenge, then, is to develop twins on which people can depend, or at least for which we know the limits of dependability.

In this introductory chapter, we take a tour through aspects of DT Engineering that are explored in greater depth in the remainder of the book. We begin with scope by introducing Cyber-Physical Systems (CPSs) as combinations of computing and physical processes, and then outlining further the concept of the DT (Sect. 1.2). Thinking about the challenges in engineering DTs for real CPSs, we consider the foundations, processes and methods that are needed, the fields of engineering that are involved and the elements that they bring together in a DT (Sect. 1.3). Finally, we consider the ways in which DTs might be in current and future products (Sect. 1.4).

---

[1] `https://www.cdbb.cam.ac.uk/DFTG/GeminiPrinciples`

[2] See `https://tinyurl.com/yc86c53v`.

[3] See `https://tinyurl.com/y5wkfewe`.

[4] See `www.gartner.com/en/documents/4011590`.

1.2 Cyber-Physical Systems and Digital Twins

This book concerns DTs of *CPSs*, and so in this section we first clarify what a CPS is (Sect. 1.2.1) before considering the exciting but challenging features of DTs (Sect. 1.2.2).

1.2.1 Cyber-Physical Systems

Informally, a *system* is a collection of components that work together to exhibit behaviour that the individual components do not. A CPS is a system that includes computational and physical components that may interact with one another, and the CPS as a whole interacts with an environment. Examples include: the battery management in a smart phone or an electric vehicle; vehicles with digital assistants from cruise control to self-parking; industrial robots controlled by computer systems to automate manufacturing processes; wearable health monitoring devices that collect physiological data and transmit it for analysis; and smart home appliances connected to the Internet for remote monitoring and control.

A CPS will typically include wired or wireless connections between cyber and physical components (Fig. 1.1). Information about the state of the environment can either come indirectly (e.g., through the friction experienced by the wheels of a vehicle) or directly from sensors which could be placed on the CPS (e.g., for measuring temperature) or may imply represent another information providing system (e.g., a weather prediction system).

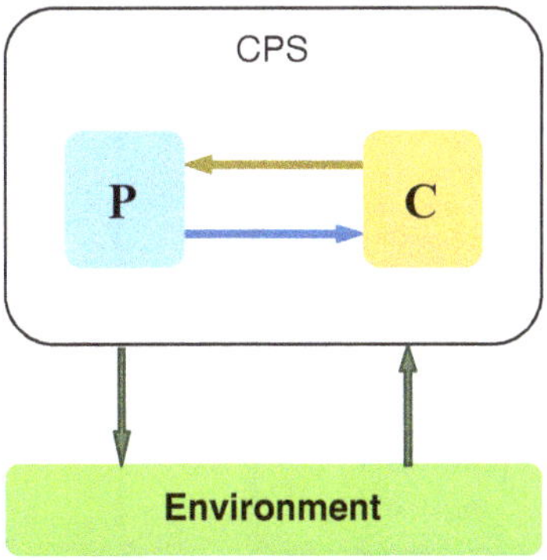

Fig. 1.1: A CPS composed of Physical (P) and Computational (C) components interacts with its environment.

1.2.2 DTs of Cyber-Physical Systems

Various definitions of DTs have been proposed, e.g., [23, 32, 35]. In this book, we will use the following working definition:

> A DT is a digital representation of a real-world entity which we call the *Physical Twin (PT)*. The DT and PT are connected by a *communications infrastructure* which allows the DT to maintain a known level of fidelity to the PT it represents. A DT offers its stakeholders a range of *services* that add value to the PT without unduly compromising the PT's operation.

Several observations flow from this working definition.

- First, it only makes sense to consider a DT in conjunction with its PT; otherwise it is not clear what the DT claims to be a twin of! Consequently, we consider the tasks of DT engineering to really be the engineering of *DT-Enabled Systems*, composed of the DT, PT and communications infrastructure, as outlined in Fig. 1.2.
- Second, the DT is a representation built from assets which include models of the PT and data derived from the PT via the communications infrastructure. It

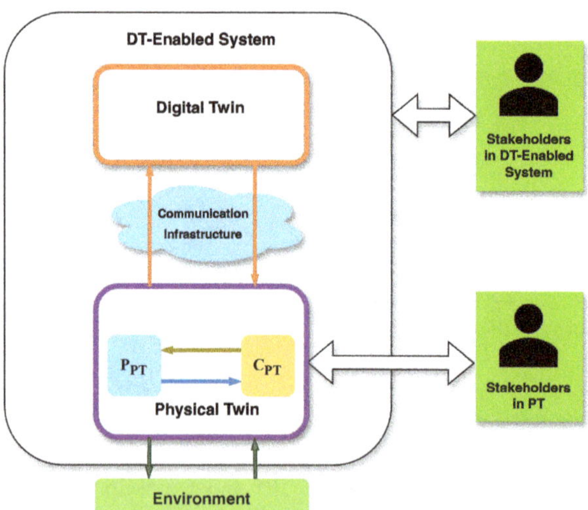

Fig. 1.2: Overview of a DT-Enabled System concept. Note that there are stakeholders in the DT-Enabled system itself, as well as existing stakeholders in the PT.

is the responsibility to curate these models as the PT changes over time. We will say that a DT is *model-centric*.

- Third, our working definition refers to a level of *fidelity* between the DT and PT. In practice this means that we have an understanding of how faithful the DT's representation is to the PT's reality. This will never be completely accurate because of the limitations of sensors, communications and computation in the DT, but we need to know how accurate is "accurate enough".
- Fourth, the DT builds on its assets to offer services that stakeholders desire, such as visualisation, preventive maintenance, failure detection, decision support and predictive analysis. It should offer these services without compromising the PT's ability to meet the demands made on it. This is particularly significant where DTs are being developed for existing PTs such as infrastructure systems on which we rely. It is unlikely that a DT can be added entirely without disruption, but this should be acceptable to the PT's stakeholders. We call this property the DT's *additionality*.

The services that a DT offers to its stakeholders are often focused on supporting decision-making. For example, a DT's visualisation and simulation can support "what-if" analysis of the consequences of environmental changes on a PT. This leads in turn to changes in the real world, closing the PT-DT loop [14, 39, 40, 1]. Tekinerdogan and Verdouw cast a light on the level of automation in the transfer of data between the PT and DT [38]. They reserve the "DT" designation for cases in which there is a high degree of automation in the link both ways, and use the term *Digital Shadow (DS)* to refer to systems in which updates by the DT to the PT are done largely manually. In our experience, the majority of current DT applications are actually DSs, focusing on offering DT stakeholders valuable insights on the state and history of the PT, rather than automatically reconfiguring it, and so our definition is agnostic about this level of automation.

Finally, we do not confine ourselves to DTs that represent objects: we consider DTs that cover processes as well. For example, depending on its purpose, the DT of a production line might have the physical machinery in scope as well as the current throughput rates of machines, and the shift scheduling of the human operators, and the digital system that is responsible for the scheduling.

1.3 Aspects of DT Engineering

How do we go about creating, maintaining and evolving systems that incorporate DTs? There is as yet no established discipline of DT engineering. However, our starting point in this book is that the foundations and methods exist, albeit in diverse and often separated areas of systems, software, mechatronic and other forms of engineering [12]. Our goal in this book is to give an overview of the diverse range of interesting and inspiring technical challenges in bringing these aspects together to deliver dependable systems enabled by DTs.

We put this in context by considering the business motivations for DT engineering (Chap. 2), laying conceptual foundations (Chap. 3) and outlining the technical processes involved in engineering a system incorporating a DT (Chap. 4).

In this section we outline some aspects of DT engineering as it is gradually emerging. We start from the principle that DT engineering is a model-centric discipline (Sect. 1.3.1) which integrates diverse types of models to deliver a DT that stays well enough aligned to the PT to add value to a PT without unduly compromising its operation. In Sect. 1.3.2, we consider the conceptual architecture of a DT in terms of layers of assets and functionality. Finally, we look outside the DT to consider how, as DT engineering develops, it will impinge on many different fields (Sect. 1.3.3). Finally, as DTs emerge as an active area of engineering, standards are beginning to be developed on which development processes can be built. For reference we list some current standards initiatives in Sect. 1.3.4.

1.3.1 Digital Twins are Model-Centric

In Sect. 1.2.2 we gave our working definition of a DT, and discussed key properties of model-centricity, fidelity and additionality. Here we dig a little more deeply into what these properties mean for DT engineering processes when the PTs are cyber-physical in character.

At the heart of a DT is a collection of models describing aspects of the PT. In practice, a comprehensive DT might manage many models, each created for different purposes. For example, a DT of a production line may contain models of the mechatronics governing the machine processes alongside very different models covering digital process control and even operator scheduling.

CPSs combine computational and physical processes and consequently their models are built from quite diverse formalisms, some describing discrete digital processes and some describing continuous flows, for example. The coupling of such models requires a standard in order to ensure interoperability. In this book we primarily use the Functional Mockup Interface (FMI) de facto standard. At the time of writing, FMI has had three major releases of which the most established is probably 2.0 [2], and its successor 3.0 [26] was released in May 2022. Different principles for creating digital models, scaling up for faster simulation and combining simulation units using co-simulation are described in Chap. 5.

Alongside model-centricity, DT engineering deals with *fidelity*, by which we mean the management of the difference between the DT's perception of the state of the PT and the reality. Techniques for aligning and realigning DTs with their PTs are considered in Chap. 6 on calibration, but one of the most interesting areas of DT engineering is understanding the bounds on how much drift and uncertainty between the DT and PT is acceptable for a given application.

The value of the DT comes from the range of services it offers stakeholders to analyse the state of the PT without interfering with it. The key characteristic here is *additionality*: the DT should create new capabilities that the original PT does not have on its own, and should not compromise the PT's behaviour. Crucially, these might include the capacity to modify or reconfigure the PT autonomously.

The DT and the PT, when coupled together with communications infrastructure, form a *DT-enabled* system. The DT engineer's goal is to create and maintain a DT that ensures fidelity and provides additionality. Fig. 1.2 illustrates the DT-enabled system concept by showing a system boundary around these elements. Notice that the stakeholders with an interest in the PT are not necessarily the same as those with an interest in the performance of the DT-enabled system as a whole: for some users of the PT, for example, the DT may not be visible at all.

These characteristics of model-centricity, fidelity and additionality, make DT engineering both exciting and challenging. Fidelity brings with it the need to harness collections of models that may derive from very different disciplines and be expressed in highly diverse notations. Maintaining a level of fidelity means bridging the "reality gap" between the world as it is as the world as the models see it, and additionality brings with it the need to support dependable decision-making, whether human or automated.

1.3.2 Inside the Digital Twin

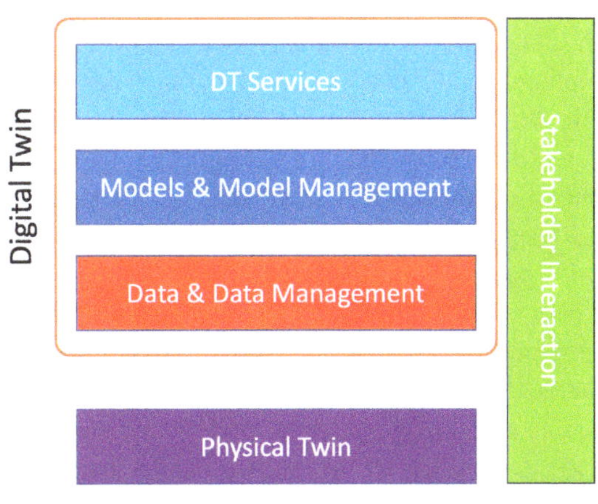

Fig. 1.3: Elements of a DT (data, models, and services).

At a very simple level, the DT can be thought of as composed of data, models, and services that are built on them (Fig. 1.3). We draw data from physical entities that are outside the DT, and within the DT we store and manage the data. Diverse models describe aspects of the physical entities in which we are interested, and these models themselves are part of the DT, which is also responsible for their management as they evolve. On these models and data we build DT services that add value, for example by tuning models on the basis of observed data, identifying potential failures before they arise, or supporting visualisation. Some services will interact directly with stakeholders in the DT-enabled system while others may be enablers, performing functions that are used by the stakeholder-facing services. We expand on this to consider models, data assets and their management within this overall conceptual structure in Chap. 3.

A DT could be realised in many different ways. It could be implemented as a software system on dedicated computing hardware, entirely virtualised over a flexible cloud computing structure, distributed computing nodes in an Internet of Things (IoT) structure or some combination. It is also worth noting that this structure is not static in the sense that a DT may allow stakeholders to explore possible functionality that eventually shifts from the DT into the physical entities that it originally twinned.

1.3.3 Fields Related to Digital Twins

Although the DT is a relatively new concept, it closely relates to numerous other technical fields. This section provides a concise overview of the most significant relationships in this context, expressed informally in Fig. 1.4.

The DT concept originated in Product Lifecycle Management (PLM), but it has been advanced particularly rapidly in the manufacturing sector, and DTs are to be found now in many other application domains [22]. Even in the software industry, the concept of Development and Operations (DevOps) includes the effort to keep track of software systems through the life cycle [20], although the potential predictive power of DTs has yet to be realised in that setting.

IoT provides one of the most prevalent settings for DT applications [16, 24, 18]. However, while IoT research primarily centres on distributing computation and managing communication between PTs and DTs, DTs encompass broader functionalities. Many DT frameworks originate from corresponding IoT frameworks, as elaborated in Chap. 11. Communication aspects of DTs are discussed in Chap. 7.

Marketing material for modelling and simulation tools can often refer to DT concepts. However, it is worth noting that models and simulations alone do not constitute a DT [45]. Digital models are indeed fundamental, but only if they can incorporate data from PTs as input for simulations. This holds true for Hardware-in-the-Loop (HiL) simulations, where outputs are not typically fed back to PTs in a

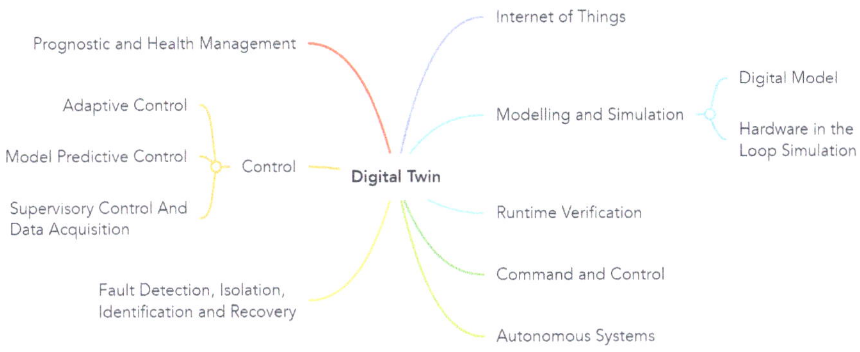

Fig. 1.4: Fields related to Digital Twins.

deployed setting. The notion of a tracking simulator closely aligns with the idea of a DS in this context [28].

Another relevant field to DTs is run-time verification [10, 6, 5]. This involves monitoring a system's behaviour at runtime to ensure adherence to specified properties or constraints. These techniques apply quite naturally to DTs as a form of monitoring [15]. More detail about monitoring will be provided in Chap. 9.

Command and control systems, such as those deployed in civil or military operations such as emergency relief, closely relate to DTs, where a digital representation with situational awareness (kept up to date) is used to assess predictions about how a situation may evolve. Despite similarities, they differ in predictive power, human involvement, and the automated transfer of data from the PT. Nevertheless, research has begun exploring the integration of DT technology to enhance such systems [44].

Supervisory Control And Data Acquisition (SCADA) systems, primarily used for real-time industrial process monitoring and control, share goals with DTs. Despite their distinct focuses, SCADA systems and DT technology can be complementary. SCADA facilitates real-time monitoring and control, while DTs provide virtual replicas for simulation and analysis, aiding predictions. These technologies can be combined for a comprehensive approach to industrial process management [33, 4].

Prognostic and Health Management (PHM), which predates DTs, closely aligns with them [36, 37, 3, 19, 41]. DTs have been employed to provide PHM functionality, particularly through predictive maintenance services. The control engineering field also existed long before DTs. While similarities exist, control engineering provides more precise assumptions about uncertainties and delays, ensuring safety and robustness. Adaptive control engineering techniques are also leveraged for autonomy in DT settings [46].

The autonomous systems field also intersects with DTs [11, 25, 9, 27]. The notion of autonomy is pivotal in the twinning aspect of DT engineering. The rapid diffusion of these systems to various industries and society at large is fueled by the swift adoption of modelling, simulation, and data acquisition. We consider autonomous reconfiguration enabled by DTs in Chap. 14.

Fault Detection, Isolation, Identification and Recovery (FDIIR) systems are used to monitor complex systems, detect abnormalities or faults, identify the root causes of these faults, and potentially take corrective actions to recover or mitigate the impact of the faults. These systems are widely used in various industries to ensure the reliability, safety, and performance of critical systems and processes, and resemble DTs, albeit with a sharper focus on faulty behaviour [8].

So, evidently, the component parts of the DT concept are not completely new. What is truly exciting is the rapid proliferation of DTs across various industries and society, catalysed by the swift adoption of modelling, simulation, and data communication technologies. Together, these create the need for methods, tools and experience supporting the delivery of model-centricity, fidelity and additionality required by modern DT applications.

1.3.4 Emerging Digital Twin Standards

Since DTs have become a hot topic, different standardisation activities are underway in order to promote collaboration. Since this tendency is likely to continue evolving in the future, limited information will be provided in this book. However, the accompanying web page[5] will provide pointers to the development of DT standards. At the time of writing, promising developments include:

ISO/IEC JTC1 (Information Technology):

- Progress Report of ISO/IEC JTC1 AG11 Digital Twin
- ISO/IEC JTC1/SC41 – Internet of Things and Digital Twin
- ISO/IEC WD30173 Digital Twin-Concepts and Terminology
- ISO/IEC WD30172 Digital Twin-Use Cases
- PWI JTC1-SC41-5 Digital Twin-Reference Architecture

ISO23247-1:2021: Automation systems and integration: Digital twin framework for manufacturing:

- Part 1: Overview and general principles
- Part 2: Reference architecture
- Part 3: Digital representation of manufacturing elements
- Part 4: Information Exchange
- ISO/TC Ad-hoc group on the Digital Twin

ISO/TR 24464:2020: Automation systems and integration: Industrial data – Visualisation elements of digital twins

[5] https://digital-twin-book.org/

IEC 63278-1:2023: Asset Administration Shell for industrial applications - Part 1: Asset Administration Shell structure (more parts are likely to come in the future)

1.4 The Transition to Digital Twins

To conclude this introduction, we consider aspects of the business context for DT technology. We first consider ways in DTs may be created for pre-existing PTs (Sect. 1.4.1). We go on to consider a future in which DTs are more routinely part of a product and in which there are more opportunities for federated DT-enabled systems (Sect. 1.4.2).

1.4.1 Digital Twins for Existing Physical Products

The developer, manufacturer or operator of a cyber-physical product or system may want to consider whether it is worth adding DT-enabled services. This may often be the case with larger and longer-lasting infrastructure systems. We will not delve into business models for implementing DTs [42] but throughout the book we do consider technical aspects, including the challenges around the acquisition and management of data derived from an existing PT.

The services that are required of the DT will tend to determine the data acquisition characteristics such as frequency of data transfer and the tolerances on timeliness, precision and accuracy. Challenges with this include detecting when the PT changes mode and starts to deviate from DT predictions. In this context, uncertainty about the PT's environment and the way in which the PT's users operate will influence the need for re-calibration of the DT's models (discussed in Chap. 6).

At a basic level, the DT services should provide additional value to that already afforded by the PT. This added value can be for both the PT's developer/manufacturer and for the PT's users. For the developer/manufacturer, the value may include intelligence about how products are used (e.g., which features are being used most) and how and when errors arise, influencing and reducing time to market for new releases. For the PT's user/operator the value may come in the form of updates to their data or software, optimisations to the PT's performance, the ability to train people in a virtual setting or services that enhance decision making. In supporting decision-making, finding the optimal way of presenting data is significant (discussed in Chap. 8). At the same time, personalised DTs can create privacy and security concerns, described further in Chap. 13.

1.4.2 A Federated Future

As new products and systems are developed with cyber capabilities, it is possible to consider a longer-term vision in which most CPS products will come with DT-based enhancements as a bundle with the product or for an additional fee. This is already supported for some consumer products such as fitness-focussed smart watches which have DT-like features allowing the user to push their recorded data to the cloud. The user of the PT is thus also stakeholder in the DT-enabled system. Added value from the DT includes the ability to analyse and track their performance, gamify a health or fitness regime and potentially share data with others. Different models inside a DT may be used for different purposes, and the DT will be able to select and use the most appropriate model depending upon the circumstances. In the Business-to-Consumer (B2C) market, products may use such digital services as a differentiator and attractor. For the developer/manufacturer benefits include richer data on how their product is used and how well it performs.

For a Business-to-Business (B2B) product such as a wind turbine, a paid DT service brings benefits such as improved performance and reliability through better scheduled maintenance. For the supplier, the DT can become the means for offering long-term services that could prove more profitable than the physical product. Today this kind of digital ecosystem tends to be limited to individual suppliers of physical products (and typically based on SCADA solutions). Eventually, however, a rise in interoperability between DT-enabled systems delivered by different providers could enable the composition of DTs, and hence of DT-enabled systems.

In this context, composition of DT-enabled systems, becomes an exercise in Systems-of-Systems (SoS) engineering. An SoS is a system that contains some pre-existing systems that may have a degree of operational and managerial independence. In thinking about DTs at the SoS level, it is necessary to extend the conceptual groups of DT components identified in Fig. 1.3 to add a layer of Twin Management services that update and orchestrate the twins involved in this SoS setting, as Zheng et al. [47] have done in defining a reference architecture for "cognitive" DTs. We will return to how this layer can be realised in Chap. 11.

In the future there will also be more DTs where the human is taken out of the loop when sufficient trust is put in the decisions taken by the DT. However, such autonomous behaviour can create challenges from the perspective of safety. This is treated further in Chap. 14.

A discipline of DT engineering for CPSs is in its infancy. In this book we hope to help it on its way by showing how disciplines, tools and techniques might work in concert to create DTs that have some of the capabilities we require of them. In the remaining chapters in this first part of the book, we examine the opportunities and demands made of DTs, lay the foundations for the DT concept, and cast DT engineering as a systems engineering process.

References

1. Attaran, M., Celik, B.G.: Digital Twin: Benefits, Use Cases, Challenges, and Opportunities. Decision Analytics Journal **6**, 100165 (2023). DOI: https://doi.org/10.1016/j.dajour.2023.100165
2. Blochwitz, T.: Functional Mock-up Interface for Model Exchange and Co-Simulation (version 2.0). https://www.fmi-standard.org/downloads (2014)
3. Booyse, W., Wilke, D.N., Heyns, S.: Deep digital twins for detection, diagnostics and prognostics. Mechanical Systems and Signal Processing **140**, 106612 (2020). DOI: https://doi.org/10.1016/j.ymssp.2019.106612
4. Brosinsky, C., Westermann, D., Krebs, R.: Recent and prospective developments in power system control centers: Adapting the digital twin technology for application in power system control centers. In: 2018 IEEE International Energy Conference (ENERGYCON), pp. 1–6 (2018). DOI: 10.1109/ENERGYCON.2018.8398846
5. Cimatti, A., Tian, C., Tonetta, S.: NuRV: A nuXmv Extension for Runtime Verification. In: B. Finkbeiner, L. Mariani (eds.) Runtime Verification, pp. 382–392. Springer International Publishing, Cham (2019)
6. Convent, L., Hungerecker, S., Leucker, M., Scheffel, T., Schmitz, M., Thoma, D.: TeSSLa: Temporal Stream-based Specification Language. In: T. Massoni, M.R. Mousavi (eds.) Formal Methods: Foundations and Applications 21st Brazilian Symposium, SBMF 2018, Salvador, Brazil, November 26-30, 2018, Proceedings, *Lecture Notes in Computer Science*, vol. 11254, pp. 144–162. Springer (2018)
7. Coveney, P., Highfield, R.: Virtual You: How Building Your Digital Twin Will Revolutionize Medicine and Change Your Life. Princeton University Press (2023)
8. Ding, S.: Model-Based Fault Diagnosis Techniques. Springer (2013)
9. Esterle, L., Ejersbo, H., Frasheri, M., Gomes, C., Macedo, H.D., Larsen, P.G.: Digital twins for autonomous intelligent systems: From development to deployment. In: 2022 IEEE International Conference on Autonomic Computing and Self-Organizing Systems Companion (ACSOS-C), pp. 53–54 (2022). DOI: 10.1109/ACSOSC56246.2022.00028
10. Falcone, Y., Havelund, K., Reger, G.: A Tutorial on Runtime Verification. Engineering Dependable Software Systems **34**, 141–175
11. Feldt, J., Kourouklis, T., Kontny, H., Wagenitz, A.: Digital twin: Revealing potentials of real-time autonomous decisions at a manufacturing company. Procedia CIRP **88**, 185–190 (2020). DOI: https://doi.org/10.1016/j.procir.2020.05.033. 13th CIRP Conference on Intelligent Computation in Manufacturing Engineering, 17-19 July 2019, Gulf of Naples, Italy
12. Fitzgerald, J.S., Larsen, P.G., Pierce, K.G.: Multi-modelling and co-simulation in the engineering of cyber-physical systems: Towards the digital twin. In: M.H. ter Beek, A. Fantechi, L. Semini (eds.) From Software Engineering to Formal Methods and Tools, and Back - Essays Dedicated to Stefania Gnesi on the Occasion of her 65th Birthday, *Lecture Notes in Computer Science*, vol. 11865, pp. 40–55. Springer (2019)
13. Grieves, M., Vickers, J.: Digital Twin: Mitigating Unpredictable, Undesirable Emergent Behavior in Complex Systems. In: F.J. Kahlen, S. Flumerfelt, A. Alves (eds.) Transdisciplinary Perspectives on Complex Systems, pp. 85–113. Springer International Publishing Switzerland (2017)
14. van der Horn, E., Mahadevan, S.: Digital twin: Generalization, characterization and implementation. Decision Support Systems **145**, 113524 (2021). DOI: https://doi.org/10.1016/j.dss.2021.113524
15. de Hoz Diego, J.D., Temperekidis, A., Katsaros, P., Konstantinou, C.: An IoT Digital Twin For Cyber-Security Defence Based On Runtime Verification. In: Leveraging Applications of Formal Methods, Verification and Validation. Verification Principles: 11th International Symposium, ISoLA 2022, Rhodes, Greece, October 22–30, 2022, Proceedings, Part I, p.

556–574. Springer-Verlag, Berlin, Heidelberg (2022). DOI: 10.1007/978-3-031-19849-6_31
16. Jacoby, M., Usländer, T.: Digital Twin and Internet of Things – Current Standards Landscape. Applied Sciences **10**(18) (2020). DOI: 10.3390/app10186519
17. Jones, D., Snider, C., Nassehi, A., Yon, J., Hicks, B.: Characterising the digital twin: A systematic literature review. CIRP Journal of Manufacturing Science and Technology **29**, 36–52 (2020). DOI: 10.1016/j.cirpj.2020.02.002
18. Kaur, M.J., Mishra, V.P., Maheshwari, P.: The Convergence of Digital Twin, IoT, and Machine Learning: Transforming Data into Action, pp. 3–17. Springer International Publishing, Cham (2020). DOI: 10.1007/978-3-030-18732-3_1
19. Kibira, D., Shao, G., Weiss, B.A.: Building a digital twin for robot workcell prognostics and health management. 2021 Winter Simulation Conference (WSC) pp. 1–12 (2021). URL https://api.semanticscholar.org/CorpusID:247059772
20. Klein, B.T.: The DevOps: A Concise Understanding to the DevOps Philosophy and Science (2021). DOI: 10.2172/1785164
21. Kritzinger, W., Karner, M., Traar, G., Henjes, J., Sihn, W.: Digital twin in manufacturing: A categorical literature review and classification. IFAC-PapersOnLine **51**, 1016–1022 (2018). DOI: 10.1016/j.ifacol.2018.08.474
22. Lim, K.Y.H., Zheng, P., Chen, C.H.: A state-of-the-art survey of digital twin: Techniques, engineering product lifecycle management and business innovation perspectives. Journal of Intelligent Manufacturing **31**, 1313–1337 (2020)
23. Liu, M., Fang, S., Dong, H., Xu, C.: Review of digital twin about concepts, technologies, and industrial applications. Journal of Manufacturing Systems **58**, 346–361 (2021). DOI: https://doi.org/10.1016/j.jmsy.2020.06.017. Digital Twin towards Smart Manufacturing and Industry 4.0
24. Luan, T.H., Liu, R., Gao, L., Li, R., Zhou, H.: The paradigm of digital twin communications. arXiv preprint arXiv:2105.07182 (2021)
25. Mahmoud, E.G., Darwish, A., Hassanien, A.E.: The Future of Digital Twins for Autonomous Systems: Analysis and Opportunities, pp. 187–200. Springer International Publishing, Cham (2022). DOI: 10.1007/978-3-030-96802-1_10
26. Modelica-Association: Functional Mock-up Interface for Model Exchange and Co-Simulation (version 3.0). https://www.fmi-standard.org/docs/3.0/ (2022)
27. Mostaq Hossain, S.M., Kumar Saha, S., Banik, S., Banik, T.: A New Era of Mobility: Exploring Digital Twin Applications in Autonomous Vehicular Systems. In: 2023 IEEE World AI IoT Congress (AIIoT), pp. 0493–0499 (2023). DOI: 10.1109/AIIoT58121.2023.10174376
28. Nakaya, M., Li, X.: On-line tracking simulator with a hybrid of physical and Just-In-Time models. Journal of Process Control **23**(2), 171–178 (2013)
29. Naseri, F., Gil, S., Barbu, C., Cetkin, E., Yarimca, G., Jensen, A., Larsen, P., Gomes, C.: Digital twin of electric vehicle battery systems: Comprehensive review of the use cases, requirements, and platforms. Renewable and Sustainable Energy Reviews **179**, 113280 (2023). DOI: https://doi.org/10.1016/j.rser.2023.113280
30. Piromalis, D., Kantaros, A.: Digital twins in the automotive industry: The road toward physical-digital convergence. Applied System Innovation **5** (2022). DOI: 10.3390/asi5040065
31. Pylianidis, C., Osinga, S., Athanasiadis, I.N.: Introducing digital twins to agriculture. Computers and Electronics in Agriculture **184**, 105942 (2021). DOI: https://doi.org/10.1016/j.compag.2020.105942
32. Rasheed, A., San, O., Kvamsdal, T.: Digital twin: Values, challenges and enablers from a modeling perspective. IEEE Access **8**, 21980–22012 (2020). DOI: 10.1109/ACCESS.2020.2970143
33. Sivalingam, K., Sepulveda, M., Spring, M., Davies, P.: A review and methodology development for remaining useful life prediction of offshore fixed and floating wind turbine power converter with digital twin technology perspective. In: 2018 2nd International Conference on Green Energy and Applications (ICGEA), pp. 197–204 (2018). DOI: 10.1109/ICGEA.2018.8356292

34. Sleiti, A.K., Kapat, J.S., Vesely, L.: Digital twin in energy industry: Proposed robust digital twin for power plant and other complex capital-intensive large engineering systems. Energy Reports **8**, 3704–3726 (2022). DOI: https://doi.org/10.1016/j.egyr.2022.02.305
35. Tao, F., Zhang, H., Liu, A., Nee, A.Y.C.: Digital Twin in Industry: State-of-the-Art. IEEE Trans. Ind. Inf. **15**(4), 2405–2415 (2019)
36. Tao, F., Zhang, M., Liu, Y., Nee, A.: Digital twin driven prognostics and health management for complex equipment. CIRP Annals **67**(1), 169–172 (2018). DOI: https://doi.org/10.1016/j.cirp.2018.04.055
37. Tao, F., Zhang, M., Nee, A.: Chapter 7 - digital twin-driven prognostics and health management. In: F. Tao, M. Zhang, A. Nee (eds.) Digital Twin Driven Smart Manufacturing, pp. 141–167. Academic Press (2019). DOI: https://doi.org/10.1016/B978-0-12-817630-6.00007-2
38. Tekinerdogan, B., Verdouw, C.: Systems architecture design pattern catalog for developing digital twins. Sensors **20**(18) (2020). DOI: 10.3390/s20185103
39. Thelen, A., Zhang, X., Fink, O., Lu, Y., Ghosh, S., Youn, B.D., Todd, M.D., Mahadevan, S., Hu, C., Hu, Z.: A comprehensive review of digital twin – part 1: Modeling and twinning enabling technologies (2022)
40. Thelen, A., Zhang, X., Fink, O., Lu, Y., Ghosh, S., Youn, B.D., Todd, M.D., Mahadevan, S., Hu, C., Hu, Z.: A comprehensive review of digital twin – part 2: Roles of uncertainty quantification and optimization, a battery digital twin, and perspectives (2022)
41. Toothman, M., Braun, B., Bury, S.J., Moyne, J., Tilbury, D.M., Ye, Y., Barton, K.: A digital twin framework for prognostics and health management. Computers in Industry **150**, 103948 (2023). DOI: https://doi.org/10.1016/j.compind.2023.103948
42. Trauer, J., Mac, D., Mörtl, M., Zimmermann, M.: A digital twin business modelling approach. Proceedings of the Design Society **3**, 121–130 (2023). DOI: 10.1017/pds.2023.13
43. Tzachor, A., Sabri, S., Richards, C.E., Rajabifard, A., Acuto, M.: Potential and limitations of digital twins to achieve the Sustainable Development Goals. Nature Sustainability **5**, 822–829 (2022). DOI: https://doi.org/10.1038/s41893-022-00923-7
44. chuan Wang, Y., Zhang, N., Li, H., Cao, J.: Research on Digital Twin Framework of Military Large-scale UAV Based on Cloud Computing. Journal of Physics: Conference Series **1738**(1), 012052 (2021). DOI: 10.1088/1742-6596/1738/1/012052
45. Wright, L., Davidson, S.: How to tell the difference between a model and a digital twin. Advanced Modeling and Simulation in Engineering Sciences **7**(13) (2020). DOI: https://doi.org/10.1186/s40323-020-00147-4
46. Zhang, C., Zhou, G., Jing, Y., Wang, R., Chang, F.: A digital twin-based automatic programming method for adaptive control of manufacturing cells. IEEE Access **10**, 80784–80793 (2022). DOI: 10.1109/ACCESS.2022.3195905
47. Zheng, X., Lu, J., Kiritsis, D.: The emergence of cognitive digital twin: vision, challenges and opportunities. International Journal of Production Research **60**(24), 7610–7632 (2022). DOI: 10.1080/00207543.2021.2014591

Chapter 2
The Potential of Digital Twins: Four Industry Perspectives

John Fitzgerald, Peter Gorm Larsen, Cláudio Gomes, Rob Charlton, Klaus Kristensen, Stylianos Basagiannis and Jonas Åkeson

Abstract We report a virtual "roundtable" discussion on Digital Twins (DTs) involving experts from four engineering sectors covering consumer electronics, the built environment, aerospace and infrastructure. Topics covered include the vision, benefits and risks for DT technology in each sector, approaches to DT engineering and the role of dependability. Themes identified by our contributors include the need for DT engineering life cycles that deal with the creation of DTs for both new and existing products, competencies in ensuring that engineering disciplines work together, and the future need for DT integration.

2.1 Round Table Discussion Structure

The benefits claimed for DT technology are often described abstractly, but what do features like predictive maintenance, autonomy and decision support mean in practice? Why might a business want to invest in DT technology in order to deliver these features and what engineering methods and tools are needed? To gain insight into these issues, we held a "roundtable" discussion with four experts from companies that work in diverse sectors but who see DTs as key to future business.

Sect. 2.2 introduces our experts and the types of system in which, for them, DT engineering has a role to play. Sect. 2.3 to Sect. 2.9 discuss facets of DT technology including the strengths and some potential risks as our experts see them, and considers how these are delivered in each domain. In Sect. 2.10 we identify some themes from the roundtable discussion, setting the scene for the DT engineering technology explored in greater depth in subsequent chapters.

J. Fitzgerald et al. (eds.), *The Engineering of Digital Twins*,
https://doi.org/10.1007/978-3-031-66719-0_2

2.2 Introductions

Hi Everyone! Could you introduce yourselves?

Rob: Rob Charlton. I'm an architect and CEO of Space Group located in the UK. We are four businesses – Space Architects, BIM.Technologies, bimstore and Twinview – which share a single ethos with a simple strategy of Digital Synergy[1]. We offer a total approach to designing, delivering and operating buildings to achieve better outcomes for people and planet.

Stelios: My name is Stylianos (Stelios) Basagiannis and I am a Senior Principal Researcher at Collins Aerospace – Applied Research & Technology (ART) Centre located in Ireland. I am also a faculty member at the Dept. of Computer, Informatics and Telecommunications Engineering, in the International Hellenic University in Greece.

Klaus: I'm Klaus Kristensen and I am a Senior Research Specialist and Senior Software/System Engineer with Bang & Olufsen (B&O) in Denmark.

Jonas: Jonas Åkeson and I am newly appointed as the Vice President of Digital Development with Grundfos in Denmark, but before that I have spent almost 20 years in different roles at Ericsson, ending with a role as head of Artificial Intelligence (AI) in automation.

2.3 Businesses

How would you describe the area of your business that we're going to talk about?

Rob: I am interested in how DTs can improve the outcomes of the built environment. We have developed Twinview, a DT platform designed specifically for buildings.

Space Group was established in 1957 as an architectural practice in north-east England. We have always been committed to technology and were among the first UK companies to use the Integrated Graphic Design System (IGDS). In 2001 the company invested in Revit, an early software product supporting parametric modelling. By 2008 the whole practice was using 3D software.

We developed our digital skills and identified an opportunity to use data to improve outcomes across the built environment. We launched a digital consultancy in 2010 called BIM Technologies, which provides specialist advice to clients about their

[1] `https://www.twinview.com/`

digital strategies. In 2011 we launched bimstore, a library for digital components used in the design of buildings.

Over the next few years, we established our digital strategy, called "digital synergy", which identified the value of collecting and using data from the building life cycle's design, construction and operational stages. In 2017, we developed a browser-based viewer following a request from a client who could not access their data. Over the next few years, this evolved into a DT, and in 2020 we launched Twinview.

Today Space Group is four businesses using the data from the building life cycle to improve building performance and reduce carbon embodied in structures and in their life phases of construction, design and decommissioning.

Stelios: Collins Aerospace is a Raytheon Technologies subsidiary and one of the world's largest suppliers of aerospace and defence products. The company is engaged in designing, manufacturing and servicing systems and components for commercial aviation, business aviation, military and defense, helicopters, space, airports, and other industries. There are six strategic business units: Aerostructures, Avionics, Interiors, Missions Systems, Connected Aviation Solutions, and Power & Controls. Collins has actually been using DT technology for quite some time. For example, engineers are using aerospace test rigs with the capability of simulating the full aircraft and all of its systems. That allows us to take equipment and put it into the aircraft, and through digital technology simulate the rest of the aircraft.

Klaus: B&O is a consumer electronics high-end luxury audio brand. We produce interoperable smart home Audio/Video (AV) systems that embody the highest engineering and design values, so our customers expect key emergent properties to be maintained as their product collection evolves. Our product portfolio is built around three types of use cases (see Fig. 2.1):

- **On-the-go** products that are mobility-based, built around nomadic use cases, including both immersive and social experiences;
- **Flexible Living** products for consumers who want flexibility in use and placement in domestic settings; and
- **Staged** products used in stationary settings for immersive listening or viewing experiences.

The use cases that first spring to people's minds concern specific products offering audio or video playback. However, B&O also delivers system-level use cases such as spatial audio, wireless multi-room, surround sound and Internet of Things (IoT) home automation. These introduce a range of challenges from the world of distributed systems. For example, system-level audio means addressing the need for global consensus among various devices, and ensuring communication that works to tight time deadlines but is also fault-tolerant.

The AV smart home market is rapidly growing, both financially and technologically, so B&O must successfully align its brand values, sustainability vision and products to the demands of the smart home market, such as device interoperability and feature extendability, that are needed for future growth. The global smart home

Fig. 2.1: B&O product use cases (l-r) on-the-go, flexible living, staged (Copyright 2021 by Bang & Olufsen A/S reprinted with permission)

market size was valued at USD$ 79.16 billion in 2022 and is expected to grow at a compound annual growth rate of 27.07% from 2023 to 2030.

Jonas: Grundfos is the largest pump manufacturer in the world, based in Denmark, with more than 19,000 employees globally. Grundfos started producing pumps in 1945 and has had several levels of innovation to pioneer solutions to the world's water and climate challenges and improve quality of life for people. It is now divided into four divisions: Commercial Building Services, Domestic Building Services, Industry and Water Utility. These jointly form one Grundfos, united by our purpose and values. In the engineering of our products first principle models have been used for a long time with different purposes.

Going forward there is a desire in all four divisions to consider how digital services can be constructed on top of the physical infrastructure. Here DTs are clearly going to be a part of the solution both at the edge as well as at the cloud level. The key for Grundfos will be to keep the domain expertise such that this knowledge can form the basis for such digital services. That will be the reason why we will be able to provide better support for our pumps than any other organisation in the world.

2.4 Where are you thinking of targeting DT technology?

What are the systems of interest, i.e., where are you thinking of applying DT technology? Can you outline an example to ground our discussion?

Rob: We appreciate that DTs can be used across many assets and sectors. Twinview was developed to address specific requirements within the construction and property sectors. Buildings contribute to 39% of greenhouse gas emissions, so improvement can significantly impact global warming. Also, with increasing energy costs building users want to understand their buildings and have control over their performance.

Twinview is a Software-as-a-Service platform. It was designed this way so it was an "out of the box" setup. With training, building owners can set up and manage their own twin. However, we have found that most building users would prefer us to do it for them. It started from our experience in BIM Technologies, a tech consultancy that had built fantastic 3D models with lots of assets and data. For example, one of our 3D models of a building would go down to the detail of a valve and would tell you what that valve is, what it weighs, what its performance is like, where it's made, etc. We made all that detail, but it was never being used!

In a perfect world, clients would come to the end of construction with a BIM model created to ISO 19650 and then just upload that to Twinview [2]. That gives you a cloud-based view of the BIM model. The sensors that are hopefully in the building will connect to the Building Management System (BMS). Twinview produces end points through the API and connects to the BMS's API and we therefore have a connection between the 3D model which is a digital representation of the physical, and the existing IoT devices that are in there. The client can then configure bespoke dashboards using widgets. That's the perfect world, but we can provide the services to help them if they don't have 3D models, or just 2D or no BMS.

We find we are competitive because Twinview is a platform with standard modules and tools which can be switched on or off. What we found is that, now, our clients are early adopter organisations who own and operate buildings. For example, we're working with a government body who will own and operate their archive building, or a project in the Middle East on a big tower that will be owned and operated by the company themselves, and we're close to working with a major airport. So these are generally big organisations, quite advanced in their thinking, confident and often intellectually driven. They're willing to take a chance because they have an interest in moving the innovation curve. We're very much in the early adopter/innovator stage, so it's a leap of faith.

Our other market is associated with digital consultants like information managers or BIM coordinators. If we can get them to understand the value — and they do because we all want to be at the forefront – they are then promoting DTs to their clients, who may be owner/operators. We initially thought commercial office operators would be a good place for Twinview, but often the people who design, construct and develop the buildings sell them on, so they have less interest in the building's performance through life.

Stelios: DT technology has been applied to a variety of aerospace systems. Collins Aerospace with its dedicated Digital Thread & Digital Twins Technologies (DT3) team quickly adopts and adapts architectures, technologies and methods to improve digital continuity across the product life cycle (as-built vs. as-designed / as-flown vs. as-maintained)[2]. Application domains may include:

- embedded systems and networks, using DT environments to constantly evaluate connected, distributed devices at the system and at the edge;
- advanced manufacturing and supply chains employing process DT technologies in an attempt to improve throughput, quality and process qualification for new products;
- additive manufacturing and materials using DT for material analysis with respect to geometry design and material validation;
- cyber systems, using DT technologies for the evaluation of cyber-attacks in aircraft systems and networks;
- new emerging aircraft technologies, motivated by ACARE Flightpath 2050 initiative (CO_2 reduction) where a DT can enable the evaluation of autonomous aircraft operations and new engine designs such as hybrid-electric or hydrogen engines [4].

Due to the evolution of the IoT, systems connectivity and software technologies, it is an important milestone to mature and foster DT technologies to improve system design and maintenance in all the aforementioned domains while drastically reducing development and testing costs.

Klaus: It's widely understood that the more you pay for a product, the less tolerant you are to failures. B&O being an luxury brand means that our customers have high expectations of our products and services. So product quality and service capability become key design parameters for us. It's also expected that luxury products are designed and built to last, and so must remain relevant to the customer as they evolve. This means that product monitoring and updating operational capabilities are crucial.

High expectations make this an exciting but demanding domain for systems development. Our hypothesis is that applying DT technology will help us to address some of the limitations in current development and service methodologies, realising the link between the B&O customer's expectations and the brand's core values such as quality, lifetime product warranties and feature scalabilities.

An example of a system-level use case is the SoundZone system. This consists of several speakers and microphones, a signal processing unit, sensors for user input, and units for visual output. The system is customisable and scalable and can be tailored to factors such as the size of the room, number of possible zones, level of sound separation, sound quality, user control and feedback. To create a sound zone from an array of speakers or soundbars, the audio sent to the speakers needs to be processed so that particular acoustic effects are achieved at different locations in the room. Dynamic sound zones rely on a large number of speakers. These need to be

2 https://www.collinsaerospace.com/what-we-do/capabilities/technology-and-innovation/applied-research-and-technology/areas-of-expertise/digital-thread-and-twins-technologies

wireless to ease installation and portability. However, we need precise control of the signal at each speaker but wireless network transmission of audio is highly error-prone. So new network and audio signal coding algorithms need to be developed.

Jonas: I think that we will be targeting a hybrid solution, partly at the edge level and partly at the cloud level, just as I experienced in the telecom industry. I think that it will start already in the processes we use for the development of our pumps, where the use of modelling and learning from data (both from previous versions and from physical prototypes of new products) will be used to a much larger extent.

Today our development processes are too static when we explore alternatives, and they are based too little on lessons learned. The whole DevOps initiative has changed the way software is delivered to its customers in the IT sector and we too will need to adapt our development to a more dynamic world. We need to get to a stage where virtually all physical assets are continuously connected to our software.

At the development stage we need to combine pumps with their DTs on a global scale so we will be prepared for the different kind of usage of the physical products. At the manufacturing stage we also need to update so we are prepared to deliver Service Level Agreements (SLAs) to our customers since that will be part of the digital services offering built on top of our physical products. This will be necessary where our pumps are included in critical infrastructure supplying clean water to people. I expect that we will see reinforcement learning, because of its ability to operate without having to label the data, along with retraining during operation in order to calibrate when changes appear in the real world.

So what do you feel DTs could bring to these kinds of system?

Rob: In the past 25 years, we have focussed on improving the design and construction process through innovation. More recently, we have invested in developing DTs to reduce the operational impact of buildings. For example, our clients, who are mainly building owners/operators connect the DT to their BMS to give them a better understanding of how their facility operates, helping them to optimise performance and reduce costs.

Stelios: Next generation transportation systems and products will be more connected and more integrated, as well as more electric. Due to their complexity, it is important that DT technology brings smart, easy-to-use automated algorithmic solutions that accelerate complex system design. Challenges may be driven by multi-disciplinary engineering tasks primarily for connecting distributed, heterogeneous (online and offline) data, models and processes in a unified body of knowledge democratising System-of-Systems (SoS) engineering processes. This is the exact point where DT technologies can accelerate a more rigorous, efficient and safer SoS design and development activities raising awareness in a multi-disciplinary engineering context.

Understandably for a domain where safety, security and energy usage are all crucial, DT technology can enable AI-based optimisations applied to new aircraft systems and designs. Automating the number of simulations, without the presence of an engineer, empowered by DT solutions could rapidly accelerate new system validation based on data originated from the physical side, reducing engineer time and cost in the complete product life cycle. For example, when new real data are being fed into the DT, automated processes could take over and re-run system simulations based on the new data, without the presence of an engineer inspecting the simulation. End results from the DT could yield potential improvements to the new (to be designed) system or highlight new system configurations that can be applied to the existing ones. It is thus crucial to expand upon new DT solutions that will allow automated system parameter space exploration by adding the ability to both automatically search spaces of architectures and to test designs using a set of different scenarios.

Klaus: In the context of B&O products and systems, the DT will not be visible to the end-user. Instead it will enable monitoring and reconfiguring the real system of interest – the "Physical Twin" (PT) – as it operates. The DT is not the product, but it helps to achieve the characteristics that we need from the product such as maintainability and extendability.

Jonas: In the context of Grundfos data about the pumps will be central to success. The key will be to ensure that we will be able to deliver services that have significant value to our customers. It is delicate to find the balance with how to deliver such services and be able to charge for them. There are different alternatives here, and we are still exploring the different options.

2.5 What does success look like?

What are the most important properties of these systems as you see them?

Rob: Providing visualisation and easy access to the data for the user. Having a 3D model helps to locate assets. For example with only an asset reference it is difficult to locate a valve, whereas it is easy to locate within a 3D visual representation.

The Twinview dashboards are configurable and accessible. They can be viewed by anyone on a mobile device. Typically this information is only accessible by an engineer on a complex system, usually in the plant room. Twinview also retains data meaning historical performance can be reviewed.

Virtual Reality (VR) adds another dimension. We know how to achieve this with Twinview as the technology is accessible. However, we feel it is a step to far for our users at the current time. I don't think we will progress this until headsets get smaller.

Stelios: Depending on the system design life cycle and its constituent model artefacts, properties include predominantly safety, reliability and performance requirements. Examples could be controller signal convergence within specific time boundaries or system responses when certain faults occur, thrust generation and engine efficiency respecting safety, pressure and combustion chamber temperature and so on. Challenging requirements are present as part of large system integration efforts, where the system as whole needs to be verified against safety and functional characteristics.

Klaus: Because of our brand, quality and maintainability are the most important properties of our systems and products – they enable longevity. The ability to maintain high quality while updating the feature set through life is key. B&O's heritage of offering lasting, high-quality products shows how B&O embeds longevity through the entire life cycle from the selection of raw materials and components, through the design, manufacturing and distribution of products, their use, their 'second life', and eventually, their end-of-life handling or disposal.

For B&O to maintain its leading audio market position and align it with the longevity strategy, the company must constantly improve development, end-user services and tools processes. In order to achieve this, DT technology should support:

- specifying and quantifying needs specific to the B&O AV domain, such as quality, maintainability and reconfiguration;
- designing, reasoning about, validating, and simulating flexible candidate architectures for services, systems, products and hardware/software platforms;
- validating and verifying the correctness of complex system(s) and product(s) using co-simulation and models to provide the ability to control environments and dependencies for testing and experimentation;
- learning about the reality of how products and systems are used in practice compared to design assumptions, with a view to improving maintainability; and
- proactive product and system health monitoring.

The DT development and maintainability benefits will impact B&O further innovations to:

- support and develop relevant B&O product lines, recognised as high-quality and sustainable products;
- expand the company's product, system and services strategies, include more premium audio products and more pro-active fault- and issue-detection services; and
- maintain market leadership through state-of-the-art competitive audio products and services.

Jonas: This is about becoming predictive and preventive. It is important to derive information about the PT and then not just tell someone about it, but take action on it autonomously. The transition towards this level of autonomy will be gradual. Initially it will be necessary to enable a human to take action, but eventually trust in DT predictions will rise to a level where the human will no longer wish to be included in the loop.

Significant time and cost are associated with sending people into the field to deal with changes. In order to reduce this, it is necessary that the DT has cognitive capabilities that are able to predict situations that one would like to avoid or prevent. Over time this will build up to form a valuable knowledge base that can be applied to increasingly complex situations, using inductive and deductive power. This will enable Grundfos to increase value at an integrated system level, instead of just supporting individual pumps. Grundfos needs to make use of the already existing domain knowledge, since that is really the main asset that we have, and this can be used constructively in our digital services.

What does engineering success look like in your domain?

Rob: We are at the beginning of the DT journey in the built environment. There is little historical data relating to how buildings perform. We are collecting operation information and beginning to analyse the outcomes. Success is visibility and understanding of all systems in a building and, therefore, the ability to optimise its performance. As well as physical systems such as heating and cooling, human interaction with buildings has an impact. We want to improve user experience by understanding how and when spaces are used.

Predictive maintenance and autonomous building operation is the next step for DTs in buildings, and it will have a significant impact on the performance. Understanding our buildings better – and most importantly being able to optimise them – will save tons of carbon across the world. Any improvement in energy use will help to achieve the 1.5-degree target for the rise in global average temperature.

Stelios: Delivering state of the art quality and airworthiness systems and products exhibiting customer requirements satisfaction with respect to safety, security and performance properties based on sound, tool-aided evidence, will compose a successful engineering story. The latter will be extremely useful in cases where requirements can be characterised as inter-dependable (e.g., safety and security in next generation autonomous systems); in this case an integrated requirements and verification analysis will be needed to resolve conflicting cases often driven by different domain-experts developing independent – yet related – system models. DT technologies could provide a "bridge" to the aforementioned challenge by connecting previously independent model representations, identifying conflicting corner cases. Through a DT early dependability verification results towards certification credit can be crucial to transition new, robust systems to the market.

Klaus: Unsurprisingly, given the values I mentioned, success is delivering a quality and maintainable product/system, where the designs and implementations are based on the right decisions, given all the stakeholders' requirements and constraints. As an example of this improvement goal, the longevity needs are captured in the non-functional requirements implicit defining the solution architectural designs to support evolving requirements and environments. DT concepts such as design space exploration at design time and what-if-analysis after deployment of the product could help reasoning about requirement completeness in both the development and maintenance stages of system life.

Jonas: We need to be able to create an architecture which allows data to be generated and stored so it can subsequently be accessed. This shall be augmented with synthetic data that can be created as well as crowd data. In the past it was possible to rely on a few goodwill customers that was able to share data. Instead it is necessary to do as the main IoT suppliers, in order to gather a sufficient amount of data.

2.6 Why Digital Twins?

So – building on that – what got you thinking about DTs?

Rob: DTs are the next step in the construction and property sector. The sector has undergone a digital transformation over the last 20 years. Both design and construction have adopted many digital workflows, enabling digital representations of the completed building. It's logical to use this information in the operational stage of the building life cycle. At the same time, energy use and costs are increasing. Also, with increased hybrid working, buildings need to be more flexible, and systems must align with the evolving workplace.

Stelios: DT technology will enable easier, and more frequent, full system analysis by embracing the multitude of tools, formalisms, and abstraction levels needed in future aircraft systems design. It can offer an engineering environment open to the introduction of a wide range of tools supporting models expressed in formalisms that include physical phenomena (continuous models) sourced from engineering disciplines such as electric, control, mechanic, hydraulic, pneumatic, thermal and fluid, and discrete events (state-based models) typically used to describe embedded systems, computing infrastructure and communications. Availability of data through the continuously evolution of IoT together with a more-connected aircraft ecosystem are some of the main reasons to adapt DTs. The latter, enabled the analysis of interactions between diverse models managed through DT and co-simulation solutions will enable faster system multi-disciplinary design and optimisation of a full range of design parameters. On top of that, recent advances in AI and Machine learning (ML) techniques will improve system design analysis and diagnostics, debugging system multi-models much more faster reducing engineering costs. All of this technology will inevitably advance system efficiencies, working with teams to improve process velocity, predictability, agility, and end-product quality. All these enhance existing DevOps practices and lead to the creation of new ones.

Klaus: We got thinking about DTs because B&O has successfully applied model-based approaches such as co-simulation and formal verification in architectural design and implementation [3]. However, the needs for quality and longevity run from development into the in-operation life, so we need a unified methodology that works across *all of the life cycle*.

DTs are model-based and therefore are able to deal with complexity through abstraction, which also aids communication during decision making. DT co-simulation and runtime verification concepts will provide capabilities through the life cycle, from design to implementation and integration validation in the development stages, and maintainability activities in the operational stages. The DT monitoring concepts bring benefits in areas such as fault detection in the deployed products and systems, where during development we could only make assumptions about the space of possible states of our devices and networks, and the ordering of events in traces.

Jonas: Here I need to go back to my experience at Ericsson where we could demonstrate that it is now possible to manage complexity that previously was not humanly possible. It is now possible to optimise complex challenges with an enormous number of parameters extremely fast. If you look at this from a pump infrastructure at a worldwide scale with a cognitive layer inside a DT, and use the data derived from all these installations it will be able to learn things that was never known in advance. This will form the basis for optimisations of full cities so from a societal perspective this will change the world for example with what-if capabilities enabling exploration of changes to the environment in a pure virtual testing sense. For Grundfos the entry ticket to this vision is the know-how about the pumps and their infrastructure. It is impossible to predict who will be the main operator of such a system, but Grundfos will be a part of it if we take the right steps going forward regarding our digital services.

2.7 Stakeholders, Developers and Users

Now we are thinking about DTs, let's consider the DT's stakeholders - who are the owners of DTs in your setting? Who are the developers? Who are the users?

Rob: Across the building life cycle, there are several stakeholders. The data starts at the design stage, growing through construction. Ultimately the DT is used in operation by the building user, owner or operator as they can generate value from the data and better control. Little data relating to buildings' operation is available, meaning data analysis is currently difficult.

As the amount of data increases in the years ahead, there will be an opportunity to use ML to predict maintenance and performance requirements.

Stelios: Modern aircraft CPSs have physical components that are monitored and controlled by distributed computing and communication systems. Development of these systems or products requires a multi-disciplinary approach where engineers can collaborate across disciplines, organisations and borders. Product stakeholders may involve program managers and project leads, systems and control engineers, modelling and verification experts, software and embedded systems engineers.

It is also important to highlight the heterogeneity of the systems being built, primarily from an ownership perspective, as the system may compose a series of products and services originated by different entities (aerospace companies and Original Equipment Manufacturers (OEMs)) that enforce various IP-flavoured constraints during system integration and testing. Up until now, aircraft OEMs and engine manufacturers are using similar Model-Based Design (MBD) techniques and approaches but in an independent and decoupled manner without exchanging engineering information during their Model-Based Systems Engineering (MBSE) activities. Collaborative MBD and DT technology is a promising approach to such multi-disciplinary development, where digital representations of individual system components (constituent models) are combined and analysed in a virtual setting. On top of that, realising the vision of a European aerospace community that "leads the world in sustainable aviation products and services" strengthens the need for new engineers (e.g., hydrogen engines will bring process or chemical engineers) utilising improved and more open-ended toolchains and methods that support collaborative design within global supply chains, with "design-for-excellence" embedded in the aircraft's life cycle.

A DT's stakeholder could span from engineers who are part of a CPS design activity (system, electrical, modelling, verification, software, hardware, integration) as well as project managers tracking down resources (labour, costs, timelines), supply chain managers and maintenance engineers and end-customers or certification authorities willing to overview methods, evidence and end-results produced from the DT.

Klaus: As I said earlier, the intent is to use DT technology in all stages of development and operation, so the DT's stakeholders would correspond to the stakeholders in each of the stages. This unified intent exposes a well-known system engineering challenge regarding the need to deal with (in)consistencies between the requirements of different stakeholder groups.

The main technical DT stakeholders would be the product engineering team and the service organisation, corresponding to the PT and DT relation. The DT's end-user would be the service departments using DT for pro-active maintenance and monitoring. Another end-user would be the engineering team using data and simulation for improvements and decision making. The DT development team would need to be cross-disciplinary because of the stakeholders come from different domains.

Jonas: With the vision I presented earlier the IP of our products actually gets centralised inside a DT capability, where this information in the past was placed in a de-centralised manner, so there is a security concern that needs to be raised here. We will have a number of proprietary solutions, that we will keep closed, but we will choose areas where we expose our pumps capabilities in an SoS setting. This will be different in the different divisions, so the target role Grundfos will wish to play and the stakeholders to collaborate with will as a consequence also differ. In short, Grundfos wish to be the main supplier supplying pumps into this agenda. It is a matter of ensuring that we can deliver business value on top of our physical offerings (possibly together with others in some kind of eco-system). This will enable others to provide innovation on top of our offerings, and we wish to monitor that continuously.

2.8 How would you expect to develop DTs?

Do you envisage developing DTs alongside physical systems, or is it more about constructing DTs for pre-existing physical systems (or a bit of both)?

Rob: DTs are at the earliest stage of the innovation curve within the property sector. To move up the curve, they need to be able to demonstrate value.

As the amount of data increases from operational buildings, we will begin to know how buildings perform and be able to optimise their performance to improve user experience whilst reducing waste and carbon.

DTs are often "applied" to the completed building. This means the systems have not been set up to allow granular access to performance data. For example, metering may not align with floors or spaces, and controls may not connect to systems. Sensors may not yet be installed, or may be installed in the wrong location. By contrast, the most effective twins are built alongside the physical asset to ensure identical replicas. At the outset, there also needs to be an understanding of how the twin will be used and what data is required.

Stelios: Determining the required level of abstraction and model representation of physical phenomena interacting with the system is crucial for the success of DT solutions. Physics-based simulations in general (e.g., Finite Element models) are front and centre in any aerospace research program since these are crosscutting technologies that impact all speed regimes and all vehicle classes. They deliver superior understanding and insight into the critical physical phenomena limiting component performance, thus opening new frontiers in aerospace vehicle design. For example, the use of Computational Solid and Fluid Dynamics (CSFD) in commercial aircraft is well documented, with particular success in the design of the high-speed wing (cruise shape) and its close integration with the engine, dating back to the Boeing 737 Classic of the late 1980s.

Finite element analysis tools describing physical interactions and impact on certain parameters, are not easily integrated with other (time-variant) simulation tools, due to large computational demands [1]. It is therefore pivotal to consider developing DTs alongside the physical systems whether new or legacy. From an engineering perspective the benefit of this will be improved productivity in the engineering workflow, such that the engineers can focus on the model development, while the adaptation to DT co-model is handled in another module, which will be available as a standard library (e.g., Read-Only Memory generation) for the most relevant cases. At the same time, valuable knowledge (standalone results, constraints) can be made available for improving/building on top of DTs of existing systems in a more agile, incremental manner.

Klaus: Given B&O's large portfolio of products that are already released and used, and the desire to maintain and extend these products, DTs will be developed for pre-existing physical systems.

Jonas: We will definitely need to establish both DTs for existing products as well as for new products. For the existing products we really need it in order to gather enough data about how our pumps are used and how we can learn from this. I think that the latter is actually a more substantial change all the way from how we develop both our pumps (the PTs) along with their DTs in a more dynamic manner as described earlier. It will become essential to consider the existing installed pumps in order to also contribute to the societal agendas, including the need to reduce CO_2.

How about composing DTs and their corresponding physical twins together?

Rob: We see this as essential to the success of DTs in buildings and in some respects we are already doing this. For example a lift will have its own system which pushes information back to the OEM. We connect into their twin and pull the information into our platform. Twinview is intended to be a single window which provides an overview of all building systems.

The dashboards can be set up so you have system information at both a macro and micro level. For example you could have a dashboard for the AV controls across the building.

Stelios: It would definitely be of interest to move into system design cycles that allow the parallel development of a system and its corresponding DT. Much of the benefit would be realised in the integration and validation phases of the design cycle that would take increasingly take account of the DT, enabling faster error-correction activities and fault diagnosis while predicting additional system responses through AI/ML features of the DT.

Klaus: There are a couple of aspects to this. Products are created from hardware and software platforms, so a product's DT will likely include DTs of the constituent platforms. There's a logical progression from developing DTs of pre-existing physical systems to composing DTs and their corresponding PTs in a consistent SoS methodology.

Jonas: In order to be able to properly support the composition it is essential for us to get started with our existing pumps because of the uncertainties, so we can learn more about the accuracy we will be able to provide in our predictions. There will clearly be challenges here, but it is essential to get started with the SoS composition. Actually I see the challenges as a positive thing, because it means that those companies that are able to manage to solve this, will clearly have a competitive advantage.

Do you see DTs monitoring PTs and reacting to observed changes?

Rob: The autonomous adjustment of building controls is fundamental in demonstrating value within the Built Environment. BMSs are typically based on logic – for example, if a fire is detected, close the fire dampers. A DT can be more intelligent. For example, if it knows how many people are on a particular floor, ventilation can increase. It has a memory and can learn, so it knows that on Mondays in February, for example, only 50 people use the building so only one space needs to be available.

Stelios: Within aerospace engineering it is essential to predict, maintain and evaluate system's health and readiness based on monitoring data. A DT attached to the real system will trigger important services to the system's owners with respect to safety, availability and performance while it can also be instrumental to its supply chain, based on up-to-date information for parts timely delivery and integration.

Klaus: Monitoring PTs and reacting to observed changes is one of the key enablers for product improvements and maintainability. Given the complexity of Validation and Verification (V&V) for the kinds of systems B&O deals with, it's not possible to engineer a pre-release V&V strategy that absolutely ensures the qualities required. The monitoring capabilities of DTs can provide real-life data, confirming or refuting assumptions made during design, implementation and operation stages.

B&O doesn't currently have a way of recording reactions to observed changes in the PT. B&O only implement the concepts issue tracking by manual analysing product states using cloud-based interfaces accessing log files.

Jonas: There is clearly a risk that uncertainties and noise will hamper our predictions, but actually we also have human errors today. The main question is what is good enough. This will depend on the different use cases because the requirements will be very different.

Are you thinking of DTs as ways of enabling people to visualise and analyse the real state of a system?

Rob: Currently, there is little performance information about the operation of a building. A DT will provide a view of the data not previously seen.

Stelios: Enabling people to visualise and analyse the real state of the system is the vital entry-point of DTs in decision-making processes. System state visualisation through DTs could potentially unlock design-driven monitoring for the real system's performance and robustness. However, there are challenges to be addressed around data acquisition and processing through third-party components, and data ownership. A crucial part, though, is the automation of the monitoring because the volume, velocity and variability of data may require customised DTs per system's design and implementation parameters.

Klaus: Yes, enabling people to visualise and analyse the real state of a system would be very beneficial for enabling sound reasoning about the system. Graphic models of transition system, and parameter value graphs would provide needed abstraction for handling the complexity of the PT.

Jonas: This is naturally a bit of speculation from my side, but I think that a lot of the DT technologies in the future will be like hidden heroes meaning they will operate without the public knowing it. Visualisation will be important for the public to get an understanding and visibility when they have a need to interact with the DT environments. AR/VR is one way to do this and there will probably be new 3D googles, head-up displays in cars etc that will make this possible in a much more seamless way than today.

So what are you expecting to sense? How?

Rob: Typical data might be room temperature and what impacts upon this. For example what impact does occupancy have or weather conditions? When is a building used, what rooms are occupied or what is the air quality at certain times of the year through to day?

What are the lighting levels and does this impact on where people sit in an office environment? How can the building energy use be optimised: is it best to pre heat a building of heat quickly when the spaces are used.

We do not have much information about the quality of the data as it is early days. However we do see errors and rogue readings when reviewing system performance. The information which has been gathered in recent years has rarely been used and therefore not checked for accuracy.

Stelios: Currently there are Prognostics and Health Management (PHM) components in large SoS solutions (e.g. fuel, pressure or combustion chamber temperature). DT sensing should not attempt to replicate those systems but rather compliment and improve their functionality. DT analysis should be able to reach system parameters that are considered to be key factors to control and manage constituent components from a design-view analysis. Those parameters, although hidden to the whole multi-disciplinary team, may yield important information about the system's anchor points preventing the latter from reaching erroneous states not predicted by PHM subsystems. Thus, it is important for the DT to monitor the PT in a governing manner, allowing a more thorough (offline) system inspection cross-correlating real and analysis data provided by the DT.

Klaus: As an simple starting point, I'd expect to sense the execution path and trace parameter anomalies. These detections need to be enabled by DT-to-PT supervised learning methods during the development stages. In the PT operational stage, monitoring will enable reasoning about learned behaviour and actual behaviour. The next step would be reacting on detection of faulty actual behaviour.

In the context of trace parameter anomalies, often this type of fault relates to deployment environments, product and system configuration issues. Because of the parameters' possible value space, it's not always possible to identify default configuration settings during development that will work in all environments. As a response to this, the DT may simply be able to reconfigure the PT. The DT reconfiguration capabilities require the DT-to-PT supervised learning model to support DT retraining based on parameter value domain(s) and value domain outlier fault classification.

Jonas: This mainly depends on a combination of factors, including how quickly you need to close the loop and how expensive it is to communicate and store data from the PT. If the customer starts asking for SLAs, this will also influence what we need to sense. In general what we sense will depend upon what makes financial sense as well, since both communication and storage of data both have associated costs.

Do you know what levels of accuracy and precision you would need to support that?

Rob: In the built environment the accuracy of data is not that critical as the sector has little knowledge of building performance. Anything is better that what is currently available. Also very few organisations are currently using DTs in our sector across the world. I would guess there are fewer than 50 true building twins in the world.

Stelios: This is entirely driven by the applications. For example safety and security critical systems will aim at high levels of accuracy and precision resulting into different DT variants.

Klaus: This really depends on the use-case requirements, which will define the accuracy and precision needed. In some cases we can accept a more relaxed state consistency model as a simple 'happen-before' causality model, but in other cases as in the audio domain where the systems operate under time constraints, the DT accuracy and precision needs are bounded by the time constraints.

Jonas: Clearly for the pump there are different data that are important to deduce how the performance is, such as vibration and flow, but probably the trend is even more essential. Here Grundfos will use its domain expertise to determine both what are the most essential parameters to use and what accuracy is necessary in order to be able to make proper predictions from it.

If we think about how faithful a DT must be as a representation of the PT, could you characterise this requirement from your viewpoint?

Rob: From a safety perspective it is important that the DT is an accurate representation of the PT. For example, fire performance of DT elements must be accurate. Spaces must represent the physical to allow evacuation monitoring and simulation.

Stelios: I would re-emphasise the term 'DT faithfulness' with an overarching system requirement that should describe the reasons and the level of precision its DT should have during the system's execution. Again depending on the application, it should be determined what is the required level of trustworthiness of the DT running (e.g., abstractions or data filtering) let aside the bidirectional validation processes that needs to be executed in order assure the DT's readiness.

Klaus: As with the accuracy and precision levels, the faithfulness of the DT depends on the use-case requirements. As an example let's define some DT faithfulness requirements for the anomaly detection use-case. For detecting execution path and parameter anomalies:

- the DT must represent the complete state- and state transition space of the PT;
- the DT must report the state transitions according to a sequential consistency causality model;
- the DT must represent the parameters sets and value domains of all traces of the PT; and
- the DT must report the parameter value transitions of for all traces according to a sequential consistency causality model.

Jonas: What is the most important thing here is to establish a triangle that is able to combine domain experts and data scientists such that they can jointly incorporate their knowledge into a common overall solution for a DT-enabled system. If that triangle is established we will have lived up to that requirement.

Do you have established ways of recording the required reactions to observed changes?

Rob: In Twinview currently significant physical changes to the PT such as walls being moved must be updated manually in the DT so a new 3D model needs to be uploaded. However, if a door type is changed, we can update it, but it is unfortunately a manual process.

Stelios: The Collins Aerospace DT3 team quickly adopts and adapts architectures, technologies and methods to improve digital continuity across the product life cycle (as-built vs. as-designed / as-flown vs. as-maintained). By connecting distributed, heterogeneous (online and offline) data, models and processes in a unified body of knowledge, it is possible to democratise the engineering processes.

Klaus: B&O used an ML-based process for monitoring the products' health. The ML models are trained from release test data, marked as correct behaviour. The ML models and running in the cloud and can be manual, or automatic trigged for validation of an deployed product(s). This monitoring method can detect the probability of faults, but have no reactions built in. Fault classifications made by the ML models are delegated to the engineering teams. This is the same reporting process used by the front-end service department inside B&O.

Jonas: This is in my opinion the most significant challenge with DTs and use of ML inside them because the explainability is complicated. This is clearly one of the places where industry relies on progress established by academia.

To what extent would you expect those reactions to be autonomous?

Rob: Currently the existing BMS systems have simple "if this, then that" For example if there is a fire, open the vents. As building twins develop and the amount of performance data increases there will be an opportunity to use ML to optimise performance.

Stelios: Autonomous DTs should not be considered a far-fetched goal. The evolution of connectivity in current ICT domains and the presence of robust and secure cloud-based infrastructures could enable an autonomous DT to react and respond to the real system when deemed necessary.

Klaus: The dream is full autonomy, meaning a reactive DT system preventing errors negatively affecting the end-user's experiences, and maintaining their expectations of the product. But how autonomous we can be, really depends on the DT fidelity level reached, and the characteristics of the use cases leading to the real errors we want to prevent. In some cases where product parameter reconfiguration is needed to fix a runtime issue and there is evidence for the correctness of the reconfiguration, this would justify full automation.

Jonas: I agree with Klaus that our dream also is full autonomy, but it needs to be proactive and preventive rather than reactive. We expect to get there in a stepwise manner where we initially will have humans involved in the loop. Naturally, the criticality of the application where our products are used influences the level of autonomy that will be acceptable society perspective. Traceability and the ability to have external audits are today included in virtually all control systems hand-coded and thus it is easier to explain anything where you have learned a part of the control. This naturally may have an effect on the allowable autonomy.

2.9 Do DTs help Dependability?

In your domain, what does dependability mean? Do you see DTs enhancing dependability? If so, how?

Rob: I think this is a big part of the value proposition. For example if you are able to reduce the use of a system it should last longer. At the simplest level, if you are leaving the lights on when the building is empty your lamps will not last as long. If the use of a boiler is monitored and maintained it will run more efficiently and not need to be replaced as frequently.

Stelios: Although the term could have a different interpretation based on the environment the system operates in, dependability in aerospace is a way to characterise the reliability and availability of a system's services. Inter-dependable requirements and their integrated analysis (e.g. describing safety, security and performance of a system) are becoming a necessity in systems engineering especially if we consider next generation autonomous aircraft systems. In a cyber-physical SoS scenario, a dependable system could be assessed by the fact that its services could be trusted within a specific time envelope or that the system could continue to offer its service after certain faults or attacks' occurrence (survivability). DT technology could help mature the inter-dependable analysis of system's components that have been originally developed in an independant manner.

Klaus: For B&O dependability means that product requirements are complete and the implementation is correct – and that there is evidence for this both in development and operation stages of the system life cycle. The DT methodologies define operational solutions for domain-specific V&V strategies that can provide the evidence we need both from design and in-operation phases.

Jonas: The key is to consider the customers requirements for the different cases where our pumps are deployed. One needs to ensure that the overall dependability of the full system including the human operators involved will stay at the same dependability level at least.

Do you see DTs as a potential threat to security, because of the communications involved creating a bigger "attack surface"? Any first thoughts about how you'd manage risks there?

Rob: DTs provide visibility of a building's performance, not previously seen. Typically a building's control system will be in the basement and only accessed by an engineer. However, a DT puts this information and more in the hands of everyone. This information will help improve behaviour and, therefore, identify future issues. As the amount of data increases, AI will help to predict and resolve issues.

Security is a regular concern. Clients want to know where their data is kept and how secure it is. Some have requested data to be stored on-premise, but if we don't provide this which is usually accepted. We use commercial servers and select a location together with the client. No one has raised the potential risk of access to individual systems, but as awareness increases this may be an issue in the years ahead.

Stelios: Connecting DTs directly or indirectly to the PT could potentially increase the attack surface from a cyber-security point of view. It is obvious that exposing new system information to large multi-disciplinary teams may raise concerns from data accessing and authorisation especially for critical system features. At the same time though, security solutions and primitives to secure communications and data handling have been mature enough nowadays to support secure DT application in the future. Initial thoughts could span from encrypted data exchange to models with limited interfaces and multi-factor authentication or Mandatory Access Control to manage user-access to system data, and so on.

Klaus: Because of the set-up we have made for our DT security is not really an issue. It is hidden away for everyone.

Jonas: The key will be to carry this out in a stepwise fashion so whenever you include more sensor data, the threat surface is extended because there are clearly security risks associated with that. In general, the technology trend is to move in this direction and eventually security mitigations will be established. This is also a place where we rely on academia to assist us. It is also related to the issue of trust by different stakeholders.

What do you see as the biggest challenges in engineering DTs? Thinking here about knowledge and skills as well as processes, methods and tools.

Rob: This level of technology is new in what is seen as a traditionally conservative property sector. Adoption will take time, but the technology is available today.

Stelios: By their very nature DTs bring an adoption challenge to existing engineering cycles. From an MBSE perspective, someone may argue that DTs have already been part of the design phase for decades (in the form of models). It's the DT governance for the complete life cycle of the system (during design, commissioning, operation and decommissioning) that brings most of the challenges. Some of those may be: a) dealing with the multidisciplinary character of engineering teams and the ability to collaborate across disciplines; b) the expressiveness of DT models, their composition and allowable abstractions; c) legacy subsystems and their lack of absence of accurate representations; d) DT to PT data acquisition, ownership and storage; and e) sound learning practices and predictions.

Klaus: The main challenges regarding deploying DT in our organisation are basically competencies. The DT domain defines a set of competencies which currently are not a natural part of development organisation skillsets. Beside competencies, some of desired DT enabled services might be constrained by the PTs communication links and real-time computation capabilities.

Jonas: As explained earlier I see the main challenges to get the domain experts and the data scientists to closely collaborate in order to come up with DT-enabled systems that will bring value. In addition, you need to find the best way to make business on the new digital solutions. Here there are extreme cases like the aero-engine domain where the owner of the engine is the engine producer so they simply sell flight hours. For water this cannot be realised in full since in most countries the water ownership is not something that private companies normally will be allowed to control. If fewer new products are expected in the long term future many companies will need to have a business strategy enabling services on top of the already installed products. The important thing about the DT technology is that it actually is an enabler for potentially changing the underlying business model in a revolutionary fashion.

2.10 Themes

Our experts work in diverse domains, so it is not surprising that they see different reasons to explore DT techology in their organisations. They also have different visions for the benefits that DTs can bring to their businesses and how the users of their products will experience the DT. However, their discussion raises some themes that will motivate the more detailed discussions later in this book.

2.10.1 The DT Life Cycle

Our experts see a need both for establishing DTs of existing products in order to gather data from them and to enable the creation of new digital services that add value to their products. For some organisations, the practicality of getting data from existing products may be limited by their design. Here it is important to note that it will always be necessary to find a balance between the cost of acquiring data and the value of the insight gained from it. However, in some cases it may not immediately be clear where the 'sweet spot' lies, and who the beneficiaries are. Clarifying these goals, and indeed the DT business model for an organisation, is going to be essential if a systematic approach is to be taken to the development and use of a DT.

Our experts also identify a need to ideally develop products and their DTs alongside one another. This will require new and more dynamic ways of thinking. We believe that it is important to consider the *DT-enabled system* from a systems engineering perspective and in Chap. 4 we consider how established systems engineering processes may be tailored to support DT development.

2.10.2 Model-Centric Digital Twins

Our experts' organisations have used different kinds of models for different purposes in the development of their products over some years. A motivation for the interest in DTs is that of maximising and sustaining the value of such models through each product's lifetime, not only during design but after their deployment.

Most of the models they have used in the past are essentially derived from first-principle laws. However, they all identify the need to incorporate models that are derived from data gathered in the real system "as built". There are many data analytics techniques and here the experts have already mentioned ML approaches. There is little doubt that considerable predictive power from DTs will come from having such models. Any kind of cognitive layer enabled in a DT will need to come from a combination of inductive and deductive reasoning. The companies need to use a combination of their existing know-how about their products (e.g., essentially their

IP) and combine this with models derived from data about usage of the products. Different ways of defining such models are treated in more detail in Chap. 5 and the necessary calibration of such models is treated in Chap. 6.

2.10.3 Getting Data from the Physical Twin

Our experts make it clear that different PTs may have different requirements for the data that needs to be transferred to its DT. There is clearly also a scalability issue since the different companies may play a role in a larger ecosystem. Thus, it may be necessary to contribute with a DT prediction capability to someone else who integrates such capabilities from different companies in a wider ecosystem. This means that we expect to see hybrid solutions where parts of the DT functionality are distributed (e.g., in or at the edge of a cloud) but also in an SoS setting where different systems may be managed and operated by different legal entities. The transfer of data between the PT and the associated DT is treated further in Chap. 7 and the realisation of DT (also in a distributed setting) is discussed further in Chap. 11.

2.10.4 Services Supported in Digital Twins

Our experts refer to the desirable services that DTs may be able to offer. These range from visualisation and monitoring to more advanced DT services such as supporting "what-if" exploration. There is some consensus around the need for basic dashboards and, where appropriate, 3D or immersive augmented reality visualisations. There is also a common desire for DT services supporting monitoring, whether this is supporting analysis of use profiles, identifying potential maintenance issues or performance defects. We consider this range of services in Part III of the book, including visualisation in Chap. 8, monitoring in Chap. 9 and more advanced services in Chap. 10.

2.10.5 Further Research in needed on Digital Twins

There is agreement on the need for further research on delivering security and safety in the context of increasing autonomy. in some of the envisaged application domains, DTs are interacting with environments and PTs that have inherent uncertainties. If DTs are to make decisions with reduced levels of human intervention, how do we ensure that confidence can justifiably be placed in them?

Uncertainties can come from many sources, ranging from noisy sensors, delays and errors introduced in data communications, the limited fidelity of models with predictive power enabling a kind of a cognitive layer, and naturally the performance characteristics of the computational platform used for running the DT. We explore these topics more deeply in Chap. 13, Chap. 14 and Chap. 15.

References

1. Thirty years of development and application of CFD at Boeing Commercial Airplanes, Seattle. Computers & Fluids **34**(10), 1115–1151 (2005). DOI: https://doi.org/10.1016/j.compfluid.2004.06.005
2. Ajayi, S.O., Oyebiyi, F., Alaka, H.A.: Facilitating compliance with BIM ISO 19650 naming convention through automation. Journal of Engineering, Design and Technology **21**(1), 108–129 (2023)
3. Antonino, P.R.G., Oliveira, M.M., Sampaio, A.C.A., Kristensen, K.E., Bryans, J.W.: Leadership election: an industrial sos application of compositional deadlock verification. In: J.M. Badger, K.Y. Rozier (eds.) NASA Formal Methods, pp. 31–45. Springer International Publishing, Cham (2014)
4. European-Commission, for-Mobility-and Transport, D.G., for-Research-and Innovation, D.G.: Flightpath 2050 – Europe's vision for aviation – Maintaining global leadership and serving society's needs. Publications Office (2011). DOI: doi/10.2777/50266

Chapter 3
Foundational Concepts for Digital Twins of Cyber-Physical Systems

Cláudio Gomes, Bentley James Oakes, John Fitzgerald, Peter Gorm Larsen

Abstract This chapter introduces concepts that form the foundations for systematically considering the engineering of digital twins for cyber-physical systems. We introduce an example based on an incubator for growing tempeh, which serves as a running study for the remainder of this book. Based on this example, we discuss basic system concepts and core features of the models that are at the heart of any digital twin. We then consider the structure of a digital twin as an assembly of services, built upwards from models and data to the delivery of advanced analytics, visualisation and "what-if" analysis. This involves the management of these services, their communication protocols, data storage, model management, and other crucial aspects inherent to distributed software systems implemented within a service-oriented architecture. The rest of the book gives details about how to realise the different services and their management.

3.1 Introduction

In this chapter we introduce technical concepts that underpin the methods, tools and techniques for *Digital Twin* (DT) Engineering described throughout the book. These concepts help explain how a DT adds value to the *Physical Twin* (PT), by giving stakeholders access to services that deliver the benefits that we explored in Chap. 2. This includes the ability to visualise the PT, explore "what-if" scenarios, preempt, detect and compensate for faults. With this view in mind, two features of DTs motivate the choice of concepts covered here. First, a DT has at its heart one or more *models* of the thing that it twins – its PT. Second, stakeholders' needs are not static and neither is the environment in which the DT and PT operate. Consequently, the DT must evolve alongside the PT and environment, continuing to add value in a dependable way [7].

J. Fitzgerald et al. (eds.), *The Engineering of Digital Twins*,
https://doi.org/10.1007/978-3-031-66719-0_3

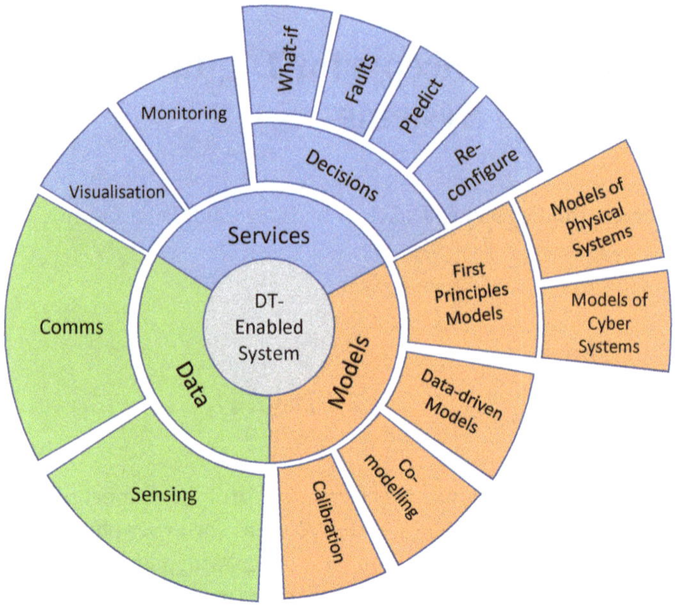

Fig. 3.1: Overview of DT-enabled system concepts covered in this book. Each colour represents one part of the book.

Through the book, we will use a running example to provide intuition and illustrate concepts. The example is based on an *incubator* which is used to cultivate tempeh, a high-protein food. We introduce the incubator in Sect. 3.2 along with our fictional company Twinpeh which aims to use a DT concept to develop its business.

The rest of the chapter gives a first taste of fundamental concepts that we explore in greater depth in the remainder of the book. Fig. 3.1 is a graphical overview of these ideas. Section 3.3 introduces essential systems concepts. We build on this to introduce models in Sect. 3.4 and DTs in Sect. 3.5.

By this point, the reader may well be wondering what the inside of a DT looks like. Finally, in Sect. 3.6 we present the interior of a DT as a structured assembly of building blocks that together deliver services that meet technical and business needs.

3.2 Running Example: the Tempeh Incubator System

Our running example is based on a DT that adds value to an *incubator* for producing tempeh, a traditional food product made from fermented soybeans. This example is relatively easy to understand but has some of the features that we observe in larger industrial examples such as industrial dry ovens and organ preservation boxes [2].

3.2.1 Tempeh and how to Make it

Known for its nutty flavour, firm texture, and high protein content, tempeh is a popular fermented foodstuff that has been consumed for centuries in Indonesia and other parts of Southeast Asia [10]. Traditionally made from soybeans, it can be created using other ingredients such as black beans, chickpeas, lentils, or a combination of vegetables. These variations offer different flavors, textures, and nutritional profiles, providing options for individuals with soy allergies or those seeking diverse plant-based protein sources.

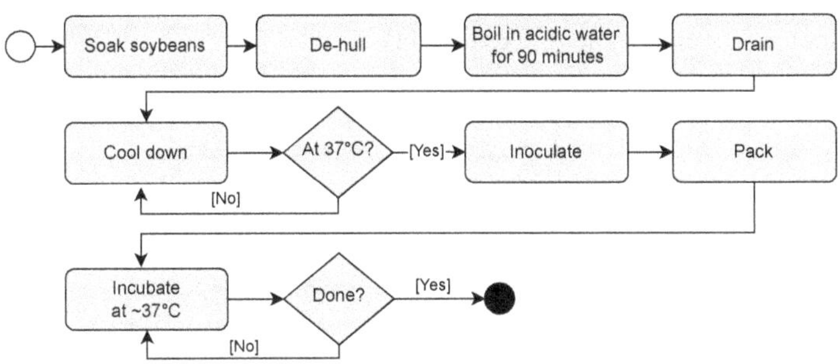

Fig. 3.2: Overview of the tempeh making process. The incubator is responsible for ensuring that the temperature is kept at 37.5°C during the incubation process.

The process of making tempeh, illustrated in Fig. 3.2, involves incubating or fermenting whole soybeans or other legumes using a fungus called *Rhizopus oligosporus*. The soybeans are first soaked, cooked, and then inoculated with the fungal culture. The mixture is then allowed to incubate, typically for around 24 to 48 hours depending on the desired flavour and texture.

During incubation, the fungus binds the soybeans together into a compact cake-like structure, as illustrated in Fig. 3.3. This gives tempeh its characteristic firm texture and a slightly nutty taste. Initially the temperature among the soybeans grows roughly linearly as the soybeans absorb the surrounding heat. Then, after about 20 hours, there is an explosion in growth of the fungus as it consumes part of the soybeans and multiplies. At this stage, the fungus starts producing rather than absorbing heat, becoming warmer than its surroundings by as much as 7°C, as shown in Fig. 3.4. The tempeh cake is at its highest density and ready for harvesting and cooking when it is past the temperature peak.

If the raw materials are changed, adjustments have to be made to the fermentation process and conditions to accommodate the specific characteristics of each ingredient. In any case, after successful incubation, Tempeh becomes a versatile ingredient that can be sliced, diced, marinated, and cooked in various ways. It can for example be used in sandwiches, stir-fries, salads and soups. Its ability to absorb flavours and

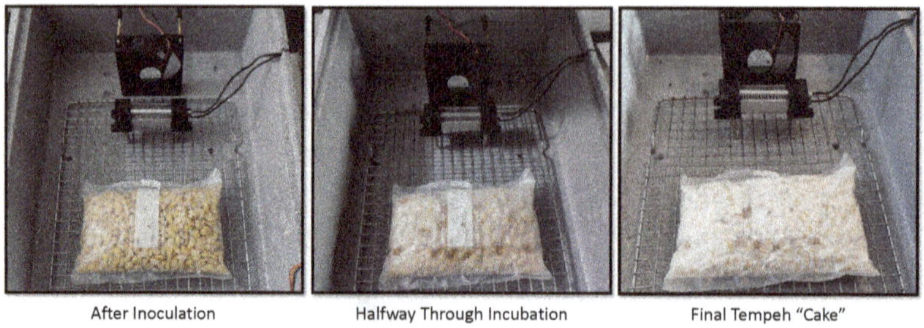

Fig. 3.3: Snapshots of the fermentation process: initial soaked soy beans (left), beginnings of the fermentation at roughly halfway through the process (middle), and final tempeh at the end of the process (right).

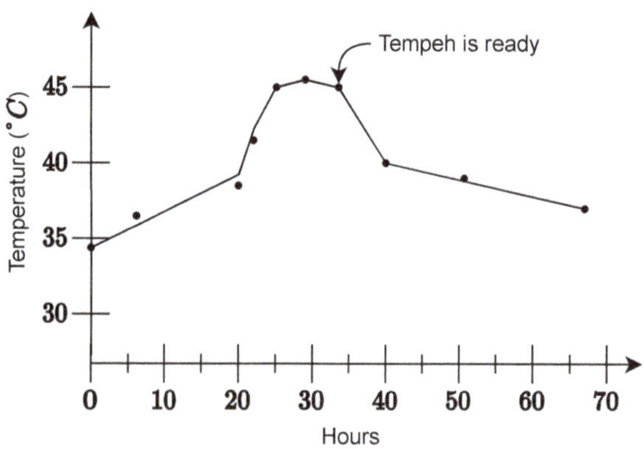

Fig. 3.4: Changes occurring during tempeh fermentation. Adapted from [10]. Note that the variability in tempeh fermentation means that this figure is merely illustrative and does not represent the exact temperature and timings involved in the fermentation process.

retain its shape during cooking makes it a popular choice as well as a good source of protein, dietary fibre, vitamins and minerals.

3.2.2 Tempeh Production

Our example is based on the idea of producing tempeh in an incubator which is a device used to grow and maintain microbiological cultures or cell cultures. It is a simple system, the goal of which is to keep the temperature inside an enclosed chamber as close to as possible to 37.5°C, to promote the fermentation.

It can be a challenge to ensure consistent tempeh quality as environmental conditions vary in different locations and climates. We imagine a company called Twinpeh which has been developing and marketing tempeh incubators. Each incubator is an insulated box containing temperature sensors, a heatbed and a fan. A temperature sensor outside the box assesses external temperature. The incubator's controller works in a simple "bang-bang" style. It receives data from sensors, using this to switch a heater and a fan on or off in order to keep the temperature at 37.5°C. A systematic diagram is shown in Fig. 3.5. Full details of one implementation of an incubator system are in a technical report [4].

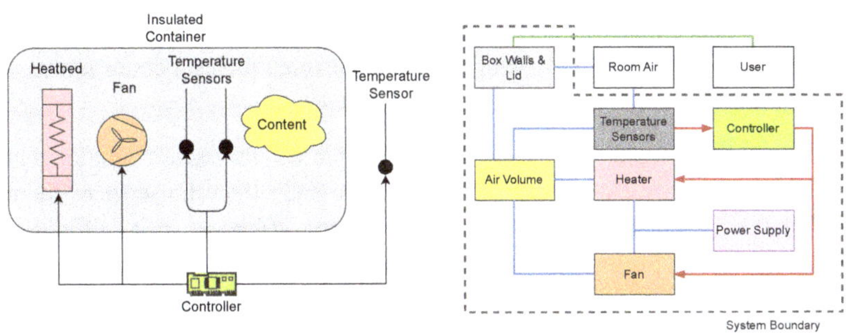

Fig. 3.5: Schematic overview of the incubator. The left is the incubator system. The right is a block diagram of the incubator. The blue connections represent energy interactions, the red connections represent digital information flow and the green connection represents the user's action by opening and closing the lid. The dashed box represents the system boundary.

3.2.3 DT-enhanced Tempeh Incubation

In our example scenario, Twinpeh's incubators are selling at an acceptable rate, but the company has noticed complaints from customers because the quality of the tempeh produced is affected by the user opening the lid of the incubator frequently to check on the progress of fermentation, as well as by the natural variability between fungi.

Given worldwide demand for alternative protein sources to meat, Twinpeh sees global potential for its incubators. They have recently heard about DT technology, so they wonder if it is worthwhile producing a DT for the next generation of their incubator product in order to achieve more dependable incubation. Part of the business need here is to gain experience with DT technology, its benefits and challenges. Twinpeh's approach is to develop a DT solution as a proof of concept, based on the current physical incubator and controller. There is not a strong appetite for making major modifications to the basic incubator, so new and expensive sensors are out of the question, for example. To an extent, this corresponds to the situation described in Sect. 1.4.1.

Twinpeh hopes to develop a DT offering that can provide the following features to users:

1. Present historical data about the incubation process. Twinpeh's rationale is that they will ultimately be able to help the user spot anomalies and adjust operations and raw materials in order to improve quality.
2. Energy saving by adjusting the controller settings when the user has the lid open.
3. Signalling that the tempeh is ready for cooking, reducing the likelihood of users opening the incubator and disturbing the controlled internal conditions.
4. Signalling that the incubator needs maintenance.

The idea would ultimately be for Twinpeh to ship a DT for each incubator. The DT can run on a device such as a user's smart phone and offers the services identified above. The incubator can still function without the DT because it fulfils the core requirement to keep the temperature around 37.5°C, but the DT adds value. The user is free to buy the incubator and not pay for the DT services. In that case, the user has to monitor the tempeh production unassisted, and the incubator is little different from an oven. From Twinpeh's business perspective, the value added comes from paying for the DT and leveraging the relevant models and data to deliver services.

Twinpeh also sees longer-term potential in developing a central DT that twins the entire fleet of DT-enabled incubators that it has online, giving the opportunity for users to permit sharing of data from their local DT-enabled incubator up to this central fleet-level DT. There is significant value to Twinpeh here in being able to get accurate information about real modes of use, and being able to push out updates to control parameters or changes to maintenance regimes based on learned experience across the whole fleet. However, that is for the future and we discuss it in more depth in Sect. 11.11. For now, Twinpeh's ambitions are to get a DT for each individual incubator.

3.3 Basic System Concepts

The *systems* considered in this book are often engineered by humans with the intention of delivering some desired behaviours in specific environments. Following the ISO/IEC/IEEE 15288 standard [5], we regard a *system* as "a combination of interacting elements organised to achieve one or more stated purposes". The system elements can include physical devices, natural phenomena and creatures including people, as well as less tangible things such as services or information. A system interacts with entities that are not part of the system itself; these elements form the *environment* of the system of interest. The environment will often contain entities that can interact with the system, such as users, so it can be viewed as the collection of potential external interactions. This leads to the concept of the *system boundary*: the dividing line between the elements that are within in the system of interest and the entities that are in the environment. This boundary must be carefully chosen through *systems engineering*, as discussed by Zeigler in [11].

A system has a *purpose*. Explicitly stating this purpose is often key to achieving a clear and shared understanding of the decisions about where to draw the system boundary. In an engineering project, for example, we might choose to draw the boundary around elements that the engineer has the power to specify, select or modify, treating others as essentially part of the environment.

A system's *behaviour* is the functionality that it offers to its environment. This is expressed in terms of the interactions that can take place across the system boundary and represent the way in which the system is "seen" by its surroundings. Functionality in turn results from the combined behaviours of the system elements and the interactions between them.

A system's *properties* are the characteristics of the system's behaviour. Examples include the property of only exhibiting behaviours that respect safety conditions, or performance levels. It should be possible to devise an experiment to determine whether the system satisfies a property or not. In the context of engineered systems, we are often interested in whether a system's properties include the requirements that were specified for it.

Incubator System

The components considered as part of the incubator *system* include: the incubator box itself, the object(s) placed inside (such as the tempeh), the heating element, and the controller. As mentioned above, the *purpose* of the incubator system is to maintain the internal temperature within the insulating box. This temperature value should be kept close to a given value regardless of the ambient temperature outside the box. The *environment* of the incubator includes the surroundings, including: the air around the incubator, the table on which it is placed, and so on. As the surrounding room goes through different temperature profiles, the incubator (and the interior air) is affected. Since the incubator's container is not perfectly insulated, more heat will be lost when the room is colder and the incubator's heating element will have to use more energy to maintain the target temperature. The *system boundary* is represented in Fig. 3.5.

Relevant *example properties* concern temperature values and timing for heating and cooling. For example: a) the incubator system must take at most 10 seconds to warm up the air from 20°C to 30°C, b) the interior temperature must remain between 15°C and 60°C at all times, and c) the incubator controller must react to variations within at most 3 seconds.

3.4 Models & Data

Incubator Physical Prototype as a Model

Suppose that you are given a set of properties that the incubator system must satisfy (for example, it must keep the temperature between 35 and 42 degrees, after some warm up period), and you are tasked with building such a system. While doing so, you build a prototype out of parts you have lying around the workshop, and you conduct experiments with the prototype in an attempt to check if it indeed satisfies those properties. This prototype acts as a kind of representative of the incubator system and is used for the purposes of answering questions about it. In a sense, it serves as a model of the real incubator.

Following [6, 9], a model should be based on the system it purports to represent, should reflect only a relevant subset of its properties, and may be used in place of the original for some defined purpose and within a particular environment. A model is created for the *purpose* of performing *experiments* experimenting with it, with the aim of demonstrating that the exhibited behaviour satisfies some properties of interest. Thus, the physical prototype of the incubator described above is a model.

Models need not be physical artefacts like the incubator prototype mentioned above. Indeed, in a digital twin, models are virtual, (digital) representations of (aspects of) the real-world system of interest. When we use the term "model" in this book, we are almost always referring to such virtual models, and will be clear when we are referring specifically to physical models. We introduce several forms of virtual model in Chap. 5, and also touch on the qualities and limits of models when we consider current and future research in Chap. 15.

Incubator Temperature Virtual Model

Suppose we are asked to build a virtual model for calculating the average temperature evolution (ignoring objects within the box) in the incubator. The system of interest is comprised of controller software running on a Raspberry Pi, and a plant consisting of the heating element, the air inside the box, and the box walls. The model could take the form of differential equations that describe how temperature evolves in the system [4]:

$$C_h \frac{dT_h}{dt} = H_h \cdot P_h - G_{hb} \cdot (T_h - T_b) \tag{3.1a}$$

$$C_b \frac{dT_b}{dt} = G_{hb} \cdot (T_h - T_b) - G_{br} \cdot (T_b - T_r) \tag{3.1b}$$

where T_h is the temperature of the heater (measured in °C), T_b is the temperature of the air inside the box (°C), T_r is the room temperature (°C); C_h is the heat capacity of the heater $(\mathrm{J}/(\mathrm{Kg}\cdot{}^{\circ}\mathrm{C}))$ while C_b is the heat capacity of the box including contained air $(\mathrm{J}/(\mathrm{Kg}\cdot{}^{\circ}\mathrm{C}))$. The electric power supplied to the heatbed is given by voltage times current $V_h \cdot I_h$ (W), while H_h is a Boolean variable utilised to switch electric power on and off. Finally, G_{br} is a coefficient that models heat transfer between the box and the room $(\mathrm{W}/{}^{\circ}\mathrm{C})$ while G_{hb} is a coefficient that models heat transfer between the heater and the box $(\mathrm{W}/{}^{\circ}\mathrm{C})$. Chapter 5 gives more detail.

The models discussed in this book, unless stated otherwise, describe dynamic behaviour. This means that a simulation — an experiment conducted with a model — produces a set of traces that describe the behaviour of the model over time.

As summarised in Fig. 3.6, a model contains:

inputs: traces that defined over time to enable a simulation,
outputs: traces that are computed when a simulation is conducted,
parameters: variables that must be defined before simulation is possible, but that are kept constant during a simulation, and
states: variables that change internally in the model during a simulation but that may not be directly observable through its outputs.

Finally, we use the term *formalism* to refer to the language and rules that dictate how the traces emerge from the model. In the incubator virtual model above: H_h and T_r are inputs; P_h, C_h, G_{hb}, C_b, G_{hb}, and G_{br} are parameters; T_h and T_b are states and outputs. In this example, the formalism is ordinary differential equations, or differential equations for short.

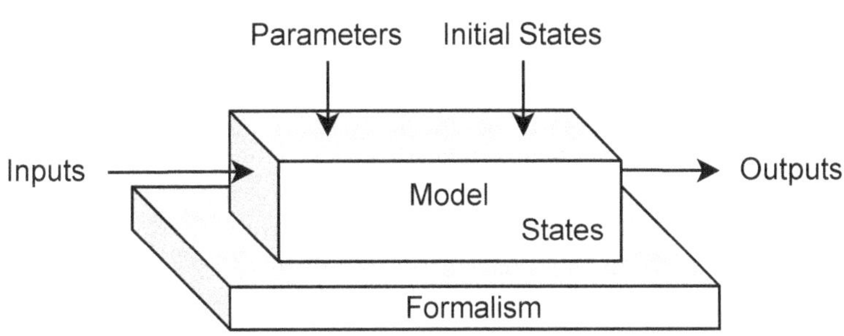

Fig. 3.6: Main concepts comprising a model.

The model formalism dictates the simulation technique used. For example, for the differential equations model in Sect. 3.4, we would use numerical simulation to produce a behaviour as shown in Fig. 3.7. The satisfaction of model properties can then be evaluated based on the simulation results.

If we wish the behaviour of the model to align with the behaviour of the PT, we need to *calibrate* the model. The calibration technique consists essentially of

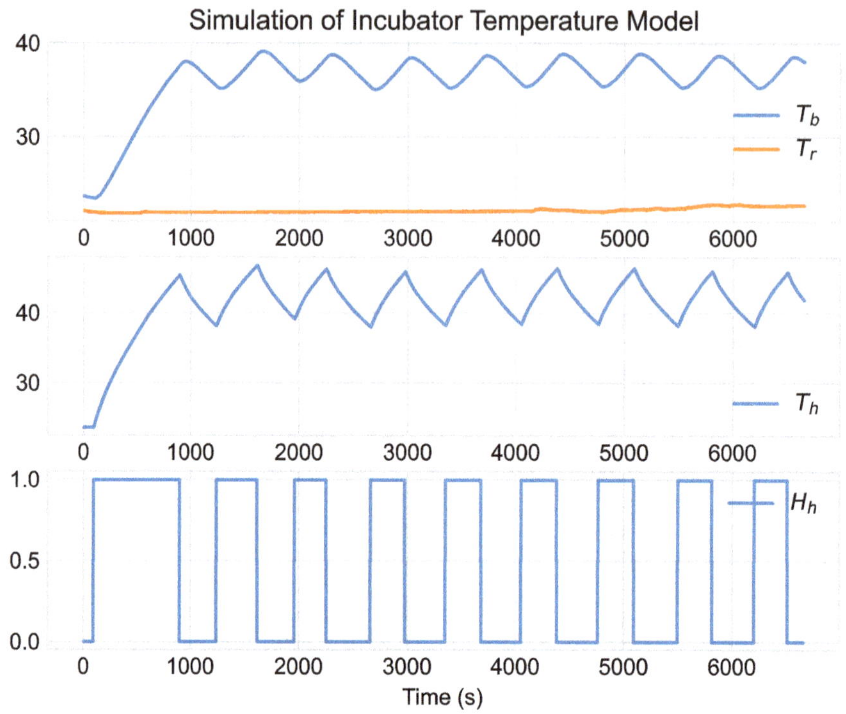

Fig. 3.7: Example behaviour of the incubator temperature model introduced in Eq. (3.1).

varying the parameters of the model as different simulations are conducted until one of the simulation results aligns well enough with the data observed from a similar physical experiment. Chapter 6 discusses calibration techniques for models specified in different formalisms, and Chap. 7 introduces some of the challenges involved in gathering the required data from those experiments, especially when it is critical to minimize delays in the transmission and processing of such data.

For our incubator model example, Fig. 3.8 shows two intermediate calibration attempts and a final successful one. The third attempt shows very good agreement with the real temperature evolution data shown in red. Note that while the physical experiment did not contain any measurements for the temperature of the heater T_h, by aligning the temperature of the model T_b with the temperature that was collected (plotted in red), we can still trust the results for the temperature of the heater because the two quantities are related by Sect. 3.4. In other words, if the model predictions for T_h were very wrong in the third attempt, then we would not observe a good agreement between T_b and the data collected.

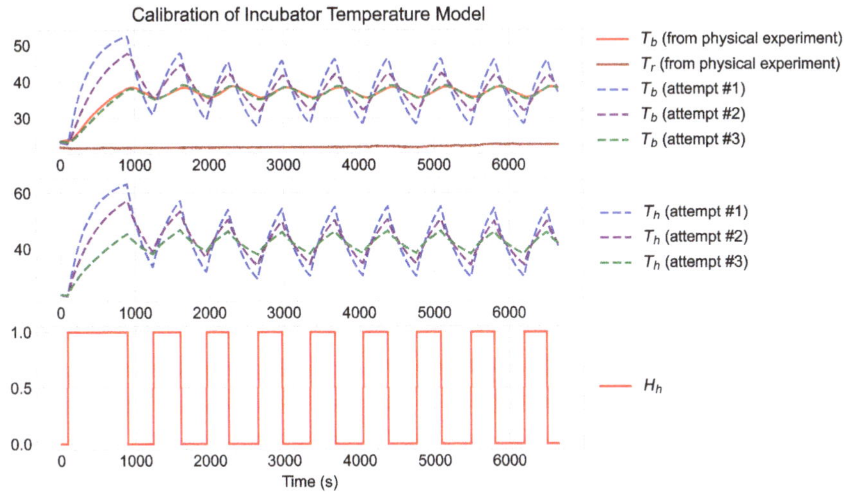

Fig. 3.8: Calibration results of Eq. (3.1). The dashed lines indicate the multiple attempts with different parameters. The third attempt represents a successful choice of parameter values. The solid lines represent the data from the physical experiment.

3.5 Digital Twin Services

Models are the basis of the analytic, predictive, monitoring or control services for which the DT is built. We discuss these services in Chap. 8 to 10, but here we introduce the main categories of DT services, with examples that have been developed for the incubator DT.

In this book, and as illustrated in Fig. 3.1, we think of DT services in three main categories: visualisation, monitoring and decision support.

Visualisation services, discussed further in Chap. 8, allow the user to consult historical data about the PT. For the incubator DT, this would likely be implemented as a web server where the user has access to multiple dashboards plotting the data, as illustrated in Fig. 3.9. These might give access to the history of temperature measured by each of the sensors as well as the control decisions made and the parameters of the controller.

Monitoring services, discussed in Chap. 9, typically observe and evaluate the behaviours of a PT as it operates [1], helping users draw insights from those evaluations. A key part of monitoring is in the definition of desirable properties, that we wish to check whether the PT's behaviour satisfies. Typically, a monitor service can be synthesised from the properties, and then deployed as part of a Digital Shadow (DS) (as introduced in Sect. 1.2.2). Monitoring can also entail anomaly detection as well as the quantification of uncertainty in the system. For the incubator example, one might develop a state estimator that helps correlate the predictions of the model

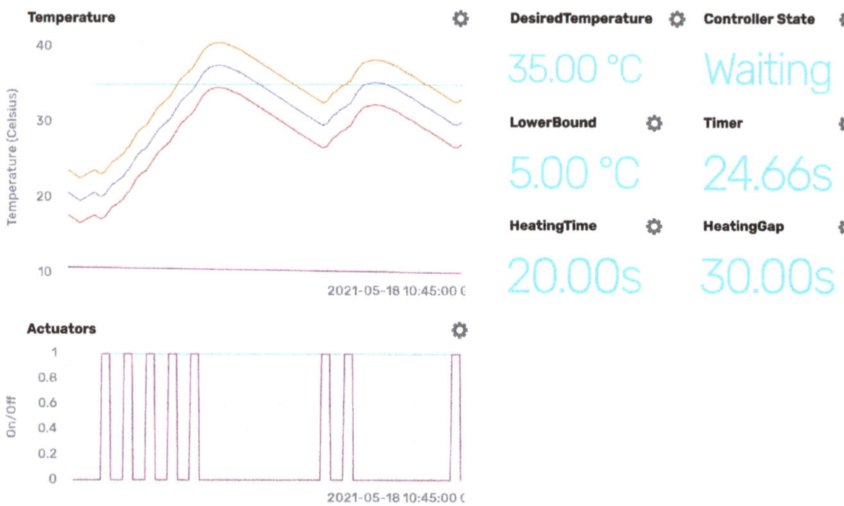

Fig. 3.9: Visualisation interface of incubator PT. The horizontal axis represents time. The plots below shows the state the heater (purple line) and the fan (blue line). The plots above show the sensor temperatures and the desired temperature. The right section shows the parameters and states of the controller.

in Eq. (3.1) with temperature measurements (detailed in Chap. 7), and provides a measure of the uncertainty that can be used as a basis for detecting that the user has opened the lid using anomaly detection techniques. this is detailed in Chap. 9. The outputs of the state estimator and anomaly detector, illustrated in Fig. 3.10, can be recorded and visualised by a visualisation service.

Finally, decision services, detailed in Chap. 10, are a broad category encompassing services that support decision making. Some of these can take automated actions to change the PT. We consider them in four categories:

What-if Analysis: services that can conduct simulations in varied scenarios using historical data or predictions in order to answer particular questions. In our incubator example, what-if simulation might be used to help determine the optimal controller heating time ahead of a physical intervention on a running incubator to swap its power supply. The intervention requires that the power is cut off for a while, so we need to determine the largest available time period for the intervention, if we allow a slight overheat of the incubator. If we resort to simulation to understand how much time is available, the results are shown in Fig. 3.11.

Fault Diagnosis: services whose goal is to classify system fault modes as occurring or not. An example for the incubator is given in Sect. 10.3.

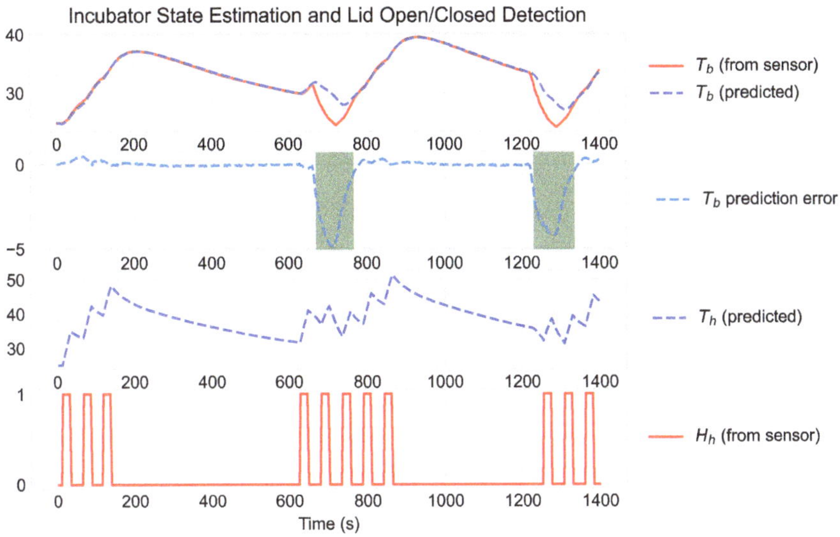

Fig. 3.10: Example showing the visualisation of the state estimation and anomaly detection services. The output of the state estimation service is a prediction for both the T_b and T_h variables, from the model in Eq. (3.1). The anomaly detection uses the difference between T_b from the sensor and T_b estimated, to detect when the lid has been opened. Its output is illustrated in green.

Predictive Maintenance: services that identify long term trends to help predict the occurrence of faults and thus prevent them with maintenance actions. An example is given in Sect. 10.4.

Re-configuration: Services whose goal is to optimise the PT when operating in a new environment. An example is provided in Sect. 10.5.

In this section, we have made a case for seeing the DT as a set of services. Each service has a clear responsibility and a clear set of inputs as well as outputs which may come from or go to other services. We have not so far specified how these services communicate (communication, data transmission, sensing, and storage are discussed in Chap. 7) or how to organise these services logically, nor have we discussed the software architecture that supports the implementation of these services. If one plots these flows between the services that make up the DT, the result is a *constellation* of data and control connections [8]. This view, however, is a purely functional and lacks other aspects such as how to handle security and authentication. In Sect. 3.6 we present one logical organisation of the DT software.

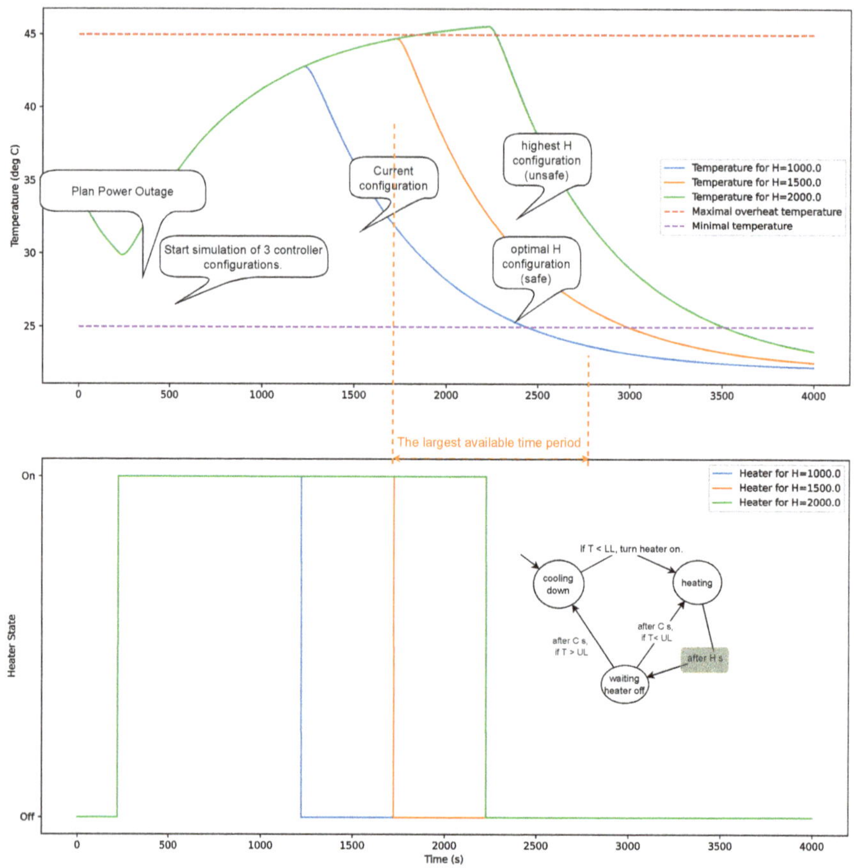

Fig. 3.11: Results of what-if simulation for controller configuration. The green line gives us the largest available time period but it overheats excessively, violating the slight overheat allowance. The blue line represents the current controller configuration. The orange line represents the controller with configuration $H = 1500s$ and yields the largest available time period. The controller model (a state machine) is on the bottom plot, for convenience, with the parameter H highlighted in green. Adapted from [3].

3.6 Digital Twin Assets and Management

In Chap. 1, we introduced at a very basic level the elements of a DT and the wider DT-enabled system, including physical entities, models, data, the user-facing services and their supporting enablers (Fig. 1.3). In this section we will look a little more deeply into these aspects as they form the skeleton on which we will build our discussion of the components of DTs in the remainder of the book. We also consider how, in our fictional case study, Twinpeh might have approached these elements. Here and subsequently, we use the term *asset* quite loosely to refer to parts of the DT that may have use beyond one particular DT-enabled system. Many assets will be models but some data and software implementing the enabling services will also be useful in many DTs.

3.6.1 Physical Entities

Our principle of additionality is that the DT adds value to the PT without compromising too much the value that the PT already provides to its users. We say "too much" here because some interference is practically inevitable. The PT will need to make data available to the DT and there may be a cost to this, at least in terms of instrumenting the PT. For our incubator example, the PT presented in Fig. 3.5, can be seen as a service offering data to the proposed DT.

If the DT is to have some capacity to update the PT, data must flow DT-to-PT and the PT may be designed or modified to enable this. For example, we may have to update the controller to receive messages from the DT and reconfigure the PT's parameters. Some of people involved in the design and manufacturing in the original PT will therefore have a significant role to play with a DT engineering team in determining how to make the PT-DT connection work for the benefit of the users of the PT as well as the users of the DT.

3.6.2 Data and Data Management

Data and data management components would address the aspects identified in Fig. 3.12. Connections to the PT could be over a variety of communications infrastructures (wired or wireless, using a range of protocols), and so adaptors and/or brokers are used to get PT data to the DT in a form that is ready for processing by enablers and top-level DT services. The data may need further processing (e.g., to remove noise) and fusing with other data. The DT would manage the storage of data, potentially in a time series where trends are to be observed or analysed. The main roles during the engineering of the data management aspects include those with an interest in sensing in the PT, the overall DT architect and requirements authority. Those involved in data engineering will benefit from functionality around data collection, storage and visualisation (note that this is lower-level than the application-specific visualisation demanded at the DT service level). It is worth noting the need to calibration in order to maintain external consistency here. These topics are treated in further detail in Chap. 7 and Chap. 11.

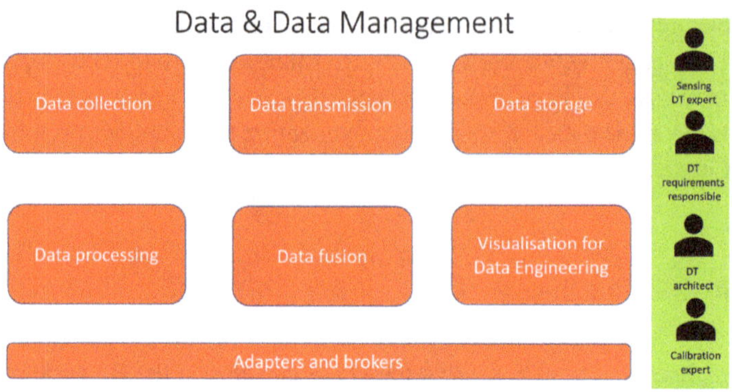

Fig. 3.12: Features of Data and Data Management.

In our incubator example, Twinpeh has prototyped a DT-enabled system in which communication is enabled by a RabbitMQ server, and the timeseries data is stored in an InfluxDB database. RabbitMQ is an open-source message broker software that facilitates communication between applications or components in a distributed system. It implements the standard Advanced Message Queuing Protocol (AMQP) for message-oriented middleware. RabbitMQ is widely used in distributed systems, microservices architectures, and scenarios where decoupling and asynchronous communication between components are essential. It is compatible with multiple programming languages and has client libraries for popular languages like Java, Python, Ruby, and .NET. InfluxDB is an open-source, high-performance, distributed, and scalable time-series database. It is designed to efficiently handle large volumes of timestamped data, making it well-suited for applications that require the storage and retrieval of time-series data, such as monitoring, analytics, and IoT applications. InfluxDB is commonly used in various domains, including DevOps and system monitoring, IoT applications, financial services, and any use case that involves tracking and analysing time-series data.

3.6.3 Models and Model Management

Models are key assets in a DT, providing a basis for analysis through a range of techniques such as simulation which allows DT users to envisage how the PT may behave under different assumptions. A DT may contain many models created for different purposes, tuned for different analyses and operating with different levels of fidelity [7]. Thus, it will be necessary for DT users, via the DT services, to select and configure models for different purposes.

The roles involved with models to be used inside a DT operates at this level, selecting, producing, analysing and maintaining models included in DTs, and so there is a need for platforms that assist with these tasks. All models need to be stored in a model repository under version control and offering traceability between the models and associated documents as well as the tools used to produce, analyse and simulate with them. There is also a need to check the consistency of the models produced, and in connection with this often knowledge graphs and ontologies are used. These topics are treated in further detail in Chap. 5 and Chap. 11 and an overview of the central concepts is presented in Fig. 3.13.

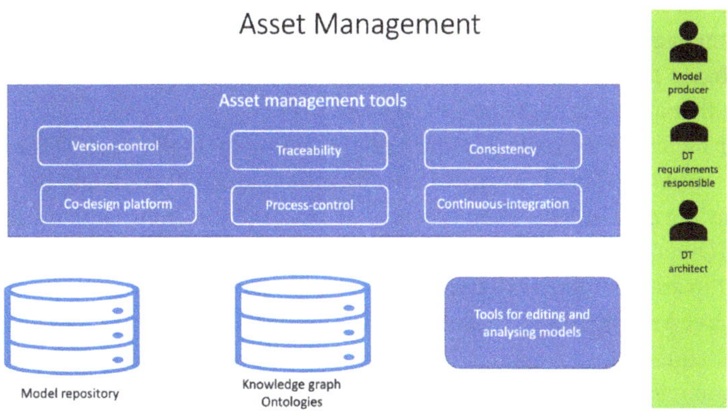

Fig. 3.13: Features of the asset management.

In the incubator example, multiple models are used by the DT services. The example model already introduced in Eq. (3.1) has been implemented in the Python programming language (which offers libraries of code for advanced numerical computation), and is deployed as a simulator service. In theory every model can be deployed as a simulator service so that other services can call upon it to conduct simulations given input, parameters, and initial states. For performance reasons, the models in the incubator are also available as library components that can be incorporated directly into other services written in Python, to avoid excessive communication overhead. An example is the what-if simulation service introduced in Fig. 3.10 in which multiple simulations need to be conducted rapidly.

3.6.4 DT Services

DT services give business value to its users. As we outlined in Sect. 3.5, these will include both the services that interact with the main users of the DT, and the *enablers* which are computational components that take models and data and support the services of the DT. Examples of enablers might include services for simulation or anomaly detection. Simulation services enable more dynamic views, potentially running models using data from the PT, permitting monitoring and anomaly detection (discussed in connection with Fig. 3.10). At the interface-facing level, one might expect DT users to begin with visualisation services that allow the presentation of selected data in forms ranging from dashboards to Virtual or Augmented Reality. We discuss such visualisation services further in Chap. 8. However, if we are to rely on simulators' predictions, some calibration services may be required (Chap. 6). At a more advanced level, we consider services that support decision-making (Chap. 10) and potentially autonomous decision-making affecting the PT (Chap. 14).

These core roles facing or enabling services that deliver the DT's functionality require management, allowing their integration, configuration and orchestration. Technologies for doing this are considered further in Chap. 11, in the context of a platform that allows users to build and share DTs.

In our incubator example, Twinpeh might choose to deploy each DT service as a *Docker container*. This is a way of encapsulating an application and its dependencies, ensuring consistency when executed across different environments. Indeed, the RabbitMQ and InfluxDB servers may also be deployed as docker containers. Service startup and shutdown is handled using the Docker Compose utility which uses a YAML (Yet Another Markup Language) file to configure the services, networks, and volumes required for an application, allowing developers to define complex, multi-container environments. They might also use Kubernetes, an open-source container orchestration platform designed to automate the deployment, scaling, and management of containerised applications, for the automatic handling of service upgrades.

References

1. Bartocci, E., Deshmukh, J., Donzé, A., Fainekos, G., Maler, O., Ničković, D., Sankaranarayanan, S.: Specification-Based Monitoring of Cyber-Physical Systems: A Survey on Theory, Tools and Applications. In: E. Bartocci, Y. Falcone (eds.) Lectures on Runtime Verification, *Lecture Notes in Computer Science*, vol. 10457, pp. 135–175. Springer International Publishing, Cham (2018). DOI: 10.1007/978-3-319-75632-5_5
2. Buhagiar, A.J., Freitas, L., III, W.E.S., Larsen, P.G.: Digital twins for organ preservation devices. In: T. Margaria, B. Steffen (eds.) ISoLA 2020, *Lecture Notes in Computer Science*, vol. 13704, pp. 22–36. Springer (2022). DOI: https://doi.org/10.1007/978-3-031-19762-8_3
3. Feng, H., Gomes, C., Thule, C., Lausdahl, K., Iosifidis, A., Larsen, P.G.: Introduction to Digital Twin Engineering. In: Proceedings of the 2021 ANNUAL MODELING AND SIMULATION CONFERENCE. IEEE, Virtual Conference (2021). DOI: 10.23919/ANNSIM52504.2021.9552135
4. Feng, H., Gomes, C., Thule, C., Lausdahl, K., Sandberg, M., Larsen, P.G.: The Incubator Case Study for Digital Twin Engineering (2021). URL http://arxiv.org/abs/2102.10390
5. IEEE: International Standard ISO/IEC/IEEE 15288:2023(en), Systems and software engineering — System life cycle processes. ISO/IEC and IEEE Computer Society (2023)
6. Kühne, T.: What is a Model? In: Language Engineering for Model-Driven Software Development, vol. 04101. Internationales Begegnungs- und Forschungszentrum für Informatik (IBFI). DOI: 10.4230/DagSemProc.04101.15

7. Oakes, B.J., Gomes, C., Larsen, P.G., Denil, J., Deantoni, J., Cambeiro, J., Fitzgerald, J.: Examining model qualities and their impact on digital twins. In: 2023 Annual Modeling and Simulation Conference (ANNSIM), pp. 220–232. IEEE (2023)
8. Oakes, B.J., Parsai, A., Meyers, B., David, I., Van Mierlo, S., Demeyer, S., Denil, J., De Meulenaere, P., Vangheluwe, H.: A Digital Twin Description Framework and its Mapping to Asset Administration Shell. In: Model-Driven Engineering and Software Development, vol. 1708, pp. 1–24. Springer Nature Switzerland. DOI: 10.1007/978-3-031-38821-7_1
9. Stachowiak, H.: Allgemeine Modelltheorie. Springer-Verlag
10. Steinkraus, K.H., Hwa, Y.B., Van Buren, J.P., Provvidenti, M.I., Hand, D.B.: Studies on tempeh. An Indonesian fermented soybean food. **25**, 777–788. URL https://www.cabdirect.org/cabdirect/abstract/19611403557
11. Zeigler, B.P.: Theory of Modelling and Simulation. New York, Wiley (1976)

Chapter 4
Digital Twin Engineering Processes

John Fitzgerald, Ken Pierce and Klaus Kristensen

Abstract We outline the technical processes defined in the systems engineering standard ISO/IEC/IEEE 15288, considering how each of them plays out when engineering a system enabled by a Digital Twin (DT). We consider how these processes, and the teams working on them, might be tailored to the challenges of engineering DTs, including the roles and competencies needed. A DT-enabled tempeh incubator system is used as a running example.

4.1 Introduction

If our industry experts highlighted the potential of DT technology in Chap. 2, our discussion of technical foundations in Chap. 3 hinted at the challenges in realising their vision. How can we create and maintain a DT that links models expressed in diverse formalisms with data from different sources, offering services that a range of stakeholders can value? In this chapter, we begin to bring order to this challenge by adopting a Systems Engineering (SE) approach.

In Sect. 4.2 we introduce SE as an approach that applies scientific, technological and management methods to the challenges of developing, deploying, maintaining and disposing of engineered systems. It is transdisciplinary because the needs and requirements that we have for such systems are rarely expressible in one formalism alone or capable of being met by just one type of engineer.

Thinking about DT engineering in these terms allows us to organise a DT project into *processes* addressing parts of the challenge. Some processes are primarily technical, and we introduce these in Sections 4.3 to 4.6 as a prelude to considering them in more depth in Parts II to IV of the book. We discuss technical processes in quite general terms, but in Sect. 4.7 we emphasise the need to tailor them to the shape of each DT project.

J. Fitzgerald et al. (eds.), *The Engineering of Digital Twins*,
https://doi.org/10.1007/978-3-031-66719-0_4

The technical processes that we describe are necessary to delivering dependable DTs for Cyber-Physical Systems (CPSs), but they are not sufficient. In Sect. 4.8 we briefly discuss the business, technical management and enabling processes that are critical to success. Equally importantly, we consider the competencies and roles of engineers carrying this out.

> If you have a background in SE, you may well be familiar with each of the engineering processes outlined in this chapter, based on ISO/IEC/IEEE 15288 [7]. However, we also consider how each process is impacted by the inclusion of a DT in the system of interest, so we still recommend reading through the chapter, treating each section as a provocation to thought.

4.2 DT Engineering as Systems Engineering

DT engineering is a systems discipline because, as we argued in Chap. 1, a DT connects and integrates models, data, communications and services that work collectively to add value to a PT without interfering excessively in the PT's business. The DT, PT and the communications that link them, together form a DT-enabled system. Our interviewees in Chap. 2 identified the value in developing a concept of *DT Engineering* which gathers methods and tools for creating and maintaining such DT-enabled systems. In Chap. 3 we introduced foundation concepts for DTs themselves and elaborated one way of viewing the constituent parts of a DT as services built on models and data.

The DT engineer's goal is to apply the methods and tools of SE to create and maintain a DT that adds value to the PT but does not compromise its operation. The technical approaches used to develop, deploy, maintain and evolve DTs will vary from project to project, influenced by the needs of their domains or stakeholders. We focus at this stage on the key processes that will frequently form a part of DT engineering, highlighting in each case features that are particularly interesting or challenging when a DT is under development.

Key decisions in the design of the DT are influenced by the rest of the DT-enabled system. For example, choices about the data used by a DT are constrained by the communications infrastructure and by the capabilities of the PT itself: there is no value in delivering a DT that relies on data that is unobtainable from the PT! This in turn depends on the extent to which the DT developer can influence the PT's owners or operators: technical decision-making is bound up with business and management processes.

SE processes are described rigorously in the international standard ISO/IEC/IEEE 15288 [7], but the International Council on Systems Engineering (INCOSE) SE Handbook [11] provides an accessible guide. Technical processes are defined in the standard as follows:

> The Technical Processes are used to define the requirements for a system, to transform the requirements into an effective product, to permit consistent reproduction of the product where necessary, to use the product, to provide the required services, to sustain the provision of those services and to dispose of the product when it is retired from service.

Each technical process expects a set of input artefacts and delivers output artefacts that embody the goals of the process and are often inputs to downstream technical processes. They support traceability from abstract needs to realised systems and enable communication across the domains and teams involved. The standard identifies 14 such technical processes. In this chapter we briefly summarise them, considering how they might play out in the context of DT engineering and in our incubator case study.

In line with ISO/IEC/IEEE 15288, we consider the *life cycle* of a system, product, service, project or other entity, to be its evolution through time from conception to retirement. We consider the technical processes in four groups covering phases of such a life cycle:

- Understanding, clarifying and articulating stakeholders' needs (Sect. 4.3).
- Defining the requirements and architecture for the DT-enabled system (Sect. 4.4).
- Realisation: the design, systems analysis and implementation of a solution meeting these requirements (Sect. 4.5).
- Life after implementation, considering the operation, maintenance and ultimately disposal of the solution (Sect. 4.6).

It is important to consider these technical processes even in a relatively simple or small-scale development because they provide a checklist. For example, some processes may require enabling tools, training or resources, so it is valuable to identify these and acquire them in time for use. The same is true for planning stakeholder engagement in some activities such as business or mission analysis, needs and requirements definition, transition, validation, optimisation and maintenance.

DT engineering practices can be as diverse as the range of DTs themselves. The technical processes that we describe here might be organised concurrently, iteratively or recursively in whatever way meets the stakeholder needs, for example to address short development times in one case or the need for evidence-based safety cases in others. In Sect. 4.7, we suggest considerations to be borne in mind when adapting these processes to a particular DT engineering challenge.

4.3 Stakeholders' Expectations, Needs and Requirements Processes

4.3.1 Business or Mission Analysis

The purpose of Business or Mission Analysis is [7]:

> . . . to define the business or mission problem or opportunity, characterise the solution space, and determine potential solution class(es) that could address a problem or take advantage of an opportunity.

A disciplined approach to this process will start from an organisation's vision, concept of operations and other strategic goals from which management defines a business or mission need supported by preliminary concepts for major phases in the system's life: acquisition, deployment, operation, support, and retirement. Business needs are then elaborated as business requirements.

In DT engineering, the goal is to create a DT-enabled system that adds value to the PT without compromising the latter's operation. This makes it essential to understand the business needs of *both* the PT's and the DT's stakeholders. The deployment status of the PT is significant in this.

- **If a deployed PT already exists,** the DT might be thought of as an add-on to the PT providing the enhanced capabilities demanded by the business. Effort is devoted to understanding the PT-side stakeholder needs as they exist and the DT's added value. This may also bring constraints: for example, it may not be possible to implement a change in the PT to accommodate the data gathering that is needed for the DT to meet its goals.
- **If a new PT and DT are being developed as a pair,** this may allow both DT and PT business stakeholders' needs to be defined together, and greatder flexibility in defining the functionality embedded in the PT and DT.

It may be that DT technology is seen as an enabler for developing products that meet business needs, as in the case with Bang & Olufsen (Chap. 2), where those business needs demand high product quality, high maintainability and development process improvements. This would scope the business case for applying the DT in both the Development and Operation stages of the product life cycle.

Incubator Business Analysis

Considering the Twinpeh scenario introduced in Sect. 3.2, the company's business mission is defined by the commercial potential of tempeh incubators. However, the qualities of the product in terms of the consistent production of good tempeh from diverse raw materials might limit the opportunity. The input to the Business Analysis process are the needs described in Chap. 3, principally the need for dependable incubation.

Chapter 3 already offers some scoping of the solution space, and so the DT may, at least initially, be considered an add-on product to an existing PT, although future incubators may be shipped with DTs with which they were specifically designed. A candidate technical solution for the proof of concept DT has already been proposed, with the working assumption that DT engineering methodologies will provide solutions for both development and maintenance processes.

4.3.2 Defining Stakeholder Needs and Requirements

Successful projects meet stakeholders' needs and requirements throughout the life cycle. The standard [7] describes the purpose of the Stakeholder Needs and Requirements process as:

> ...defining the stakeholder requirements for a system that can provide the capabilities needed by users and other stakeholders in a defined environment.

A stakeholder is any individual or organisation with a legitimate interest in the system. This would typically be a very broad group, including users, operators, parties to agreements, regulatory bodies, support organisations and society at large (within the context of the business and proposed solution).

Starting from the business needs specified in the Business or Mission Analysis, this process determines the groups of stakeholders involved, expressing their requirements and those of any enabling systems, products or services. These needs might be communicated as scenarios, defining interaction needs, implicitly defining a needed system structure and interactions with other stakeholders. The products of the process should be needs defined as verifiable and traceable requirements.

A DT may not be mandated as a requirement at this point. Instead, the stakeholders may define implicit needs for services such as visualisation, monitoring or preventive maintenance, which in a systems requirements process would be mapped to DT concepts as one possible solution to those needs. In practice, however, it may be that a DT is a desired solution. In either situation, the activities in this technical process produce the requirements specification needed for appraising the completeness of candidate solutions against the stakeholder requirements.

Incubator Stakeholder Needs and Requirements

Recall from Sect. 4.3.1 that Twinpeh wishes to develop a proof of concept DT that could be worked up to a full product rolled out with its next generation of incubators. The focus is on individual DTs for each incubator PT, although there may be a longer-term goal to develop a corporate DT for the whole fleet. The incubator works satisfactorily without the DT, but the DT adds value. This means that Twinpeh does not wish to add a lot of extra technology to the physical incubator product just to support the DT, although this could change over time as the added value of the DT becomes worth the technology investment.

Twinpeh would identify stakeholders who have an interest in the solutions. Main stakeholders and their needs would include:

Twinpeh. Driven by patterns of customer use, Twinpeh as a company has a business interest in enabling an advanced set of services to incubator users, driven by the need for dependable incubation outlined in Sect. 4.3.1.

Incubator users. Users (owners/operators) of incubators have an interest in tempeh quality and the production process, as well as reliability, warranties, safety and other facets of the incubator itself. Such users may be driven by same need for dependable incubation identified by Twinpeh, and they will also expect to benefit from DT services that can be accessed, understood and used easily.

In our example, Twinpeh has a presumption in favour of applying DT technology. This leads to stakeholder requirements for the four key features (listed in Sect. 3.2.3) that Twinpeh wants to supply to the stakeholder group of incubator users, in addition to the requirements for successful incubation that are already met by incubator PT. As a reminder, these features include:

1. Presenting historical data about the incubation process.
2. Adjusting controller settings when the user has the lid open to save energy.
3. Signalling when the tempeh is ready.
4. Signalling that the incubator needs maintenance.

4.4 System Requirements and Architecture Processes

4.4.1 Defining System Requirements

The purpose of the System Requirements Definition process is [7]:

> ... to transform the stakeholder, user-oriented view of desired capabilities into a technical view of a solution that meets the operational needs of the user.

The process marks a move from requirements being expressed in terms relevant to stakeholders towards requirements on the technical system of interest, constraining the system's boundaries, elements and interactions. It is important to recognise that system requirements definition takes place alongside the other technical processes, often iteratively. For example, the process of defining system requirements often reveals and clarifies stakeholders' requirements and even business needs.

The transformation to system requirements will constrain the DT-enabled system which includes the PT, DT and communications infrastructure. Factors such as the deployment status of the PT can expose tensions. For example, consider a DT that needs bidirectional communication with an existing, deployed PT. The requirement to provide data and control interfaces may conflict with communications performance or security on the PT. We discuss means of addressing such tensions in Chap. 7 and Chap. 13 respectively, but the freedom to apply them might be limited by the extent to which the PT is open to modification.

A key activity in system requirement definition is the development of the operational concept and other life cycle concepts. This is often done by identifying operational scenarios and associated capabilities, behaviours, and responses of the system and environments. Scenarios help to identify requirements that might otherwise be overlooked, and may raise requirements that would not have been described directly by the DT stakeholders. For example, the need to monitor performance of the DT itself and detect and recover from DT faults might feature at this level.

Incubator System Requirements

The stakeholder needs and requirements defined previously in Sect. 4.3.2 are transformed into requirements on a DT-enabled system with the PT and the DT are identified as entities interacting with an environment in various operational scenarios. For example, we might at this level expect to see many requirements, including some derived from the business need for dependable incubation:

SR1: The DT must offer users an interface showing historical performance of the incubation process.

SR2: The DT must adjust the PT to limit energy consumption when the incubator lid is open.

4.4.2 System Architecture Definition

As described in [7], the purpose of the System Architecture Definition is:

> ... to generate system architecture alternatives, select one or more alternative(s) that address stakeholder concerns and system requirements and express this in consistent views and models.

The process takes system requirements as the main input, as well as the concepts, constraints and other artefacts developed from the stakeholder engagement. It results in a description of the fundamental properties and concepts of the system to be delivered, including its key components and their connections, and their expected life cycles. The architecture is the result of a creative exercise of proposing and evaluating design alternatives, in an effort to satisfy as far as possible the problem or opportunity expressed by models and views for a set of stakeholder and system requirements (traceable to business/mission requirements, as applicable) and life cycle concepts. The system architecture description is not a static artefact, but a model that evolves as expectations and understanding develops.

Outputs from the System Architecture definition would normally be expected to identify (at the top conceptual level) the core elements of a DT and PT, with potentially a distinct communications component. It will be important in this context to provide traceability information to the stakeholder requirements identified in the upstream development phases. An architectural description would also include a rationale indicating why specific decisions have been made, particularly around any modifications to already deployed PT elements.

Incubator Architecture

At a high conceptual level, the structure of a System Architecture for the incubator DT-enabled system might appear roughly as shown in Fig. 4.1. The bi-directional DT-PT communication links represent well-defined DT-to-PT system-level architectural interfaces scoping the contracts between the DT and the PT for enabling the system requirements. The communications infrastructure might be shown as a separate component at this level or contained within one of the defined blocks. Communications interfaces to stakeholders such as DT and PT users and Twinpeh could also be shown at this level. For example, SR2 requires interfaces, data and function definitions for communications to allow the DT to detect the lid being open, and from DT to PT to allow the reconfiguration of the incubator controller in the PT. At

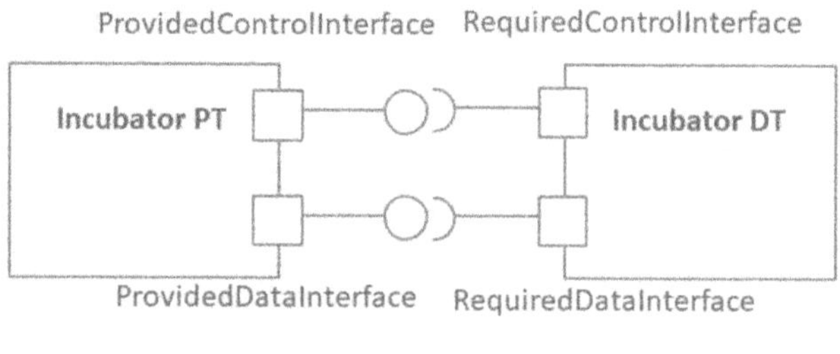

Fig. 4.1: System Architecture model, showing the two main incubator PT-to-DT communication interfaces needed for meeting the system requirements.

a lower level, the DT-enabled system architecture would have features similar to that of Fig. 12.1, in which data from the PT feeds into the DT. The DT maintains models and data, using them in enabling services such as dashboard generators to supply user-facing services such as real-time visualisation and anomaly detection.

4.5 Realisation Processes

4.5.1 Design, Systems Analysis and Implementation

The Design, System Analysis and Implementation processes take outputs from the Requirements and Architecture process and realise the elements of the actual system. The Design process takes the architecture and requirements and produces detailed data about how the system elements should be realised, and how each of those fits within the architecture and realises the requirements. This process also establishes traceability from each requirement into the design elements. Traceability continues through the subsequent processes through to the implementation.

System Analysis process provides a "rigorous basis of data and information" [7] to help with decision making and technical understanding. There are a wide range of analysis techniques that can be deployed with various levels of formality and rigour, and their use will depend on the needs and importance of the system elements and the available time and resources. These range from expert judgement and analysis of historical data to running simulations and creating prototypes. As with the design process, traceability of the analysis data is also established in this phase.

An Implementation process realises system elements using appropriate techniques and media such as software development, machining or manufacturing or adapting and integrating pre-existing system elements. It is important to record objective evidence that shows that a given implementation meets its requirements. Test results, both positive and negative, are linked to earlier engineering processes, allowing problems discovered in implementation to be linked back to earlier activities, and for iteration through these processes to resolve them. As with the other technical processes it is important to establish traceability from each of the implemented system elements to the design (and ultimately to stakeholder needs and requirements).

In the design of the DT and communications infrastructure, detailed decisions will be made about the components of the DT through consideration of a range of alternatives. Approaches to making many of these decisions are described in depth in later chapters. They include: the types of models to be included (Chap. 5) and how they are to be calibrated against the PT (Chap. 6); data and data transmission techniques and protocols (Chap. 7); the design and implementation of visualisation interfaces (Chap. 8); the approach to monitoring (Chap. 9) and its use to detect faults and support predictive maintenance (Chap. 10). The decision might be made to use a platform for realising the DT such as that discussed in Chap. 11.

Incubator Design, Systems Analysis and Implementation

The outcomes of a real incubator design and implementation process are described in some detail in Sect. 12.3.

4.5.2 Integration

While the Implementation process realises individual system elements, the Integration process synthesises these elements into a system that satisfies the system requirements. This may be iterative or proceed in stages as artefacts are realised and the system becomes more and more complete. This process requires interfaces to be activated and connected, and interoperation to start both between system elements and with external enabling systems. This typically proceeds alongside verification activities which increase confidence in the correct joint behaviour of the elements.

Integration arises in several contexts. For example, there is a need to consider combining the DT's own internal components to create an overall architecture like that described in Sect. 3.6. If the DT engineering organisation has sufficient knowledge of and control over the DT's components, this may be a reasonably standard integration activity. Then, at the level of the DT-enabled system as a whole, it is necessary to plan the systematic integration of the DT with the communications infrastructure and the PT to deliver the stakeholder needs and requirements for the DT-enabled system as a whole.

Incubator Integration

In our Twinpeh proof of concept incubator example, we assume that a PT containing the incubator and its software controller has already been built, so that the bulk of the integration activity is formed from connecting the models, data and services that make up the DT, following the architecture described above, followed by the progressive link-up of the PT with communications infrastructure and DT, enabling the top-level services implementing the system requirements to be verified.

4.5.3 Verification

Verification aims to provide evidence that the system elements and their integration produced in the processes described above fulfill the specified characteristics. Verification may identify anomalies in the system requirements, architecture description, designs, or implementation. This may be supported by the development of an assurance case, and should link results into traceability. Where anomalies are detected, the process should provide sufficient information to correct these by feeding back to the previous processes.

The first step in the Verification process is to select appropriate methods to provide the necessary evidence. These include both demonstration and testing, but also include modelling and simulation, which are a key element of the SE process in the DT engineering context. It is important to identify the constraints and objectives based on the Verification strategy, which include the limitations on the accuracy and uncertainty in measurement, and repeatability of the tests. Chapter 6 provides guidance on the predictive power of calibrating digital models in relation to real data. As with the testing in the Implementation phase, results generated during Verification should be recorded, including any potential anomalies encountered. The Verification process continues the chain of traceability back to requirements by recording requirements satisfaction, or any anomalies/deviations, in the digital model, highlighting deviations from requirement satisfaction.

Incubator Verification

Verification of the DT-enabled incubator system is an ongoing stepwise process that takes place alongside integration, as discussed above. We first consider confirming that the PT, DT and communications behave to specification, and then that the composition does. Verification plans and outcomes form a part of the traceable connection back to system requirements. The requirements on the DT-Enabled System, including improvement of the process of data, need more longitudinal study to verify.

4.5.4 Transition

The Transition process aims

> ... to establish a capability for a system to provide services specified by stakeholder requirements in the operational environment.

In practice this covers the process of installing a system in the environment and business for which it was intended.

The process of transition to a DT-enabled system is particularly interesting if the PT is already operational. In this case, transition involves the integration of an existing system, on which some reliance is placed by stakeholders, with a possibly new communications infrastructure and the DT. This may introduce the potential for disruption caused by commissioning the communications infrastructure and the DT. In some application domains, there may be almost no tolerance for this, so that transition has to be effectively invisible to many of the PT's stakeholders.

Transition may be best seen as a critical project in itself. It is worth mapping out the process in terms of the existing environment, the changes needed to allow the DT-enabled system to function, the enabling systems required for managing the transition, planning, resourcing and scheduling the transition, installing and activating the DT-enabled system as well as ensuring the capability to maintain it.

It is clearly important to manage the results of transition. As the INCOSE guidance points out, there is a strong link to Verification and Validation processes here in confirming the integration with the target environment [11]. This is likely to lead to identification of gaps in requirements and design, driving further change. At one level this change might be as simple as undertaking (re-)calibration of a DT in situ; on the other hand, newly identified requirements could entail modelling, development and operational activities taking place in parallel while the new DT-enabled systems is 'broken in'. We consider modelling in depth in Chap. 5 and calibration in more detail in Chap. 6.

Incubator Transition

Transition is straightforward for Twinpeh's proof-of-concept DT, given that it is a relatively small system on the laboratory bench. However, at a larger scale, the successful transition to DT-enabled incubators needs significant planning. For Twinpeh's user community, this is about ensuring the ordered roll-out of DT-ready incubators, then the communications infrastructure and the DT. We can certainly envisage situations in which the transition to the full-scale deployed DT solution leads to reconfiguration. For example, suppose the incubator is deployed in a corner of a room, with walls adjacent to the incubator on several sides. The DT-enabled system will require to recalibrate after some time, with new parameters of the plant model.

4.5.5 Validation

Validation provides objective evidence that the DT-enabled system once in use in its intended operational environment, meets its original objectives. The process is often undertaken for specific artefacts such as software, requirements or models, as well as for an implemented system. Its main output is evidence to support claims that the right system or element has been constructed to deliver the intended performance in the intended context. As so often stated, validation answers the question "Did we build the right thing?". This contrasts with verification (Sect. 4.5.3), which answers the question "Did we build this thing correctly".

It is worth thinking of validation in terms of a well-designed campaign that starts by defining the scope of validation in terms of the artefacts being validated and the stakeholder requirements against which they are to be assessed, for example, those defined during the definition processes for Stakeholder Needs and Requirements Definition (Sect. 4.3.2) and System Requirements (Sect. 4.4.1).

In the DT context, a validation campaign might be mounted to gain confidence that, for example, the DT-enabled system offers analysis services that meet the stakeholder needs. To address the requirements, a range of validation activities are possible, offering different qualities of evidence. They include inspection of models, through running test scenarios or simulations, to formal proofs. Given the fact that a DT-enabled system is often embedded in a complex environment, the engineers' ability to use some of these will be constrained by factors including cost and feasibility. For example, the closed Intellectual Property (IP) of some of elements of the DT-enabled system can restrict the range of validation actions that could be carried out. For each validation activity, it is essential to define the criteria against which success will be judged. There will of course be resource limits and an understanding of how much risk is acceptable, that together determine how much validation activity can be undertaken.

Validation

Twinpeh might anticipate using DT services to identify whether a specific update is needed to incubator controllers in the deployed products, and expect the DT to help determine the likely outcomes of a particular controller version on, for example, on heater and fan remaining useful life. This would be designed as a "what-if" analysis service in the DT. A validation campaign might be undertaken to increase confidence that a prototype service satisfies the user requirements here. The scope of the validation may be the service in the context of the DT, and the activities undertaken might include inspection of the design, including mock-ups on interfaces, and potentially playing through test scenarios.

4.6 The DT-Enabled System in Operation

4.6.1 Operation

The Operation process covers the use of the deployed system for the benefit of the stakeholders in line with the needs and requirements that they have articulated in earlier technical processes. In the context of a DT, this entails using the DT's services to add value to the PT. The DT is likely to operate in the context of a range of other systems in its environment, and it is important to ensure that any changes that have taken place during DT development have been taken into account, and continue to be monitored as use is made of the DT's services. Part III of this book covers the DT domain-specific services that can be used during the operation process.

Preparation is needed by the stakeholders to ensure alignment of the DT with their needs and requirements, and the concept of operations, all within the regulatory, ethical, environmental and governance constraints. There is a need to plan for ramping up the ability of the stakeholders to use the DT services, including training, and sustaining this support through operations. Operation includes confirming successful transition to the use of the DT's capabilities, ensuring operation remains within the procedures defined, detecting and recovering from abnormalities. In the context of the DT development at a technical operations level, the preparation activity of this process would most likely refer to activities such as calibration of the DT with real PT data, DT "what-if" configuration validation scenarios can used as preparation of the DT and PT for ensuring an well-defined system state before entering the operational stage.

It is worth noting that Operation and Maintenance often run concurrently, with maintenance activities focusing on ensuring that system capabilities are sustained. There is a mutual dependency here in that maintenance reports and data may affect the ways in which operation of the DT-enabled system is performed. Observations made during operation, including incidents, failures and other anomalies can lead to improvement, which entails revising system requirements and potentially stakeholder needs and requirements.

Incubator Operation

The successful operation of Twinpeh's DT-enabled incubator is the guiding aim of the SE processes described above. During operation, we would expect to see the features originally intended to achieve dependable incubation come into play, and the potential to arise for further development of the DT. It is worth noting that one of the many reasons to develop a DT in the first place is to gather and gain value from the data gathered precisely during operation, and we can expect Twinpeh to consider developing the product along these lines.

4.6.2 Maintenance and Disposal

Maintenance seeks to maintain operation of the system at a level that can sustain the services it offers. Within the process, the capability of the system to work as required is monitored by recording incidents where these are deviated from, and a variety of actions are taken to restore capabilities. Actions include prevention, correction and adaptation. Maintenance may involve replacing parts of the system, either in a proactive or reactive manner. While maintenance may be driven by failure (or expected failure) of a part through wear and tear, it may also be required in response to evolution of the system or services on which is depends and technical obsolescence of components.

The Disposal process also runs in parallel with the Operational and Maintenance process, because it covers removal or of as any waste products generated during operation, including consumables, as well as artefacts such as prototypes. The process also covers retirement of part or all of the system. Retirement may include transferring the system to a new life cycle, passing it on to a different organisation, disassembly, repurposing or recycling of components.

Maintenance of the DT-enabled system continues through the life cycle. A maintenance strategy should be defined to include provision for *preventative maintenance* (we revisit this in Sect. 10.4). The DT allows for better prediction of maintenance requirements as the digital model evolves alongside the PT and reflect the life of the system, including wear and tear and potential failures. Problems that arise and any mitigation or maintenance actions taken on the PT should be recorded, and the impact of any changed analysed. The design alternatives analysed during the Design process can help to determine the cost-effectiveness of different designs and estimate the cost of ongoing maintenance given the selected system elements and components. Ongoing analysis with the DT should identify and record trends of problems or incidents, and maintain traceability for future logistics and maintenance.

The Disposal phase of the systems engineering process may include complete disposal of the entire system, or transfer of all or part of the system a new owner who begins a new life cycle. It is important to both identify and record information about the Disposed system or system element to permit future system creators and users to build a knowledge base from experience. DT engineering allows for ongoing traceability where Disposal involves transfer to a new owner, to permit informed decision making on replacing or repurposing of the system or system elements.

Incubator Maintenance and Disposal

The physical components of the PT are obvious candidates for predictive maintenance DT-enabled incubator system. Data collected by the DT could be used to detect a reduction in performance of heating and/or cooling suggesting potential failure of the heat bed or fan. The controller of the DT may need to be maintained by applying security fixes and patches to library software (because the device is connected to the Internet), and the control software may be directly updated to improve performance based on insights provided by the DT.

We may consider that the Twinpeh company will upgrade and replace their incubators after a period of use. It may be possible to refurbish older incubators for use in less critical applications (e.g., for raising seedlings in a botanical setting), in which case the DT data will allow new owners to understand the previous and current performance of the incubator, and identify which components may need refurbishing.

4.7 Tailoring Processes and Teams

Each DT engineering project has its own characteristics and context. There is a world of difference between a well-established engineering organisation developing an ecosystem of DTs of its deployed products and a small startup business like our fictional Twinpeh company, with limited SE experience and capabilities, creating its first experimental DT of a new product. The former may well have a formalised engineering life cycle model with inputs and outputs to each process that are clearly defined and highly traceable through digital repositories; the latter will have a more ad hoc life cycle model, perhaps using a small development team in a prototyping process that uses iterative sprints.

In this chapter we have outlined SE processes that may be applied in the development, maintenance and operation of a DT-enabled system, and we have done so in rather general terms. We have indicated some of the ways in which those individual processes may be affected by the fact that we are engaged in the creation and operation of a DT-enabled system. In practice, however, these processes will need to be adapted to each individual engineering context and project. The guidance from INCOSE (in Chapter 4 of [11]) advocates treating this tailoring as a deliberate process, taking as inputs the strategic plan of the engineering organisation and the life cycle models being employed and generating a strategy for the adaptation of individual processes to the needs of the engineering task.

There are many diverse influences on the tailoring of SE technical processes. For example, work towards DT engineering approaches for the maintenance of deployed naval systems [1] tailors processes that are similar to those we have described here. Tailoring a technical process may involve modifying its expected inputs, activities and outputs, bearing in mind the life cycle model of which it forms a part. Some factors in this tailoring task come to the fore when the focus is on engineering a DT-enabled system. These include, but are certainly not limited to:

Stakeholders: In the DT-enabled systems context, it might be appropriate to think about a wider group of stakeholders than just those associated with the PT. Developers and users of the additional services afforded by the DT, and stakeholders in the communications infrastructure linking DT and PT are relevant here. Indeed, one may wish to re-think the engineering roles involved in the DT-enabled system's life cycle. We discuss examples of DT engineering roles in Sect. 4.8.

Tolerance of risk: In this book, we describe a range of techniques applicable in DT engineering. These are intended to help reduce the risks associated with developing a DT that cannot be depended on. However, all these techniques come at a cost, and so there is a balance to be struck between the level of rigour in the SE processes and the risks associated with not applying them. It may help here to consider the forms of evidence that would satisfy a DT stakeholder that the services offered by the DT can be relied on. In some critical applications, it will be worth a highly rigorous application of technical processes to deliver design justifications; in other situations a lower level of rigour is perfectly satisfactory.

Complexity in the DT-enabled system: Complexity may in itself justify the use of relatively formal technical processes because they give a stronger basis for managing the volume of design data. The communications infrastructure, the need to understand the environment of the DT as well as that of the PT, and the enabling systems that allow the DT's stakeholders to access its services all add to the potential complexity here.

The need for integration: As discussed in Sect. 4.5.2, in a DT-enabled system there is a need to manage the integration of DT components, as well as the top-level integration of the DT with communications infrastructure and the PT. In some cases, the PT may continue to evolve during DT development, particularly with fast-moving technologies and application domains, and in these situations it will be particularly important to perform integration processes alongside development and Validation and Verification (V&V) from early stages.

Systems Engineering Methodology: We have defined technical processes relatively independently of the development methodologies being used. In this connection, it is worth reflecting on the role of Model-Based Systems Engineering (MBSE). MBSE uses models through much of the system life cycle, and most if not all of the technical processes discussed in this chapter can operate with models of the DT-enabled system and its elements rather than documentary descriptions. Such models are intended to provide a comprehensive and interconnected view of the system, enabling better communication, collaboration and understanding among multidisciplinary teams.

At the time of writing, MBSE has a recognised potential, is seen as central to the future of the profession by institutions like INCOSE [8] and is being embraced by prominent organisations such as NASA [12] and ESA [4]. From a practical standpoint, implementing MBSE requires a rigorous framework for describing requirements and specifications for models as well as how they can be coupled or interchanged. Many notations have been proposed to achieve this goal in the context of a CPS, where coupling of a broad range of different formalisms has been always a priority. Among the best known of these notations are the the System Modelling Language (SysML) [3] and the Architecture Analysis and Design Language (AADL) [5] as well as the more software-focussed Unified Modelling Language (UML) [6].

System types: It is worth considering the extent to which the character of the PT may affect the DT engineering processes. There are many system types that may have an influence here, but we focus on CPSs and Systems of Systems (SoSs).

DT-enabled systems are inherently cyber-physical because they couple computational process in the DT and communications infrastructure with physical processes in the PT. The PT may itself combine computational and physical processes, as is the case in our incubator example. This has a bearing on the diversity of models that are to be considered, including discrete models of computational phenomena, continuous or analytic models of physical aspects, probabilistic models of environmental factors and so on. This in turn affects the range of analysis and simulation tools that are deployed to deliver the services that comprise the DT. In some application domains, DT-enabled systems may be composed of multiple independently operated and managed systems, some of which may be DT-enabled and some not. In an SoS, we are really saying that we rely on some of the properties that emerge from the collective operation of these constituent systems. For example, one might consider a DT project that aims to monitor and improve energy use across an estate of domestic dwellings and commercial premises, serviced by electricity and gas suppliers. Such a DT would need to interact with independent constituent systems, some of which may already be DT-enabled and may have competing goals. This independence and diversity of ownership and management has implications through the development process, from the identification of stakeholders to managing the permissions needed to make changes to the estate itself, e.g., as a result of preventive maintenance. Thinking in SoS terms encourages us to consider explicitly the ownership and management of the constituent systems, and the relationships between them in almost contractual terms: what must each of them rely on and what can each of them guarantee to their fellow systems? In Sect. 15.5 the challenges of establishing such ecosystems is emphasised.

4.8 Processes and Competencies

In this chapter, and in Parts II to IV of the book, we concentrate on technical processes for engineering DT-enabled systems. However, DT engineering does not take place in a vacuum, and all these technical processes will be fruitless without a disciplined approach to creating the business, setting up the infrastructure that allows creative design, implementation and test, and managing the technical project to ensure it is properly resourced and scheduled. Success will require engineers able to work successfully in a DT engineering environment in which models and data are combined, diverse formalisms are brought together across disciplines, and in which a wide range of stakeholders need to collaborate. In Sect. 4.8.1 we consider the non-technical SE processes key to DT engineering, and in Sect. 4.8.2 we consider competencies and roles needed in DT engineering teams.

4.8.1 Agreement, Organisation and Management Processes

Alongside the technical processes, ISO/IEC/IEEE 15288 describes SE-related business activities such as agreement, organisational project-enabling and technical management processes [7]. Agreement processes set terms and conditions under which a DT engineering project will operate. Organisation processes enable the DT engineering project activities and so work at the level of the whole organisation rather than just a single project. They include the enabling of life cycle models, the management of infrastructure, people quality and knowledge in the whole DT engineering enterprise. Technical management processes include the operation of the life cycle model processes covering planning, assessment and control, decisions, risk, information, measurement and quality assurance.

It is worth stressing the value of engineering involvement in all these processes, even those which at first sight appear to be the domain of negotiators and lawyers. For example, acceptance criteria agreed between parties influence directly the tests required to pass validation; engineering input is vital to ensure that the criteria are testable, and that progress and the consequences of changes can be evaluated. All of these processes are likely to be coloured to some extent by the characteristics of DT-enabled systems, particularly the need to integrate models with data, deal with the heterogeneity of engineering models and in some cases deal with the diverse owner and operator organisations forming parts of the DT-enabled system.

4.8.2 Competencies and Roles in DT Engineering

In our roundtable discussion (Chap. 2), participants considered some of the biggest challenges in DT engineering to be less obviously technical skills such as the ability to work in multidisciplinary engineering teams, or to get domain experts and data scientists to work closely together. The widely used term "competencies" refers to the skills that are required of individuals undertaking an engineering role.

For organisations wishing to adopt a sustainable approach to DT engineering, there are some frameworks for professional competencies that are worth considering. Within SE generally the INCOSE competency framework [10] identifies the need for knowledge and skills in four broad areas engineering:

- foundational engineering knowledge such as mathematical foundations and skills such as critical thinking;
- technical competencies which include the skills required to undertake the processes outlined in this chapter;
- management competencies such as planning, configuration management; understanding;
- competencies such as effective communication, team working, and ethical conduct.

Here again, working on DT-enabled systems tends to lead to highlight particular abilities, notably the capacity to work across disciplines and organisations, demanding fluency in models, data and communications.

Studies in the aerospace industry [2] have identified a model of competencies based on the T-shaped principle. "Vertical" competencies reflect the deep knowledge and skills of a home discipline and specific domains. "Horizontal" competencies reflect cooperation at the technology, systems and business levels. Deep knowledge about software verification tools might be seen as a vertical competency in this framework. For example, an appreciation of how software verification integrates with other verification activities, such as equipment testing, forms a basis for multidisciplinary cooperation at the technology level. Understanding the overall DT-enabled system architecture is a horizontal competency at the systems level. The ability to work influentially and self-critically with colleagues in a multi-discipline team are also horizontal competencies.

There appears to be a paucity of work on DT-specific competencies. The UK's National Digital Twin programme has proposed an extensive competencies framework for information sharing in the built environment, embracing groups of business and digital skills [9]. Business skills include championing the role of the DT, as well as communication, collaboration, adaptability, commercial mindset and business analysis. The digital skills grouping deals with data, life cycle assurance and quality management, data modelling, analytics, application, security and ethics.

Stakeholders for a DT-enabled system include the stakeholders in the original PT, who may expect service to be almost as good as usual once a DT and its communications infrastructure have been added. To this group, it is important to add those with a stake in the capabilities enabled by the DT. However, the developers and maintainers of the DT-enabled system's elements are stakeholders too, and so it is worth considering the roles that one might define in a DT engineering team.

Just as SE processes may be tailored to the needs of DT-enabled CPS engineering, it is worth considering how practitioner roles and capabilities might also adapt. Activities involving the identification, design and implementation of the core model, data and communications elements of the DT-enabled system will demand particular skills profiles. Some modified capabilities and roles might be as follows:

DT requirements management: Requirements for the DT-enabled system are devolved onto the DT, in turn influencing the choices about the abstractions embodied in the models at the core of the DT as well as the data gathering and communications needs. DT requirements management thus involves an appreciation of the potential allocations of responsibility to these elements of the DT.

Model management: The core models in a DT can be highly diverse, describing physical or computational phenomena, and can be based on first principles or be driven by data. In practice, there may be several diverse models with shared elements that need to be used in a coordinated way. In Chap. 5 we give a flavour of this variety of models. In the most straightforward scenarios, there may be a model of the PT that is ready "off-the-shelf", that is well understood by the DT engineering team, and on which it is possible to build the services needed to deliver the DT requirements. In many situations, however, several models will be needed, and so an understanding of the semantic links between models, the need for consistency, and the strengths and limitations of each model type will be needed. The ability to marshal a range of models is rare, given the focus of much engineering training in a single discipline.

PT design and manufacturing: In certain circumstances, the design and production of the PT might hardly be affected by the fact that a DT is being developed. It is, however, likely that an existing PT will be augmented with a data-gathering capability, or the capacity to have DT-controlled reconfiguration. Over time, an increasing tendency to develop and deliver DTs alongside physical products is likely to mean that PTs are designed specifically for DTs to add value, as is already the case in some sectors like healthcare. Chap. 7 introduces a range of technical issues addressed in this role, including important decisions around sensing characteristics such as sampling frequency and data compression.

PT maintenance: As was noted in Chap. 2, predictive maintenance of the PT is often identified as a benefit of the monitoring capabilities that come with a having a DT. Further, the information gathered by the DT about the use, performance and dependability of the PT as built and deployed in its real environment, has great potential in helping to identify design upgrades that may be applied to the PT over time. We explore these monitoring capabilities and the capacity for effecting change in the PT in Chap. 9 and Chap. 10. The capacity for PT maintenance, whether done through the normal front-line customer-facing services or by a third party, potentially assumed greater significance in the DT-enabled system.

DT architecture: The DT needs to be deployed somewhere and here there are multiple options including edge and cloud solutions. This role needs to be familiar with the needs on a timing perspective depending upon the purpose of the DT. The aspects relevant for this role are described further in Chap. 11.

Calibration: In case the DT needs to be calibrated, typically because of high variability of the environment the DT is operating in this role needs the ability to determine how the DT can be calibrated during its operation. The aspects relevant for this role are described further in Chap. 6.

References

1. Bickford, J., Van Bossuyt, D.L., Beery, P., Pollman, A.: Operationalizing digital twins through model-based systems engineering methods. Systems Engineering **23**(6), 724–750 (2020). DOI: https://doi.org/10.1002/sys.21559
2. Delicado, B.A., Salado, A., Mompó, R.: Conceptualization of a T-Shaped engineering competency model in collaborative organizational settings: Problem and status in the Spanish aircraft industry. Systems Engineering **21**(6), 534–554 (2018). DOI: https://doi.org/10.1002/sys.21453
3. Delligatti, L.: SysML Distilled: A Brief Guide to the Systems Modeling Language, 1st edn. Addison-Wesley Professional (2013)
4. European-Space-Agency: ESA's Technology Strategy. https://technology.esa.int/page/esa-technology-strategy (2022). Version 1.2, Last access December 2023
5. Feiler, P., Gluch, D., Hudak, J.: The Architecture Analysis & Design Language (AADL): An Introduction. Tech. Rep. CMU/SEI-2006-TN-011, Software Engineering Institute, Carnegie Mellon University, Pittsburgh, PA (2006). URL http://resources.sei.cmu.edu/library/asset-view.cfm?AssetID=7879
6. Fowler, M., Scott, K.: UML Distilled: A Brief Guide to the Standard Object Modeling Language. Addison-Wesley (2003)
7. IEEE: International Standard ISO/IEC/IEEE 15288:2023(en), Systems and software engineering — System life cycle processes. ISO/IEC and IEEE Computer Society (2023)
8. INCOSE: Systems Engineering Vision 2035: Engineering Solutions for a better World. https://www.incose.org/2023_redesign/publications/se-vision-2035 (2021)
9. Plummer, D., Kearney, S., Monagle, A., Collins, H., Perry, V., Moulds, A., Mroczka, M., Robertson, J., Smith, T., Trewavas, S., Lewis, C.: Skills and Competency Framework - Supporting the development and adoption of the Information Management Framework (IMF) and the National Digital Twin. Tech. rep., Centre for Digital Built Britain, Cambridge, UK (2021). DOI: 10.17863/CAM.65347
10. Presland, I. (ed.): INCOSE Systems Engineering Competency Framework. International Council on Systems Engineering (2018)
11. Walden, D.D., Shortell, T.M., Roedler, G.J., Delicado, B.A., Mornas, O., Yew-Seng, Y., Endler, D. (eds.): Systems Engineering Handbook. A Guide for System Life Cycle Processes and Activities, Version 4.0., 5 edn. Wiley (2023)
12. Weiland, K.J.: Future Model-Based Systems Engineering Vision and Strategy Bridge for NASA. Tech. Rep. NASA/TM-20210014025, National Aeronautics and Space Administration (2021)

Part II
Models and Data

At the heart of every Digital Twin (DT) is the link-up between models and data. Data gathered from the real-world Physical Twin (PT) give essential clues as to the state of the PT and its environment. Models tell the DT how the PT might be expected to behave in response to these conditions and thus support decision-making whether by humans or by autonomous computing systems.

This part of the book tells the story of models and data. It discusses the first principles and data-driven models that can be used inside DTs of Cyber-Physical Systems (CPSs), how they can be produced, combined and used in a coherent way (Chap. 5), and how they calibrated to the needs of a given PT (Chap. 6). Chap. 7 discusses how data sensed from the PT can be communicated to the DT in a sufficiently accurate and timely way. Entire books have been written about models, calibration and sensing, so we focus on how these topics influence the engineering of DT-enabled systems. The contents of this part are shown in overview in the figure below.

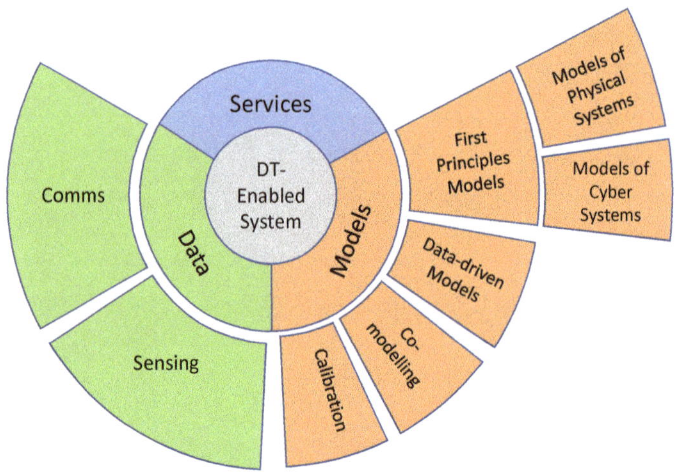

Topics related to models and data in DT-enabled systems.

Chapter 5
Modelling for Digital Twins

Giuseppe Abbiati, Cláudio Gomes, Michael Sandberg, Zahra Kazemi, Simon Thrane Hansen and Peter Gorm Larsen

Abstract At the heart of a digital twin is one or more models of the real-world system that it aims to describe. There are many facets to real-world systems, and many types of model available to describe them, each with different strengths. We provide an introduction to a range of model types. These include physics-based models founded on first principles laws, data-driven models derived by machine learning from experimental data, and formalisms rooted in discrete mathematics and logic to describe computational processes. We describe approaches to harnessing these diverse types of model together, in particular co-simulation in which models are integrated in a principled and coordinated execution, and the formal integration of discrete and continuous behaviours in hybrid automata. The practical use of some of the most common formalisms has been showcased using an incubator example.

5.1 Introduction

At the heart of a Digital Twin (DT) is a set of models describing the system that it represents – the Physical Twin (PT). These models will reflect different facets of the PT. For example, one might describe its energy consumption; another might describe its functional performance. These models, though describing the same system, may be exploited in different ways to deliver the services demanded of the DT by its stakeholders. They will each have been developed for a specific purpose and within a budget, and they have to be usable within the available computational resources, so they all trade off qualities like precision and accuracy against cost and speed.

A key choice at the heart of the design process (Sect. 4.5.1) is the selection of "the right" models to use within the DT, weighing up factors that include their levels of accuracy and precision, the speed with which they can be utilised, and the costs of using them to meet stakeholders' needs within the constraints of a DT-enabled system.

J. Fitzgerald et al. (eds.), *The Engineering of Digital Twins*,
https://doi.org/10.1007/978-3-031-66719-0_5

An important characteristic of these models is their heterogeneity. Models describing different system aspects are likely to be developed by specialists from different disciplines. The formalisms for expressing such models may vary as widely as the engineering disciplines that produce them. For example, finite-element models are popular in mechanical engineering (for example for a model expressing physical stresses) whereas statecharts are used by engineers from a computational background (for example, expressing functionality of a software controller). Reciprocal understanding of each other's formalism may be limited. Nevertheless, while crafting a DT, at some point, such models need to be combined.

Against this background, what are the modelling options facing the engineer designing a DT for a CPS? The range is wide, and in this chapter we showcase modelling formalisms that are particularly useful to represent *cyber* and *physical* systems, and we will discuss how models can be coupled to represent systems. The incubator case study is used as a running example through the entire chapter, and so will not be separated out into break-out boxes. Through the chapter, we outline *what* is needed to produce models for DTs and *why*. The reader interested in the details of *how* this is done will guided to the relevant literature as the chapter unfolds.

We first introduce some common modelling paradigms and link them to the incubator example (Sect. 5.2), discussing the different formalisms of the incubator models in Sect. 5.3. We examine models of physical processes represented as Ordinary Differential Equations (ODEs) or Partial Differential Equations (PDEs) in Sect. 5.4. We are seeing a dramatic rise in the use of Machine Learning (ML) to produce data-driven models, and these are outlined in Sect. 5.5. We move on to consider models of computational system elements in Sect. 5.6 and finally introduce approaches to the combination of diverse models in Sect. 5.7.

5.2 Overview of Modelling Formalisms

DT engineers face a choice from a very wide range of formalisms for modelling the PT. Some have developed over centuries, while others are an outcome of the rise of digital technology and ML in the last 80 years. In this section we briefly describe some of these approaches as an indication of the choices to be made. We will consider three types of models: *physics-based models*, *data-driven models* and *models for computer-based systems*. Each group is briefly introduced here and then expanded on in a dedicated section later.

Physics-based models: These models rely on "first principle laws", typically derived from scientific works and stated in terms of *algebraic equations*, *differential equations*, or a mix of the two, namely *differential algebraic equations*. *Difference equations* introduce the notion of time in algebraic equations. *Differential inclusion* extends the concept of differential equations to systems characterised by quantities that follow probability distributions. In our incubator example, the use of physics-based models is particularly indicated for describing heat transfer. Some of the most relevant physics-based models are outlined in Sect. 5.4, which focuses on differential equations and their numerical solution.

Data-driven models: These models reproduce the input-output relationship based on sampled data. The simplest possible example of such model is a *linear regression*. The *machine learning* community has developed the concept of regression and extended its application to a very broad range of models and areas of ever greater complexity. The common ground of all such models is that they are characterised by equations and parameters that do not reflect information on the inner workings of the system they reproduce but only the input-output behaviour that one can observe from data. More recently, researchers have started to add more information on the physics of the system into such models, leading to development of *physics-informed* ML. Owing to their versatility, *neural networks* have emerged as a central building block to construct data-driven models. We discuss data-driven and in particular neural network models in Sect. 5.5.

Models for computer-based systems: These models have been developed to describe the behaviour of both hardware and software components of computer-based systems. In such systems, discrete logic describes the rules governing the transition of the system from one state to the next. To represent such systems both graphical and textual *specification models* have been developed. More about these models are given in Sect. 5.6.

A model intended for use in a DT is described in a formalism that defines some aspect of the PT's behaviour. The model is characterised by a *state*, which is a collection of variables that form the model's *memory*, that is, what you need to know about what condition the system reached at some point in time. For example, think of a model describing the motion of a simple spring-mass system. This is a physics-based model typically represented using an Ordinary Differential Equation (ODE). The state variables that describe the system might include the *position* and *velocity* of the mass. The same model can be used to describe several systems that have similar behaviour by setting some specific parameter values. In the spring-mass system model, these may include the mass, stiffness and damping.

The properties of the state are useful as a means of grouping modelling formalisms into clusters that help in making modelling choices. For example:

State form: In principle it is possible to define models characterised by states of infinite dimensions, but in practice the need to represent a model in a computer means that implementations are typically limited to models with states of finite dimension. State variables may be numerical or symbolic, or both. Both cases conceal several possible options.

State evolution: The state can vary continuously over time (termed a *continuous* time model), through discrete jumps (a *discrete* time model), or a mix of the two, that is, continuously with localised jumps (termed *superdense* time).

Table 5.1: Characteristics of a some common modelling formalisms.

Formalism	State form	State evolution	Model behaviour
Difference Equations	Numerical	Discrete time	Deterministic
Ordinary Differential Equations	Numerical	Discrete time	Deterministic
Partial Differential Equations	Numerical	Continuous	Deterministic
Stochastic Differential Equations [47]	Numerical	Continuous time	Stochastic
Differential Inclusions [63]	Numerical	Superdense time	Deterministic
Finite-state machines	Symbolic	Discrete time	Deterministic
Hybrid Automata [40]	Both	Superdense time	Deterministic

Model behaviour: The model behaviour may be *deterministic* or *stochastic*. If the state evolution is deterministic, models experience the same exact trajectory if subjected to the same input. If the state evolution is stochastic, the model experience different trajectories if exposed to the same input.

These criteria do not aim to provide a rigorous framework for classifying all possible models but are useful to grasp the essence of each formalism. By way of example, Table 5.1 applies the criteria to some common modelling formalisms.

There are any more formalisms than are mentioned in this list. Each specific formalism embodying specific assumptions about the behaviour of the real system being modelled. Crafting the model requires finding the best balance between accuracy and computational cost for the model's intended use. Finding such a balance requires a sound understanding of the physical system as well as pragmatism and creativity. We see modelling as a *craft*, and developing skill at the craft requires experience gained over many different modelling challenges from many different angles. Within the confines of a book such as this, we can only aim to give a flavour of this experience by showing one path through modelling and simulation for our incubator example.

5.3 Models for the Incubator Example

Fig. 5.1 shows the incubator without its lid, making it possible to distinguish its main components, namely the Styrofoam™ box, the heater, the fan, and two temperature sensors. Two critical element in the figure are not visible – the air volume inside the box and the controller. As introduced in Sect. 3.2, the purpose of the incubator is to expose its content to a controlled temperature and its functioning can be explained in few sentences. The controller receives temperature measurements and decides whether to turn on/off the electrical heater. The later injects thermal power into the air contained within the box. As a consequence, the temperature of the air volume enclosed by the box increases. The operating temperature range of the incubator is between 20 and 60 °C and the incubator is placed in a room with sufficient ventilation.

Fig. 5.1: Close-up view on the incubator without the lid.

So one can reasonably assume that both temperature and pressure of the air outside the box are at constant ambient conditions (i.e., 20 °C, 100 kPa). The controller adjusts the electric supply of the heater based on the air temperature measurements averaged over two thermocouples installed in the box. An electrical fan is installed in the box to ensure a uniform temperature distribution.

In order to avoid heat loss due to air flow exchanges with the outside, the incubator operates with a closed lid. Air, as a gas, experiences pressure increase as temperature increases while the volume is constant. However, the lid is not perfectly sealed. In addition to that, the thermal deformation of the box is too small to affect the air volume. Therefore, air pressure inside the box can be assumed to be constant.

Given this understanding of the flow-thermal behaviour of the incubator it is possible to show several modelling formalisms that are useful to craft a DT. Specifically, physics-based models are presented to describe heat transfer within the box starting from "first principles" like energy conservation (Sect. 5.4). Data-driven models are presented as an alternative approach to describe heat transfer based on previously measured data (Sect. 5.5) rather than pre-existing equations. Finally, models for computer-based systems are outlined as a tool for describing the incubator's temperature controller (Sect. 5.6).

5.4 Physics-based Models

Differential equations are a mathematical tool for representing classical physics-based models, suchas the description of heat transfer in the incubator. This section presents both a simplified model based on Ordinary Differential Equations (ODEs) and a more advanced model based on Partial Differential Equations (PDEs).

> Simulation of physical systems like the physical incubator system is typically conducted by means of *first principle* models. Such models are based on mathematical equations (differential and/or algebraic) that embed some basic physical principles, such as energy or momentum conservation. For engineers, it is often natural to adopt such models as they reflect the inner workings of the physical system. As of today, the market offers many software packages for the simulation of physical systems using first principle models. 3DS[a], Ansys[b], Comsol Multiphysics[c] or Altair[d], allow engineers to craft advanced simulations. However, in order to have full control of such models, a coarse understanding of the methodologies behind these software packages is needed, which is the aim of this section.
>
> [a] `https://www.3ds.com/`
> [b] `https://www.ansys.com/`
> [c] `https://www.comsol.com/`
> [d] `https://altair.com/`

5.4.1 Ordinary Differential Equations

ODEs are very useful for formulating relatively simple yet effective physics-based models starting from basic conservation laws (e.g., energy or momentum). For an introduction to this subject, we recommend [18]. In order to describe the heat transfer dynamics of the incubator, the following ODE is formulated,

$$C_h \frac{dT_h}{dt} = H_h \cdot V_h \cdot I_h - G_{hb} \cdot (T_h - T_b) \tag{5.1a}$$

$$C_b \frac{dT_b}{dt} = G_{hb} \cdot (T_h - T_b) - G_{br} \cdot (T_b - T_r) \tag{5.1b}$$

where T_h is the temperature of the heater (measured in °C), T_b is the temperature of the air inside the box (°C), T_r is the room temperature (°C); C_h is the heat capacity of the heater ($\mathrm{J}/(\mathrm{Kg} \cdot °\mathrm{C})$) while C_b is the heat capacity of the box including contained air $(\mathrm{J}/(\mathrm{Kg} \cdot {}^\circ\mathrm{C}))$. The electric power supplied to the heater is given by the product of voltage and current $V_h \cdot I_h$ (W) while H_h is a Boolean variable used to switch electric power on and off. Finally, G_{br} is a coefficient that models heat transfer between the box and the room (W/°C) while G_{hb} is a coefficient that models heat transfer between the heater and the box (W/°C).

Eq. (5.1) is an instance of the first law of thermodynamics, which states energy conservation. In fact, the term $G_{hb} \cdot (T_h - T_b)$ is the heat flux exchanged between heater and air whereas $G_{br} \cdot (T_b - T_r)$ is the heatflux exchanged between air and external environment (the room). Accordingly, $C_h \frac{dT_h}{dt}$ and $C_b \frac{dT_b}{dt}$ are the instantaneous variations of thermal energy stored in the heater and the air.

In order to simulate an ODE in the form of Eq. (5.1), it is necessary to define the initial values of the state variables T_h and T_b. If the incubator is heated from room temperature, it is correct to set the initial value of both state variables to the room temperature:

$$T_h(0) = T_b(0) = T_r = 20\ {}^\circ\mathrm{C} \tag{5.2}$$

Models in the form of Eq. (5.1) are called *lumped-parameter* models. Lumped-parameter models assume all model components are concentrated at a single point. For instance, Eq. (5.1) assumes that the air temperature is the same for the full air volume (close or far from the heater), which is not true.

It is important to remark that physics-based models are inherently inexact as it is not practically feasible to provide a comprehensive mathematical description of any engineering system down to the sub-atomic scale. It follows that engineered systems are often designed such that their behaviour can be predicted with a sufficient degree of accuracy by using simple models. For this reason, a fan was installed in the incubator box to ensure a uniform air temperature as assumed in the model ODE of Eq. (5.1), helpng to make the model an accurate description of the temperature behaviour.

The ODE in Eq. (5.1) and the initial configuration form an Initial Value Problem (IVP). The field of applied mathematics that deals with the solution of ODEs via numerical approximation is called *numerical time integration*. In this area, one can find a broad variety of *time integration schemes*, which are formulated to solve generic ODEs of the form,

$$\frac{d\mathbf{x}}{dt} = \mathbf{f}(\mathbf{x}, \mathbf{u}, \mathbf{p}) \ , \ \mathbf{x}(0) = \mathbf{x}_0 \tag{5.3}$$

where $\mathbf{f}(\mathbf{x}, \mathbf{u}, \mathbf{p})$ is a vector function that maps the state vector $\mathbf{x}$ and the input vector $\mathbf{u}$ into the time derivative of the state vector $\frac{d\mathbf{x}}{dt}$; finally, vector $\mathbf{p}$ contains all the parameters of the model. For the specific case of the incubator ODE of Eq. (5.1),

$$\mathbf{x} = [T_h, T_b]^T \ , \mathbf{u} = [H_h]^T \ , \mathbf{p} = [C_h, C_b, G_{br}, G_{hb}]^T \ , \mathbf{x}(0) = [20, 20]^T \tag{5.4}$$

The numerical solution of such ODEs is carried out using a *time integration algorithm*, which computes the response history of the state vector for discrete time intervals $t_k = k\Delta t$, where $\Delta t > 0$ is the time step size. For this reason, time integration algorithms are also called *time stepping algorithms*. The main idea behind numerical time integration is that the state vector of the ODE and its derivative(s) are related by approximating equations and, based on that, the value of the state at a specific time step k is related to a finite number of past states (and their derivatives). The simplest possible time integration scheme is the Euler method whose approximating equation reads,

$$\mathbf{x}_{k+1} = \mathbf{x}_k + \frac{d\mathbf{x}_k}{dt}\Delta t = \mathbf{x}_k + \mathbf{f}(\mathbf{x}_k, \mathbf{u}_k, \mathbf{p})\Delta t \tag{5.5}$$

More advanced time integration schemes use more response steps in the past to predict the step ahead at t_{k+1} (multi-step methods) or more evaluation of the state transition function within the same time step (multi-stage methods). In either case, the output of the simulation is the trace of the model over time (also known as the *time history response*). Fig. 5.2 shows a zoomed in version of Fig. 3.7 where the first 30 steps of the integration scheme can be seen. The reader interested in numerical time integration can consult [16, 34].

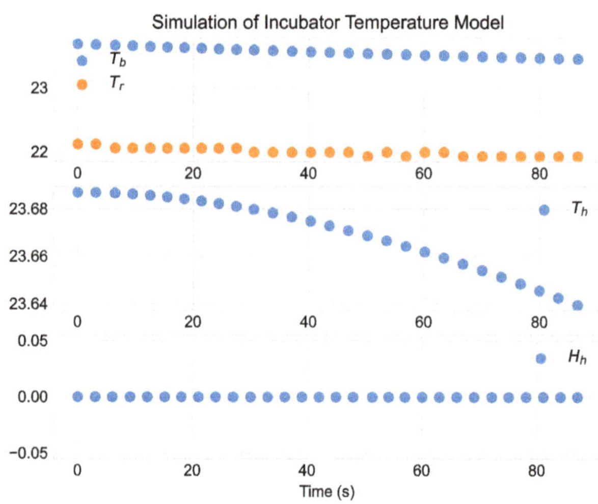

Fig. 5.2: Example behavior of the incubator temperature model introduced in Eq. (3.1), obtained by running a time integration method for 30 steps from the given initial values.

Time integration schemes are typically part of the algorithmic framework of simulation software. Each application domain has its own preferred choices. In addition to that, libraries for numerical time integration are available for several programming languages. For example, the *SciPy* library contains a number of schemes that can be used to solve ODEs in *Python*.

5.4.2 Partial Differential Equations

Like ODEs, Partial Differential Equations (PDEs) are typically used to represent first principle laws such as momentum or energy conservation. However, a PDE typically refers to an infinitesimal portion of a physical system. Accordingly, together with the PDE, a geometrical domain (i.e. portion of space) where it applies is defined and the solution of the PDE is constructed over such domain. It follows that modelling with PDEs implicitly means considering a geometrical representation of the physical system, which is overlooked in lumped-parameter models like the one presented in Eq. (5.1).

In practical terms, modelling with PDEs starts with the description of the geometry of the physical system. Geometric modelling is carried out using CAD tools, which are typically part of the *pre-processor* of the modelling environment. Note the geometry is not intended to reflect the physical system exactly. Instead, it is meant to support a sufficiently accurate mathematical description of the *physics* involved in the behaviour of the physical system. This is a central aspect, which constantly steers modelling choices.

In the case of the incubator, the starting point is to recognise that the environmental conditions outside the box are not affected by the incubator thermal response. Therefore, modelling shall be limited to the incubator and not the surrounding environment. It is important to stress that geometrical complexity in principle affects the computational cost of the resulting simulation model. The degree of proportionality between the two aspects is hard to quantify because computational cost might grow to a smaller or larger degree depending on the specific case. Based on this premise, one can reasonably construct the geometry of the incubator as the union of a cuboid representing the Styrofoam™ box and fill the hollow part with another cuboid representing the air volume. The incubator operates mostly with a closed lid. Therefore, the box volume encloses the air inside. Both box and the air are bulky and characterised by non-uniform temperature distribution. In order to represent their physical behaviour a 3D description is needed.

Modelling the heater and fan is much less intuitive. A novice modeller might be tempted to provide an accurate geometrical description of both. This means modelling the heater as a volume associated with volumetric heat source and the fan as a mechanical entity with moving parts that interact with the air volume. This is definitely not the way to go! Such an approach to geometrical modelling might be reasonable for a high-fidelity analysis of a combustion engine. In fact, the resulting simulation would be associated with a tremendous computational cost. Therefore, the heater and fan are put on hold for the moment and their modelling is introduced later in the section, after the definition of the physics. At this point, it is worth looking ahead to Fig. 5.5(b), which provides a graphical representation of the geometrical modelling described so far.

Having defined the geometry of the model, the next step is to associate some specific *physics* to each portion of the model domain. Here, the term *physics* refers to the first principle laws that describe the spatio-temporal evolution of the relevant solution *fields* of the state variables. The latter, for the incubator example, are temperature, pressure and velocity for the air, but only temperature for the Styrofoam™ box. Mathematical modelling of such physics relies on *continuum mechanics* [36], which is a branch of *mechanics* that deals with the mathematical description of materials as a continuous system rather than as set of discrete particles. In principle, mass, momentum and energy *conservation laws* together with *constitutive laws* are used to formulate PDEs that describe the *physics* of a material. For the specific case of the incubator, one has to distinguish between the box and the air volume.

The air volume is fluid so both heat (energy) and mass transport must be accounted for. The *Navier-Stokes* continuum mechanics model deals with the mathematical description of mass and energy transport in fluids [22] and therefore can be conveniently used to simulate energy and mass transport associated with the air volume inside the incubator. The corresponding system of PDEs reads,

$$\boldsymbol{\nabla} \cdot \mathbf{v}_a = 0 \tag{5.6a}$$

$$\rho_a \left[\frac{\partial \mathbf{v}_a}{\partial t} + (\mathbf{v}_a \cdot \boldsymbol{\nabla}) \mathbf{v}_a \right] = -\boldsymbol{\nabla} p_a + \rho_a \mathbf{g} + \mu_a \nabla^2 \mathbf{v}_a \tag{5.6b}$$

$$\rho_a C_{p,a} \left[\frac{\partial T_a}{\partial t} + \mathbf{v}_a \cdot \boldsymbol{\nabla} T_a \right] - \boldsymbol{\nabla} \cdot (k_a \boldsymbol{\nabla} T_a) = 0 \tag{5.6c}$$

where $\mathbf{v}_a(x, y, z, t)$, $p_a(x, y, z, t)$ and $T_a(x, y, z, t)$ are the time-dependent spatial velocity, pressure and temperature *fields* associated with the air inside the box; however space (x, y, z) and time (t) dependency is omitted for brevity. For the same reason, *differential operator* $\boldsymbol{\nabla} := [\partial/\partial x, \partial/\partial y, \partial/\partial z]$ is used to shorten the notation for spatial partial derivatives. The Navier-Stokes' equations represent an Eulerian approach in which all fields are evaluated for a fixed point in space, and not a point anchored to a fluid particle in motion. The air properties are given by the parameters, ρ_a (density $\mathrm{kg/m^3}$), μ_a (viscosity, $\mathrm{N/(m^2 \cdot s)}$), k_a (thermal conductivity $\mathrm{W/(m \cdot {}^\circ C)}$) and $C_{p,a}$ (heat capacity at constant pressure $\mathrm{J/(kg \cdot {}^\circ C)}$). The system of PDEs of Eq. (5.6) describes mass conservation (Eq. (5.6a)), momentum conservation (Eq. (5.6b)), and energy conservation (Eq. (5.6c)) for a Newtonian fluid. Note the mass conservation equation of Eq. (5.6a) indicates that air is assumed to be incompressible. Alternatively, one can extend Eq. (5.6) with a state equation that relates temperature, density and temperature fields.

The Styrofoam™ box is solid. Therefore, a heat equation, which is a simpler variant of Eq. (5.6c) without a convective term, is associated with this component.

$$\rho_s C_s \frac{\partial}{\partial T_s} t - \nabla \cdot (k_s \nabla T_s) = 0 \tag{5.7}$$

where $T_s(x, y, z)$ is the temperature field associated with the Styrofoam™ box, ρ_s $(\mathrm{kg/m^3})$ is the density k_s $(\mathrm{W/(m \cdot °C)})$ is the thermal conductivity and C_s $(\mathrm{J/(kg \cdot °C)})$ is the heat capacity parameter.

At this stage, geometry, physics and material properties of the model are specified. In order to solve for pressure $(p(x, y, z, t))$, temperature $(T(x, y, z, t))$ and velocity $(\mathbf{v}(x, y, z, t))$ fields, their Boundary Conditions (BCs) must be defined, that is, the values that such fields and some of their spatial derivatives assume at the boundary of the model domain. For the specific case of the incubator, convective heat exchange is applied to the exterior of the box so that the expression of the heat flux normal to the box reads,

$$-k_s \nabla T = h_a (T(x, y, z) - T_r) \tag{5.8}$$

with h_a (W/m^2K) is the convection coefficient and T_r is the room temperature. Finally, a forced internal air flow is modelled as a constant velocity averaged over the cross-sectional area of the fan.

Just as in Sect. 5.4, since the purpose of the model is to simulate the transient response of the incubator, Initial Conditions (ICs) shall be specified for all associated fields. In detail, initial temperature is assumed uniform and equal to the room temperature for both air and Styrofoam™ box. Air pressure is assumed uniform and equal to ambient pressure. Air velocity is assumed to be zero.

All the steps described so far are carried out within the *pre-processor* of a simulation framework. Commercial simulation frameworks like ANSYS, ABAQUS and Comsol Multiphysics, typically come with a pre-processor. Open-source frameworks like OpenFOAM typically come without a pre-processor although other packages are available, such as GiD[1].

In general, the analytical solution of a PDE, if available, is limited to a few elementary cases where the geometry of the domain is simple (e.g., rectangular or spherical), and the material properties are constant and uniform. It is clear that, such oversimplified cases are not enough to cover real engineering systems. For the specific case of Eq. (5.6b), mathematicians have neither proved that smooth solutions always exist, nor found any counter-examples. This is called the *Navier-Stokes existence and smoothness problem* [49] and the Clay Mathematics Institute in May 2000 made this problem one of its seven Millennium Prize problems in mathematics[2]. Thankfully, engineers do not need *exact* solutions and so PDEs are solved using numerical approximations.

Since computers have finite memory, and analytical solutions of Eq. (5.6) and Eq. (5.7) are typically not available, the corresponding fields (pressure $p_{a,s}(x, y, z, t)$, temperature $T_{a,s}(x, y, z, t)$ and velocity $\mathbf{v}_{a,s}(x, y, z, t)$ are approximated for a discrete space-time grid. Time discretisation is achieved using a time integration scheme as

[1] `https://www.gidsimulation.com/`

[2] A 1,000,000 USD prize is assigned for the solution of each of these problems.

described in Section 5.4.1. Spatial discretisation is performed using a completely different class of algorithms. The reason is physical. Let us imagine suddenly switching the incubator's heater off at ambient conditions. Based on a qualitative understanding of physics, it is reasonable to accept that a temperature perturbation propagates through the air volume in all directions. However, temperature perturbation propagates only forward in time (and not backward). The latter is the immediate consequence of causality. Such properties shall be preserved in the numerical approximations underlying the solution of Eq. (5.6) and Eq. (5.7). Therefore, spatial and time discretisation are pursued with radically different approaches.

Two approaches for spatial discretisation are briefly described in the following, namely, the Finite Element Method (FEM) [61] and Finite Volume Method (FVM) [66]. The precise mathematical explanation of both FEM and FVM goes beyond the aim of this chapter. However, the basic underlying idea is given for both.

The basic idea of the FVM is to divide the domain into control volumes called *cells*. Each cell is associated with volume-averaged values of the field variables. The governing equations are integrated over control volumes, resulting in a set of discrete equations. These equations relate the fluxes of the conserved quantities at the control volume interfaces to the changes in their average values within the volume. As a result, the FVM method is particularly suited for conservation laws, as it emphasises the balance of quantities across control volume boundaries must be equal to the observed changes in the cell. If one considers a uni-dimensional temperature field, the average temperature of the i-th cell reads,

$$\overline{T}_i(t) = \frac{1}{V} \int_V T(x,t) dV \tag{5.9}$$

where V is the volume of the cell. A schematic representation of the spatial discretisation of FVM is shown in Fig. 5.3.

The basic idea of the FEM is that the domain of the PDE is discretised into non-overlapping *elements*, which are interconnected through their *nodes*. This is called a mesh. Each field associated with the PDE is *interpolated* based on nodal values. A central concept of FEM is thus the *shape function*. A shape function typically consists of a polynomial that interpolates the field within each element and the corresponding partial derivatives based on nodal values. As an example, the general interpolation expression for a uni-dimensional temperature field reads,

$$T(x,y,z,t) = \sum_i T_i(t)\phi_i(x) \tag{5.10}$$

where $T(x,t)$ is the interpolated temperature within a generic element $T_i(t)$ is the temperature of node i-th and $\phi_i(x)$ is the associated shape function. Fig. 5.4 provides a graphical representation of the interpolation of a uni-dimensional temperature field.

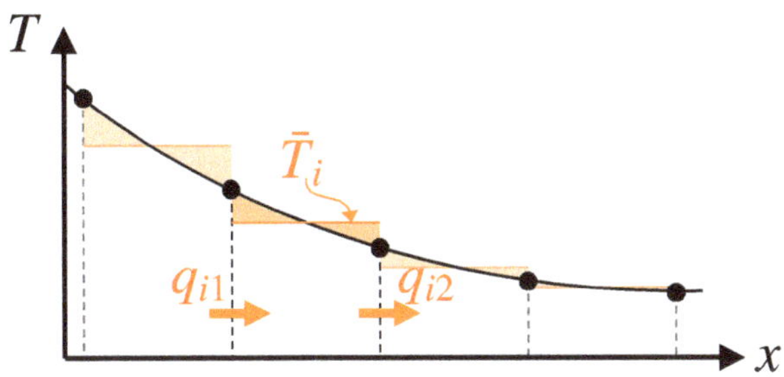

Fig. 5.3: Spatial discretisation of the FVM: the volume-averaged value of the spatial variable, $\bar{T}_i$, is determined based on the fluxes, q_i, that leaves and enter the cell. For a thermal problem, the fluxes can be calculated based on the temperature difference between two adjacent cells and the heat conductivity.

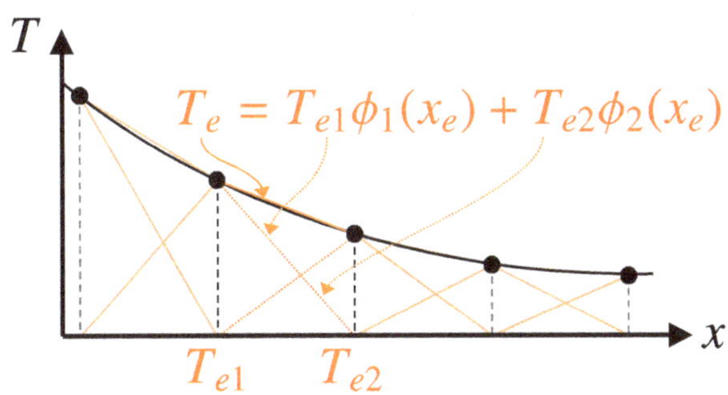

Fig. 5.4: Spatial (linear) discretisation of the FEM for an element, e, based on the shape functions, ϕ_1 and ϕ_2, and the local nodal values, T_{e1} and T_{e2} and spatial element coordinate, x_e.

After spatial discretisation, the PDEs of Eq. (5.6) and Eq. (5.7) are converted into an ODE in the form of Eq. (5.3) where $\mathbf{x} = [\mathbf{v}_1, \ldots, \mathbf{v}_n, p_1, \ldots, p_n, T_1, \ldots, T_n]^T$ is the state vector of the model after spatial discretisation using the FEM with n nodes or, equivalently, the FVM with n cells. Such an ODE is then solved using a time integration scheme as described in Sect. 5.4.1.

As an example, the three-dimensional spatial mesh discretisation of the FEM of the incubator implemented in Comsol Multiphysics is shown in Fig. 5.5(a). The corresponding heatmap of the spatial temperature distribution together with the vector plot of the velocity field can be seen in Fig. 5.5.

Commercial software packages (e.g., Comsol Multiphysics, Ansys Fluent and STAR-CCM) typically come with tools for meshing (pre-processor) and visualisation/processing of simulation results (post-processor). Non-commercial packages,

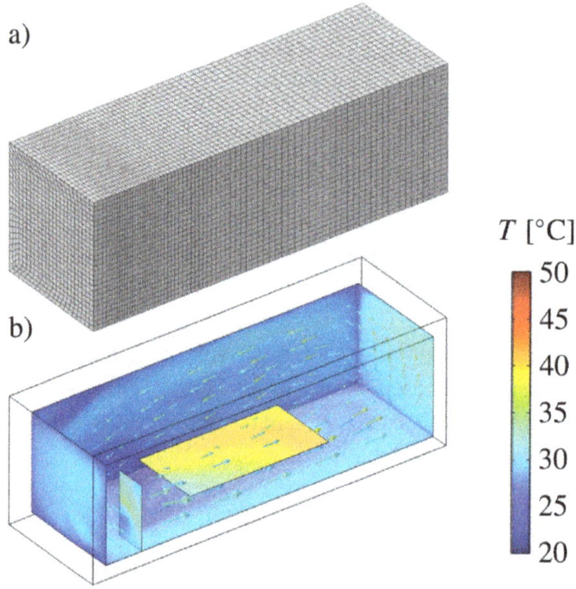

Fig. 5.5: (a) Mesh discretisation of (b) the incubator geometry, including the post-processed solution of the spatial temperature distribution and a vector field plot of the velocity components (reproduced from [25]).

typically associated with open-source repositories (e.g., OpenFOAM and Elmer FEM), often relies on third-party pre- and post-processors like GiD or Gmsh[3]. It is important to note that due to storage limitations, it is often not possible to store all possible model results and only a specific subset is selected (e.g., temperature distribution in the air volume only).

PDE-based models are typically classified as high-fidelity models, and their development is time consuming. Before starting to develop such a model, it is worth exploring the behaviour the physical model by means of simpler alternatives such as lumped-parameter models based on ODEs. Another important consideration is that complex models are more likely to contain implementation errors. It may prove useful to compare the results of PDE-based models with ODE models, and see if their results are consistent, in order to detect potential errors in the models.

[3] https://gmsh.info/

5.4.3 Model Order Reduction

Model order reduction is a multidisciplinary field shaped by diverse research communities over recent decades with the common goal of speeding up numerical simulations. Although the methodology is broadly applicable, one can identify two primary approaches: the system-theoretic and the numerical analysis approaches.

System-theoretic methods are developed for systems described by large sets of ODEs (see Sect. 5.4.1) or Differential Algebraic Equations (DAEs). Their aim is to derive a reduced-order model with fewer state variables, which nevertheless approximates the true system response well. Sometimes these reduced-order models are obtained by splitting a single model of unmanageable complexity into its components, which are reduced and re-assembled in a divide-and-conquer approach.

Numerical analysis approaches tackle computational speed-up of models based on PDEs (see Sect. 5.4.2). The strategy relies on *Galerkin projection* on finite-dimensional subspace to approximate spatial and temporal discretisation of the PDE. Projection subspaces are obtained from some snapshots of the solutions of the full-order model corresponding to a range of model parameters.

For both system-theoretic and numerical analysis methods, the final result is that the model state dimension (i.e., *system order* in automatic control) is reduced. The simplest possible example of model order reduction, which is called *Galerkin projection*, is illustrated to reduce the state dimension of a generic ODE such as Eq. (5.3). The starting point is to define a reduced-order state vector $\mathbf{q}$ with $k = \dim(\mathbf{q}) << n = \dim(\mathbf{x})$ as,

$$\mathbf{x} \approx \mathbf{V}\mathbf{q} \tag{5.11}$$

where $\mathbf{V} \in \mathbb{R}^{n \times k}$ is a constant projection matrix. If one replaces Eq. (5.11) in Eq. (5.3), after few manipulations, the following ODE is obtained,

$$\dot{\mathbf{q}} = \left(\mathbf{V}^T\mathbf{V}\right)^{-1} \mathbf{V}^T\mathbf{f}\left(\mathbf{V}\mathbf{q}, t\right) \tag{5.12}$$

which approximates Eq. (5.3) with a reduced-order state vector $\mathbf{q}$.

Intuitively, the goodness of the approximation is dictated by the projection matrix $\mathbf{V}$ whose estimation conceals the model order reduction methodology. An up-to-date account of the latter can be found in [7, 8, 9], including an overview of relevant case studies. As of today, model order reduction is a well-established methodology. However, model order reduction tools are seldom integrated in modelling environments. We foresee that the massive adoption of DTs will change this scenario.

Model order reduction is useful if one needs to solve the model in real-time or close-to-real-time. For parametric studies or uncertainty quantification analyses, non-intrusive surrogate modelling approaches, such as those based on ML models, are more appealing, and these are introduced in the following section.

5.5 Data-driven Models

In Sect. 5.4, we discussed models that rely on fundamental physical laws and mathematical equations to describe the system's behaviour, under a range of assumptions. Even though such physics-based modelling has been the cornerstone of engineering design for many decades, it has several limitations that hinder its application in DTs. Systems are becoming more complicated and interconnected, and traditional physics-based approaches may no longer be sufficient for accurate modelling of such nonlinear and complex systems with complicated time-varying interactions between different variables, parameters, and subsystems, as this requires relaxing many assumptions to make the modelling feasible. As seen in Sect. 5.4, the simplicity of the model comes at the cost of gross simplifications of the physics.

In some situations the underlying physics of a system may not be fully understood, and the required laws or equations to describe it may not be available or may not be accurate enough to produce reliable predictions [51]. Effective modelling paradigms are needed to address the complicated requirements of new systems.

Data-driven modelling embraces empirical data from the system itself to establish relationships and patterns between the system states, inputs, and outputs, and subsequently uses them for predicting or classifying the system behaviour and states or for representing its underlying dynamics. Data-driven models empower DTs with accurate and responsive representations of their corresponding PTs, but rely heavily on the quality and quantity of the data.

Contrary to the physics-based models, the development of data-driven models is relatively quick and straightforward, with the major task being the estimation of the model parameters that make the model's behaviour match the data that should be readily available. The procedure of estimating the model parameters, called *calibration*, is discussed in more detail in Chap. 6. In fact, all models, even physics-based, require calibration, with parameters being estimated from empirical experiments, so the line between data-driven modelling and physics based modelling is blurred and mostly relates to how much prior domain knowledge is incorporated into the structure of the model. This has led many practitioners and researchers to coin the term *grey box modelling*.

It should be noted that there are many models that can be used in data-driven modelling. While in this section we cover one well-known formalism (neural networks), we refer the reader to [58] for a comprehensive overview of data-driven modelling approaches. In practical terms, the choice of model type in a DT depends on the exact modelling requirements which vary depending on the application area, use cases, and available data, as discussed below.

5.5.1 Requirements of Data-driven Modelling

Finding and establishing an effective data-driven model for a DT depends on understanding several key requirements which can be considered under several broad headings:

Quality of data: The quality of training data is critical to ensuring a high-fidelity data-driven model. The data should be rich, i.e., it should be sufficiently precise, reliable, and complete to capture the full operating range of the system. According to control theory, the experiments conducted on the system to be modelled should be *persistently exciting*, i.e., the resulting behaviour should contain non-repetitive data and be able to excite and represent unique operating conditions in the system at hand. Thus, data preprocessing and cleaning might be needed to handle missing data points, repetitive data, inconsistencies, and outliers [35]. Biased or skewed data can cause poor generalisation results.

Quantity of data: The size of the dataset is also a critical requirement for data-driven modelling. Sufficient data should be available to enable full coverage of the operating range and with good resolution. This is particularly important since in data-driven modelling a good portion of data should usually be reserved for validation of the models, which should be different to the data used for model training.

Model accuracy: The accuracy needed from DT models depends on the application area and stakeholder requirements. More accurate models are often more complex and difficult to operate. Thus, it is important to define what is "acceptable" accuracy for the DT model [24]. Generally, a good data-driven model should have a strong generalisation capability to predict unseen data. Overfitting (when DT model responds well on the training dataset but performs poorly on new data) and under-fitting (when the DT model performs well with neither training nor validation data) should be minimised. Model accuracy should be evaluated based on an appropriate metric, such as an "F1" score (a common metric used in ML and statistics to evaluate the performance of classification models), mean squared error, etc.

Model Computational Complexity: The complexity of the model should meet the real-time performance requirements to assure implementation feasibility and avoid delayed runtime and synchronisation issues between PT and DT [19].

Interpretability: In some DT contexts, models might need to have interpretability, particularly where they inform critical decisions. This means that the models' sequences or hierarchy (e.g., when a mega model is formed of smaller models or interacts with other models) should be able to interact, process, and exchange information between them [19].

Scalability: The data sizes increase over time and thus the data-driven models should be scalable in order to handle larger datasets. The scalability of data-driven models is also required to ensure feasibility in real-time or near-real-time predictions [43].

These are general requirements that should be considered in the design phase of the data-driven models to be used in a DT; other specific requirements will typically be added for specific applications.

5.5.2 Artificial Neural Networks

The idea of Artificial Neural Networks (ANNs) has its origins in an understanding of the operation of the human brain, in which neurons react to the activation of adjacent connected neurons [56]. Each neuron maps an input-output relationship with a specific sensitivity depending on the strength of the input-output connection links. Thus, neurons may represent different levels of responsiveness to inputs. All existing ANN models have a common building block which is formed from a neuron and a networked interconnection structure. A widely used model for neurons has been presented by McCulloch and Pitts [30] and is illustrated in Fig. 5.6.

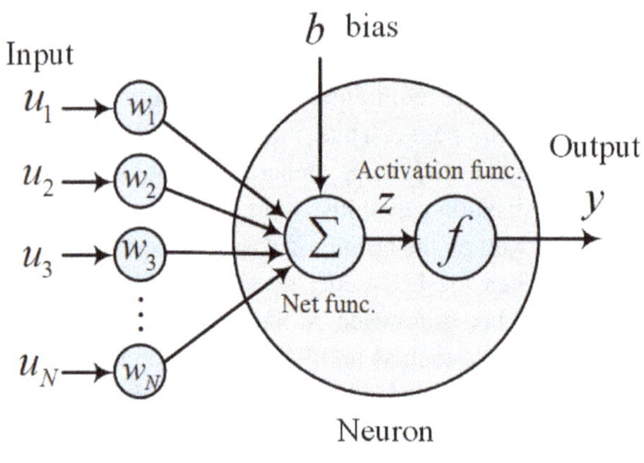

Fig. 5.6: A neuron model based on McCulloch and Pitts' work.

As shows in Fig. 5.6, each neuron is composed of two functional blocks: 1) the *net function* that determines the combination of the network inputs in the neuron and 2) the *activation function* which produces the neuron output. For the sake of simplicity, the following discussion refers to the simple case of a *linear* net function, which is defined as,

$$z = \sum_{i=1}^{N} w_i u_i + b \tag{5.13}$$

Table 5.2: Example of some well-known activation functions.

Activation function	Formula
Sigmoid	$\frac{1}{1+\exp(-z/T)}$
Hyperbolic tangent	$\tanh(\frac{z}{T})$
ReLU	$\max(z, 0)$

where z is the net function output for the input vector $\mathbf{u} = [u_1, ..., u_i, ...u_N]$ the $\mathbf{w} = [w_1, ..., w_i, ...w_N]$ are the corresponding synaptic weights and b is a bias term. Other types of net functions for the combination of network inputs may also be considered such as higher-order polynomial function ($z = \sum_{i=1}^{N} \sum_{j=1}^{N} w_{ij} u_i u_j + b$), delta function ($z = \Pi_{i=1}^{N} w_i u_i$). The activation function relates the combined network input to the neuron output y_i using a transformation function as follows:

$$y = f(z) \tag{5.14}$$

The complexity and nonlinearity that can be achieved by linear activation functions are limited compared to nonlinear activation functions. Thus, the choice of activation function depends on the type of nonlinearity of the system. Some of the most commonly adopted activation functions are *sigmoid*, *hyperbolic tangent* and *Rectified Linear Unit* (ReLU) functions, as listed in Table 5.2.

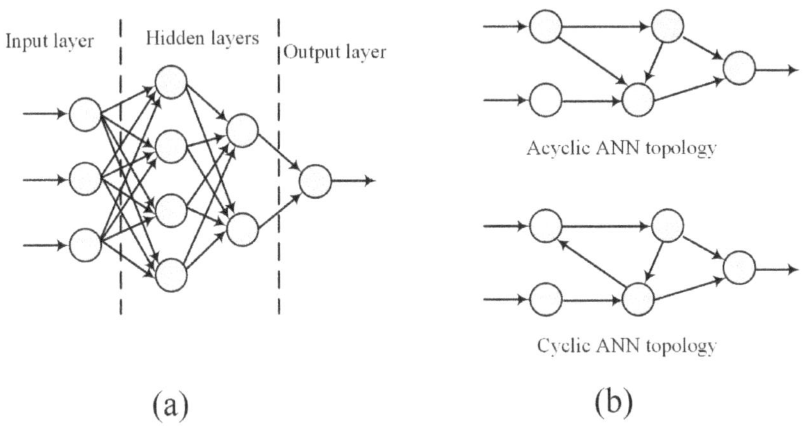

Fig. 5.7: (a) The structure of a typical ANN model, (b) different types of ANN topologies.

A combination of several neurons in a certain structure or architecture is considered in an ANN model to enable modelling of complex patterns in data. Each neuron in the network performs computations on the input data, transforming and passing information to the next layer. By having multiple neurons, the model can simultaneously process different aspects of the input, extracting diverse features and representations. Additionally, a larger number of neurons provides the network with more flexibility and expressive power, enabling it to learn and approximate complex functions more accurately [33].

The interconnection of different neurons in a typical ANN is shown in Fig. 5.7 [64]. As depicted, the graphs are composed of nodes to indicate neurons and external inputs and directed arcs to indicate the synaptic links. Fig. 5.7.(a) shows the layered structure in the ANN model. From the left-hand side, the first layer is the *input layer* and each node (neuron) in it represents a feature or attribute of the input data. The number of nodes in the input layer is determined based on the dimensions of the input data. The layers in the middle are named the *hidden layers*. One or more hidden layers can be considered in the ANN structure between the input and output layers. Adding hidden layers increases the complexity of the ANN model. The number of hidden layers and the number of nodes in each hidden layer are design parameters that should be considered based on the complexity and performance requirements of the model. Proper selection of these parameters ensures the accuracy and generalisation capability of the ANN model while avoiding overfitting (see Sect. 5.5.1). The last layer is the *output layer* which is responsible for producing output predictions. One or more neurons may be included in the output layer and this choice depends on specific model requirements.

The number of layers indicates the depth of the network. Deep networks with multiple layers can learn hierarchical representations of the data, enabling them to capture complex relationships in the data. For all cases, the generic input-output relation of the ANN model reads,

$$\mathbf{Y} = g(\mathbf{U}; \mathbf{W}, \mathbf{b}) \tag{5.15}$$

with $\mathbf{U}$ and $\mathbf{Y}$ being input (a.k.a. features) and output (a.k.a. labels) vectors of the ANN model [4] and $\mathbf{W} = [\mathbf{w}_1, ...\mathbf{w}_l, ...\mathbf{w}_P]$, $\mathbf{b} = [b_1, ..., b_l, ..., b_P]$ the parameters of the ANN model with P the number of neurons.

Incubator Next Temperature Prediction Model

Fig. 5.8 shows the inputs and outputs of an ANN whose purpose is to predict the next temperature of the box, from the measured previous temperature, the heater status, and the room temperature. This model is already discretised, and its inputs and outputs are similar to the model in Eq. (5.1) discretised as in Eq. (5.5). The difference resides in the state of the model, which is assumed to be given as input to the network. This is convenient if measurements of the temperature of the box are available, which they are. An example plot used in the dataset is shown in Fig. 5.9. As can be seen, all the required model inputs are available. Samples are equidistant

[4] Input/ouput of the ANN are indicated with capital letters $\{\mathbf{U}, \mathbf{Y}\}$ whereas input/ouput of single neuron are indicated with small letters $\{\mathbf{u}, y\}$.

according to the sample rate of the sensors, which is 3 seconds. The model will therefore predict the temperature after 3 seconds.

One might wonder what the use is of a model that predicts the next temperature value. After all, the physics-based model introduced in Eq. (5.1) is capable of predicting the temperature evolution for any given time interval. As we will show later, the model introduced here can be used for anomaly detection. In general, such models can also be used for improved control, and can be trained to predict more than one temperature reading ahead.

For improved efficiency, all the data used for training are concatenated into the input matrix with shape:

$$\mathbf{U} = \begin{bmatrix} T_{b,0} & H_{h,0} & T_{r,0} \\ T_{b,1} & H_{h,1} & T_{r,1} \\ \dots & \dots & \dots \end{bmatrix}$$

and all the outputs are assembled as

$$\mathbf{Y} = \begin{bmatrix} T_{b,1} \\ T_{b,2} \\ \dots \end{bmatrix}$$

So the neural network is just matrix operations carried out on large matrices. Modern computers have been optimised to carry out such operations.

The neural network used for the incubator data consists of a single layer with 3 inputs and 1 output, and no activation function. After training with 70% of the data as described in Chap. 6, the results can be seen in Fig. 5.9. The area in green represents input data that the model was not trained on.

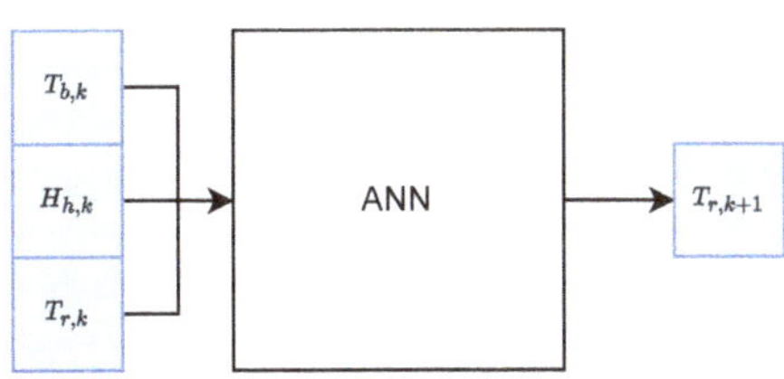

Fig. 5.8: Incubator box temperature prediction neural network models, showing the inputs and output.

Depending on how neurons are connected together, ANNs can also be classified into *acyclic* or *cyclic* graphs (see Fig. 5.7b) [64]. ANNs with acyclic configuration do not include feedback loops and are suitable for approximating *static* nonlinear input-output relationships. On the other hand, ANNs with cyclic configurations include at least one cycle composed of directed arcs. Such ANNs are also referred

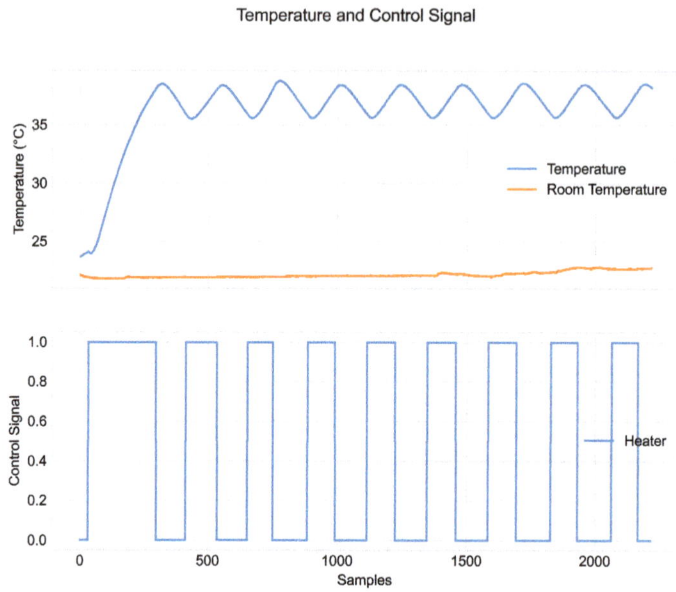

Fig. 5.9: Example dataset used for the incubator next temperature prediction model.

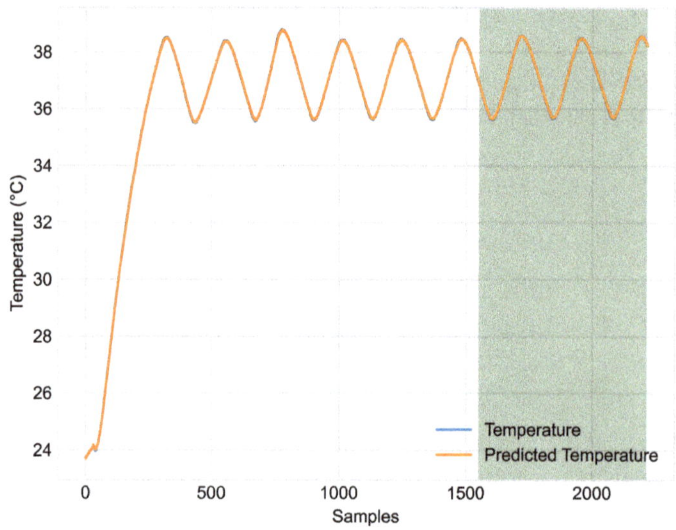

Fig. 5.10: Example model prediction results. The area highlighted in green represents the test data.

to as *recurrent* ANNs. The inclusion of a feedback loop in a recurrent ANN allows for emulating *dynamic* model with the possibility of containing internal memory. Thus they acquire a state just like other dynamical models such as differential or difference equations. In fact, the hidden state in a recurrent ANN can be seen as a continuous-time approximation of the solution to a difference equation. The recurrent connections in the recurrent ANN update the hidden state based on the current input and the previous value of the hidden state. This updating process resembles the iterative nature of difference equations, where each new value is calculated based on the previous values [44]. For example, the generic expression of a discrete time recurrent ANN reads,

$$\mathbf{Y}_{k+1} = g(\mathbf{Y}_k, \mathbf{U}_k; \mathbf{W}, \mathbf{b}) \tag{5.16}$$

where k is the sample index corresponding to the instant t_k, consistently with the notation of Eq. (5.5). More recently, *neural* ODEs [20] have been proposed to generalised recurrent ANN to dynamical models in continuous time,

$$\dot{\mathbf{Y}} = g(\mathbf{Y}, \mathbf{U}; \mathbf{W}, \mathbf{b}) \tag{5.17}$$

For both (Eq. (5.16)) and (Eq. (5.17)), $\mathbf{Y}$ coincides with the state of the dynamical system represented by the ANN. However, neural ODEs, to be trained, are coupled with a time integration scheme as such the Euler method presented in Sect. 5.4.1.

5.5.3 Learning Methods

The term *Learning* refers to the procedure for calibrating the parameters of the data-driven model, a topic discussed in detail in Chap. 6. For the ANN example used in this chapter, the parameters are the synaptic weights $\mathbf{W}$ and the biases $\mathbf{b}$. There are three main learning methodologies, namely, *supervised*, *unsupervised* and *reinforcement* learning, each of which is discussed below.

5.5.3.1 Supervised Learning

In *supervised* learning, the ANN model is trained using *labeled* data, that is, data containing both input and output variables generated by a real system,

$$\mathcal{U} = \{\hat{\mathbf{U}}_1, ..., \hat{\mathbf{U}}_s, ..., \hat{\mathbf{U}}_S\}, \mathcal{Y} = \{\hat{\mathbf{Y}}_1, ..., \hat{\mathbf{Y}}_s, ..., \hat{\mathbf{Y}}_S\} \tag{5.18}$$

with S the number of *samples* in the dataset. To avoid confusion, input/output data is indicated with the hat symbol $\{\hat{\mathbf{U}}, \hat{\mathbf{Y}}\}$ while input/output of the ANN is indicated as $\{\mathbf{U}, \mathbf{Y}\}$. The ANN parameters $\{\mathbf{W}, \mathbf{b}\}$ are tuned to map input variables $\hat{\mathbf{U}}_s$ to the corresponding output variables $\hat{\mathbf{Y}}_s$ by solving the following optimisation problem,

$$\mathbf{W}, \mathbf{b} = \arg\min_{\mathbf{W},\mathbf{b}} \sum_{s=1}^{S} L\left(\hat{\mathbf{Y}}_s, g\left(\hat{\mathbf{U}}_s; \mathbf{W}, \mathbf{b}\right)\right) \tag{5.19}$$

where $L()$ is a *loss* function, which measures the predictive performance of the ANN model [33]. Depending on the type of the output data, one can use the ANN to solve a *regression* or a *classification* problem. For example, in our incubator study, a regression problem is solved if the ANN model is trained to predict the air temperature t_b inside the box based some previous measurements of temperatures and inputs, as shown in Fig. 5.8. A classification problem is solved if the ANN model is trained to predict whether the lid is open or closed based on the same measurements. There are a range of loss functions: an example for regression problems is the *mean square error* whereas an example for classification problems is the *cross-entropy loss*.

5.5.3.2 Unsupervised Learning

In *unsupervised* learning, the ANN model is trained using *unlabelled* data, that is, data containing only input variables generated by a real system,

$$\mathcal{U} = \{\hat{\mathbf{U}}_1, ..., \hat{\mathbf{U}}_s, ..., \hat{\mathbf{U}}_S\} \tag{5.20}$$

with S the number of *samples* in the dataset. The aim of unsupervised learning is to find structures in data and not mapping input/output relationships. This may seem very abstract so consider a couple of examples. Two common applications of unsupervised learning are *clustering* and *dimensionality reduction*. Clustering is used to subdivide data into groups (or clusters) whose elements manifest some strong reciprocal similarity. Dimensionality reduction is used to identify the patterns in data that allow for representing the same information with less data.

Autoencoders are a powerful application of ANNs to perform dimensionality reduction. These ANNs are trained to reconstruct their input data by compressing it into a lower-dimensional representation and then reconstructing it back to its original form. The key idea behind autoencoders is to learn a compressed and meaningful representation of the input data, capturing its essential features without relying on explicit labels. As shown in Fig. 5.11, the architecture of an autoencoder typically consists of an encoder network that compresses the input data into a latent space representation, followed by a decoder network that reconstructs the original input from the compressed representation [5]. The training process involves two main steps, namely, *encoding* and *decoding*. In the encoding phase, the input data is passed through the encoder ANN, which learns to compress the data into a lower-dimensional representation called the latent space. The decoder ANN then takes this compressed representation and attempts to reconstruct the original input. The reconstruction is compared to the input, and the model's parameters are adjusted to minimise the difference between the two. Encoder and decoder ANNs are trained by

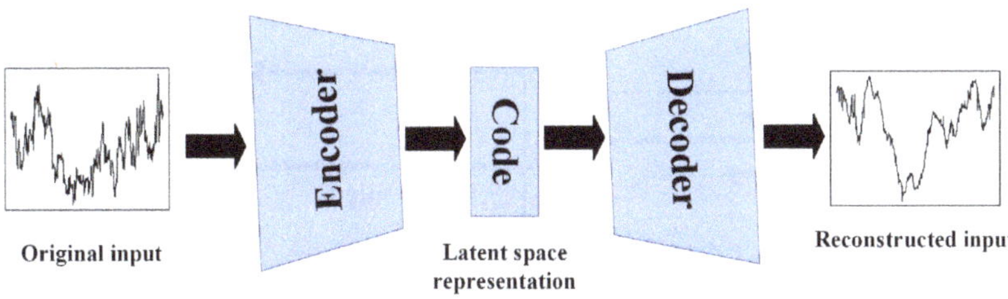

Fig. 5.11: An example of autoencoders. The input time series data is compressed into a latent space representation using the encoder and is reconstructed back to its original form using the decoder.

solving the following optimisation problem,

$$\mathbf{W}_{E,D}, \mathbf{b}_{E,D} = \arg \min_{\mathbf{W}_{E,D}, \mathbf{b}_{E,D}} \sum_{s=1}^{S} L\left(\hat{\mathbf{U}}_s, g_D\left(g_E\left(\hat{\mathbf{U}}_s; \mathbf{W}_E, \mathbf{b}_E\right); \mathbf{W}_D, \mathbf{b}_D\right)\right) \tag{5.21}$$

where the ANN parameters are grouped into $\mathbf{W}_{E,D} = \{\mathbf{W}_E, \mathbf{W}_D\}$ and $\mathbf{b}_{E,D} = \{\mathbf{b}_E, \mathbf{b}_D\}$; L is the *reconstruction* loss function which measures the distance between the output of the decoder (g_D) and the input ($\hat{\mathbf{U}}_s$) of the encoder (g_E). The reconstruction loss function is usually set to be the *mean square error*.

After being trained the ANN can be split into the encoder and decoder, each capable of performing the compression and decompression tasks, respectively.

Different regularisation techniques have been developed for autoencoders to ensure that the learned compressed representation is meaningful. Examples include the under-complete [52], denoising [67], sparse [53], and adversarial [54] autoencoders.

5.5.3.3 Reinforcement Learning

Unlike supervised learning, where labeled examples are provided, and unsupervised learning, where patterns are discovered in unlabelled data, reinforcement learning revolves around an agent interacting with an environment, and receiving feedback in the form of rewards or penalties. The key elements of a reinforcement learning method are the policy function, reward function, environment model, and the reinforcement learning algorithm (as shown in Fig. 5.12).

The policy function within the agent takes observations from the environment and maps them to actions to be applied. For instance, a controller for regulating the heater embedded in the incubator system discussed in Sect. 3.3 can be designed using reinforcement learning. In this example, the observations would be the incubator's

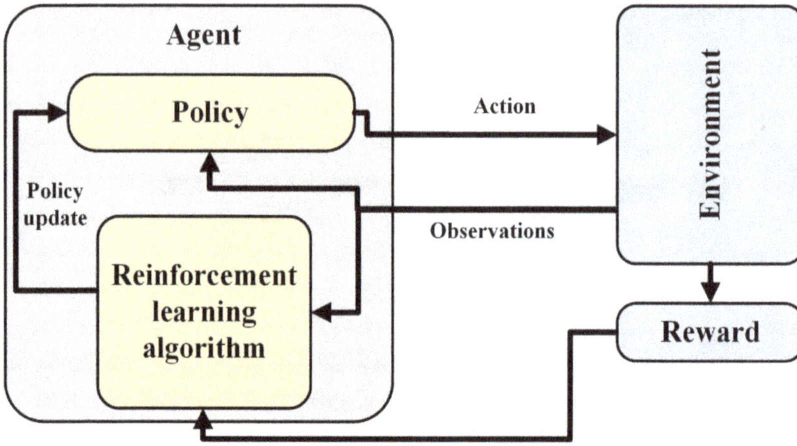

Fig. 5.12: Main elements of the reinforcement learning method. The agent comprises the policy function and reinforcement learning algorithm. The environment can be a real-world system, model, or simulation.

temperature, and the policy would use the current temperature as input and output the appropriate command to turn the heater on or off. The environment then provides a reward value to give the agent feedback on the effectiveness of the command. By considering the actions taken, observations received, and the cumulative reward, the reinforcement learning algorithm learns the best policy through interaction with the environment. The best policy ultimately generates the optimal actions to maximise cumulative reward over time [68].

The training process typically involves the agent taking actions based on its current state, receiving feedback from the environment, and updating its policy or strategy to improve future decision-making. Reinforcement learning algorithms often employ a trial-and-error approach, wherein the agent explores different actions to learn the optimal behaviour by balancing exploration (trying new actions) [48] and exploitation (repeating actions with known high rewards).

Several design parameters influence reinforcement learning [68], including the reward structure, discount factor, exploration-exploitation balance, and the choice of algorithm (e.g., Q-learning or policy gradients), among others (see [57, 68]). Popular applications include robotics, gaming, and optimisation tasks.

5.6 Models for Computer-Based Systems

The growth of computing technology since the middle of the 20th century has been accompanied by an exploration of modelling techniques, notations and tools that permit reasoning about computing systems with the same rigour that can be applied to physical processes. A vast range of formalisms has arisen for describing the behaviour of computing systems, their performance and other qualities such as security and dependability. Most of them are agnostic about the particular programming language employed, or about the specifics of computing hardware. They permit abstraction from those implementation details, instead focussing on the essence of *what* is being computed without overly detailing *how* it is to be done. For this reason we often call these *specification formalisms*. The fact that these formalisms are defined mathematically allows verification processes (as described in Sect. 4.5.3) to provide evidence of value in confirming or refuting claims about a computing system's properties.

This section showcases several models and verification methods for computing processes. Application examples refer to the temperature controller of our incubator example.

Software models can take either *textual* (e.g. VDM [11, 12], Z [69] and B [1]) or *graphical* (e.g. SDL [17], Petri Nets [42], Finite State Machines (FSM [39]), timed automata [3, 50]) forms. Textual models tend to look like programs at first sight, while graphical models have some similarity to block diagrams. Both styles are larely presentational since they have underlying mathematical semantics. Below we consider two example formalisms: the graphical Finite-State Machine (FSM) formalism, and the textual Vienna Development Method (VDM). In illustrating these approaches, we will use examples drawn from the temperature controller of our incubator study.

5.6.1 Finite State Machines

FSMs are a valuable tool for modelling software systems. An FSM represents a finite number of discrete states, each corresponding to different operation modes. These states are connected by transitions, which define the rules for moving between them. Creating an FSM begins by identifying the various system behaviours and the states associated with them. Next, you define the conditions that trigger transitions between states in response to inputs. These transitions can also include actions that modify variables or initiate processes.

Consider using this approach to model the temperature controller of the incubator as an FSM. First, we identify the different operating modes of the controller. Based on the incubator system's block diagram in Fig. 3.5, the controller receives a reading of the average temperature in the box, and determines whether to turn the heater on or off, with the goal of trying to keep the temperature around 37 °C. Consequently, the controller can be in one of two states: *Cooling* or *Heating*, depending on whether the heater is on or off.

The controller starts in the *Cooling* state, and transitions to the *Heating* state when the temperature T_b drops below the Lower Limit LL (36 °C). It transitions back to the *Cooling* state when the temperature exceeds the Upper Limit UL (39 °C). The limits of the controller are chosen to account for noise in the temperature measurement, and to avoid the heater switching on and off too frequently. This line of thinking leads to the FSM of the temperature controller depicted in Fig. 5.13.

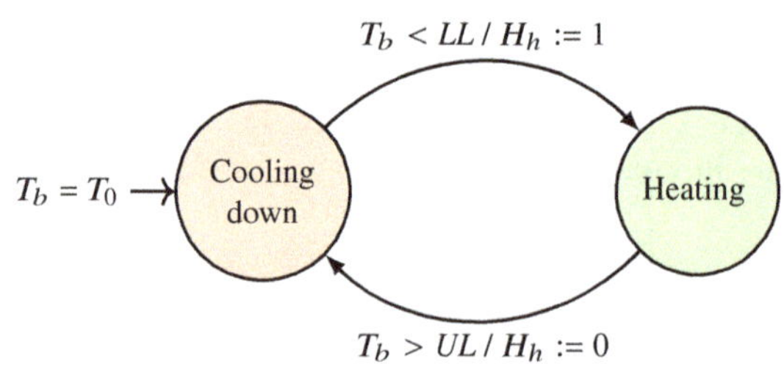

Fig. 5.13: An FSM model of the incubator temperature controller (see Eq. (5.1) to find the meaning of mathematical terms).

In order to simulate such a model, one has to provide some real or virtual temperature readings and observe the controller reactions. However, since a model of the heat transfer within the incubator is not present, one cannot see some important shortcomings associated with the physical implementation such as, for example, communication delays and measurement noise, as discussed in more depth in Chap. 7. Addressing these would require a more complex software model, and one of the drawbacks or of the visual model is that it can quickly become difficult to read if the number of states is large. By contrast a textual formalism such as the next one shown tend to scale.

5.6.2 Vienna Development Method

The Vienna Development Method (VDM) is one of the longest-established textual formalisms for developing computer-based systems and software based on specification models [26]. VDM comes with several tools for verification (see Section 5.6.3). In principle, all computer programming languages can be considered as modelling formalisms describing the desired actions of computers. In order to cover the broad range of programming paradigms (e.g., object-oriented (C++) vs. declarative (LISP)), different specialised versions of VDM have been produced.

For example, VDM++ extends the scope of VDM to object-oriented models [27]; similarly, VDM-RT allows for abstracting time, which is necessary for real-time computer-based systems [65][5].

The following gives an example VDM-RT model for the incubator temperature controller. The structure of a VDM-RT model is the definition of a *system* (i.e., computer-based system) and a collection of *classes*, which are used to define its components. Specifically, the instances of classes are called *objects* and are declared as *instance variables* within a system. Objects can be deployed to abstractions of processors (CPUs) and connected by communication buses. The VDM-RT formalism provides standardised abstractions for such CPUs and buses. Both systems and objects are instantiated by *constructors*. The VDM-RT model for the temperature controller of the incubator is presented as follows:

Listing 5.1: VDM-RT system for the incubator temperature controller.

```
class Controller -- declaration of the controller class
  values -- constants of the class Controller
  LL: real = 36.5; -- lower limit
  UL: real = 39.5; -- upper limit
  instance variables -- member variables
  current_state: <Heating> | <Cooling> := <Cooling>;
  T_in: real := 20.0; -- temperature inside the box
  T_out : real := 20.0; -- temperature outside the box
  heater_on_out : bool := false;

 inv T_in >= T_out; -- invariant

  operations -- methods of the class
  Ctrl_step: () ==> () -- method to be called at each time step
    Ctrl_step() ==
    if T_in >= UL -- too hot
      then (current_state := <Cooling>; heater_on_out := false;)
    elseif T_in <= LL -- too cold
      then (current_state := <Heating>; heater_on_out := true;)
    else if current_state = <Heating> -- if heating
      then (heater_on_out := true;)
    else (heater_on_out := false;)
    update_BoxTemperature();
end Controller
```

The VDM-RT model of the incubator temperature controller is composed of a single class called `Controller`. The class contains instance variables which are used to store the current state of the controller, the temperature inside and outside the box, and an indicator of whether the heater is on or off. The class also contains a description of functionality (a *method* or *operation*)) called `Ctrl_step`, which represents the actions to be taken at each time step. The method checks the

[5] One can find descriptions of each VDM dialect at `https://en.wikipedia.org/wiki/Vienna_Development_Method`.

current temperature and switches the heater on or off accordingly. It also updates the temperature inside the box using the `update_BoxTemperature` method, which is not shown in the listing. The `inv` keyword defines an invariant, a property is assumed to always hold. In this case, the invariant states that the temperature inside the box must be greater than or equal to the temperature outside the box.

Each class can in addition define its own values, types, functions and operations. A class can also include concurrency primitives, which are used to synchronise, coordinate, and communicate between multiple threads or processes that run concurrently.

5.6.3 Verification Methods

The expected behaviour of a model is stated in a *contractual* style. Contracts define the circumstances under which a part of the program can be executed by means of a *pre-condition* and specify properties of the corresponding output by means of a *post-condition* [62]. In the incubator temperature controller, a contract may be used to specify that the box temperature must be maintained within the range $[LL, UL]$, where LL and UL are the minimum and maximum temperatures respectively. A very simple controller model might indicate that, if the temperature drops below LL, the heater is switched on; and if the temperature exceeds UL, the heater is switched off. Verifying the model against the contract involves rigorously checking whether the model respects a given contract. This may be approached using either *reasoning methods* or *exhaustive methods*.

Reasoning methods: These methods focus on the use of formal logic and discrete mathematics. The specification model is typically transformed into a set of logical statements (*axioms*) that describe the properties of the elements in the specification model, including the operations describing functionality. The contract is expressed in the same logical form and a *conjecture* is constructed that the model satisfies the contract [10]. It is this conjecture that must be proved from the axioms. Such a proof is typically performed with a greater or lesser degree of machine support.

Exhaustive methods: These methods explore all states in the specification model to discover whether or not a given property holds. The main technique here is *model checking* [21], which typically involves translating a specification model into an FSM and then exploring the state space of the FSM to find states and transitions that contradict the contract being checked.

Both reasoning and exhaustive methods have their advantages and disadvantages. Reasoning methods are robust and scalable but require considerable specialist expertise in discrete mathematics and logic and are, therefore, more expensive to apply.

Exhaustive methods are less robust and scalable, but they are more accessible, and many toolchains exist to completely automate the process.

5.7 Coupling of Heterogeneous Models

In this book, we focus on DTs of cyber-physical systems (CPSs). CPSs involve both computational and physical processes, and of course the DT-enabled system is itself inherently cyber-physical in character. In pursuing our goal of model-based engineering for such systems, can we possibly rely on the availability of a single formalism to describe these diverse digital, mechanical, electrical and other processes? In practice, no. It will often be the case that the DT-enabled system begins from diverse models that may have been developed independently of one another and of the DT, and which clearly and accurately describe facets of the system to be twinned. There is an important challenge then, in bringing these diverse models together in a sound way. Furthermore, the models may be owned by different parties, which do not intend to disclose their models as they conceal the reciprocal industrial secrets. In response to these challenges a number of approaches have been developed to couple heterogeneous models. In the following, two among the most promising approaches for coupling models are presented, namely, co-simulation and hybrid automata.

5.7.1 Co-Simulation

Co-simulation is the coordinated simulation of an ensemble of models that are coupled through their inputs and outputs using a standardised software interface. The use of such an interface allows the individual models to keep their inner workings private. Co-simulation [32] promotes the idea that different subsystems are best modelled and simulated using specialised tools. Each subsystem is modelled using the most appropriate modelling formalism and simulation tool, which is then wrapped in a standardised interface that allows it to be integrated with other subsystems without worrying about the details of the underlying simulation tool or modelling formalism.

Interoperability between heterogeneous models is often achieved using the Functional Mock-up Interface (FMI) standard [14, 13, 45], which defines a standardised interface. This interface is supported by a wide range of modelling and simulation tools, including Simulink [55], Dymola [15], OpenModelica [31] and 20-sim [46] which all allow models to be exported as Functional Mock-up Units (FMUs). An FMU is a binary file containing the model code and meta-data describing the model structure and the interface to provide controlled access to the model it represents. The model code implements three functions: `get` and `set` and `doStep`, which make it possible to query the value of a variable, set it, or ask the model to simulate for a short period of simulation time.

The interface provided by FMI enables a tool-agnostic approach in which models may be used and combined without requiring access to or knowledge of their individual underlying simulation tools and implementations. Consequently, models can shared and used in interdisciplinary projects without requiring the different domain specialists to learn how to use each other's tools, which can be a significant barrier to collaboration.

The Modelica Association[6] continuously updates the FMI standard to facilitate the simulation of increasingly complex models and systems. Although FMI 3.0 is the latest version of the standard, our discussion will focus on FMI 2.0, as it is (at the time of writing) the most widely supported version with more than 140 tools [28]. FMI 2.0 provides two interfaces, called *co-simulation* and *model exchange* describing the functions that can be used to simulate the encapsulated model known as an FMU. The co-simulation interface is the most commonly used interface, as it is the most flexible and allows the FMU to be simulated independently of the simulation environment. It comprises a set of input and output ports that allow stimuli to be applied to the model embodied in the FMU and observe its behaviour. The content of the FMU remains private.

A system composed of multiple subsystems, such as the incubator, can be constructed from a collection of FMUs by coupling their inputs and outputs using a set of *connections*. A connection denotes a data dependency between the output of one FMU and the input of another. A collection of FMUs and connections between them is called a *scenario*.

Co-simulation of the Incubator

A scenario for the incubator is shown in Fig. 5.14, composed of three FMUs: `Plant`, `Room`, and `Controller`. Comparing it with the block diagram in Fig. 3.5, one can see that the `Plant` implements a simulation of the incubator box temperature dynamics, introduced in Eq. (5.1); and that the `Controller` implements a simulation of the state machine introduced in Fig. 5.13. Finally the `Room` just represents a model of the room temperature dynamics, which is just providing an input to the incubator dynamics. This co-simulation yields the results presented in Fig. 3.7, through the steps presented in the remainder of this section.

[6] `https://modelica.org/`

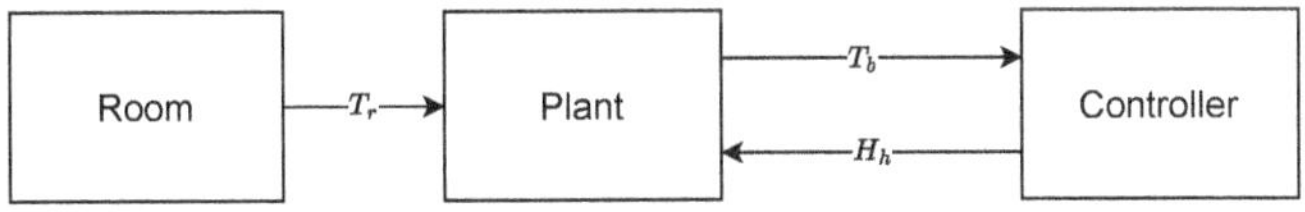

Fig. 5.14: The co-simulation scenario of the incubator.

The behaviour of the co-simulation is defined by the combined behaviour of the FMUs, which can be obtained by simulating each FMU in parallel while exchanging values on the coupled ports throughout the simulation. This proceeds as follows:

1. **Load FMUs:** load the FMUs that are part of the scenario.
2. **Initialisation:** initialise the scenario by initialising the FMUs, setting their parameter values, and exchanging the initial outputs so that all FMUs have the initial values for their inputs.
3. **Begin Simulation:** begin the simulation by setting the simulation time to the starting time and setting the simulation step size.
4. **Exchange Values:** exchange values on the coupled ports using `get` and `set`.
5. **Update FMUs:** advance state of the FMUs by calling the `doStep` function of the FMUs.
6. **Advance Time:** advance the simulation time by the simulation step size.
7. **Iteration:** Repeat steps 2 to 6 until the simulation time reaches the desired endpoint or termination criteria are met.
8. **Termination:** terminate the simulation and unload the FMUs.
9. **Output Collection:** Throughout the simulation, collect and store data about the system's states, variables, and any relevant outputs. This data can then be used for analysis and visualisation.
10. **Visualisation and Interpretation:** Visualisations might include state trajectories, changes in continuous variables, and transitions between modes.

The steps above are typically carried out by a *co-simulation engine* such as Maestro2 [38], and the algorithm to complete each simulation step (also known as the orchestration algorithm) may vary and can be optimised to specific co-simulation scenarios. For more details on co-simulation, we direct the reader to [32].

5.7.2 Hybrid Automata

The incubator is a good example of *hybrid system*, that is a system where continuous dynamics are mixed with discrete state transitions. Specifically, heat transfer within the box, which is associated with the continuous dynamics, can be either described using first principle models based on ODEs/PDEs or data-driven models produced using ML. The temperature controller, which is a computer-based system, can be efficiently represented using a FSM.

In general, the behaviour of a hybrid system can be described by alternative sets of differential equations, which can be switched on and off when certain conditions are met. Such differential equations (typically ODEs), define different continuous behaviours of the same hybrid system. Hybrid systems can be described using various formalisms, such as *hybrid automata* [40], *hybrid process algebra* [23] and *differential dynamic logic* [59, 60]. Among these, following the work of Alur & Dill [4, 2], hybrid automata have become popular.

A hybrid automaton has a set of *state variables* which may evolve continuously or discretely. This evolution is described by a control graph that consists of modes or locations linked by edges. When the system is in a given mode, the state variables evolve in accordance with a set of ODEs. The system may transition from one mode to another, meaning that a *discrete event* has occurred. Each mode (continuous evolution) is labelled by an initial condition on the state variables, an invariant which holds throughout the evolution in that mode, and a set of *flow equations* – the ODEs describing behaviour in that mode. A mode thus describes a region of the space of all possible states. The transition between locations is driven by *guards*, which as part of the transition, execute a set of *assignments*. An example of assignment is the update of the state variables. Several tools have been developed to simulate hybrid automata, including SpaceEx [29], Uppaal [6], HyTech [41], and Modest [37].

Fig. 5.15 shows an example of a hybrid automaton which describes the behaviour of the incubator, combining the ODE in Eq. (5.1) and the controller FSM in Fig. 5.13. Two locations are represented, showing the case when the incubator is warming up, and when the incubator is cooling down. The advantage of the hybrid automaton model compared to the FSM of Sect. 5.6.1, is that the temperature controller is coupled with the ODE of Eq. (5.1), which represents heat transfer within the incubator. The results are presented in Fig. 3.7.

The behaviour of a hybrid automaton can be computed by iteratively applying the following steps:

1. **Initialisation:** begin by initialising the model's initial state. This involves setting the values of both location and other variables, including differential equation states.
2. **Discrete Transition Detection:** Check if the current state of the system satisfies the conditions for any discrete transitions (mode changes). These conditions are often specified as guards associated with transitions.
3. **Discrete Transition Handling:** If a discrete transition condition is met, update the current location and any relevant discrete variables. This transition might also trigger changes in the continuous dynamics or initiate discrete actions.
4. **Continuous Dynamics Update:** Using numerical integration methods (e.g., Euler's method or Runge-Kutta methods), update the continuous variables based on their current values and the current location's differential equations. This is the simulation procedure for the continuous dynamics, as explained in Sect. 5.4, except it is carried out for a very small period of simulation time, in order to allow for discrete transition detection and handling as in the previous steps.

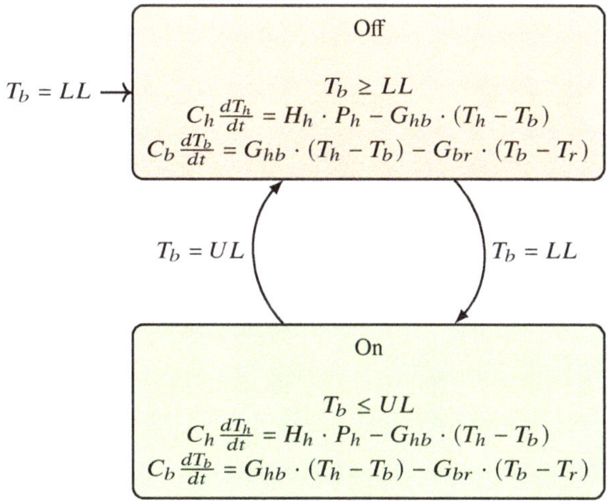

Fig. 5.15: A hybrid automaton describing the behaviour of a thermostat that controls the temperature of a room by switching a heater on (LL) and off (LU).

5. **Event Handling and Simulation Termination:** Throughout the simulation, check for events that might halt the simulation, such as reaching a predefined time limit, a specific state configuration, or other termination criteria.

Steps 2 to 5 are repeated until the simulation time reaches the desired endpoint or termination criteria are met. At each iteration, the system's mode, discrete variables, and continuous variables may change based on transitions and dynamics. Throughout the simulation, collect and store data about the system's states, variables, and any relevant outputs. This data can then be used for analysis and visualisation. Visualisations might include state trajectories, changes in continuous variables, and transitions between modes.

Hybrid automata models can be used both for simulation of the trajectory of the hybrid system and in combination with verification methods like those described in Sect. 5.6.3. This latter use is typically not possible with co-simulation. Also, in contrast to co-simulation, hybrid automata models will typically be implemented in a single tool, which may limit their applicability in multi-stakeholder CPS development.

References

1. Abrial, J.R.: The B Book – Assigning Programs to Meanings. Cambridge University Press (1996)
2. Alur, R., Courcoubetis, C., Henzinger, T.A., Ho, P.: Hybrid automata: An algorithmic approach to the specification and verification of hybrid systems. In: R.L. Grossman, A. Nerode, A.P. Ravn, H. Rischel (eds.) Hybrid Systems, *Lecture Notes in Computer Science*, vol. 736, pp. 209–229. Springer (1992)
3. Alur, R., Dill, D.: Automata for modeling real-time systems. ICALP'91 (1991)
4. Alur, R., Henzinger, T.: Realtime logics: complexity and expressiveness. Tech. Rep. STANCS901307, Departments of Computer Science and Medicine,Stanford, CA (1990). Appeared in the 5th IEEE LICS'90
5. Bank, D., Koenigstein, N., Giryes, R.: Autoencoders, pp. 353–374. Springer International Publishing, Cham (2023). DOI: 10.1007/978-3-031-24628-9_16
6. Behrmann, G., David, A., Larsen, K.G.: A Tutorial on Uppaal. In: M. Bernardo, F. Corradini (eds.) Formal Methods for the Design of Real-Time Systems, International School on Formal Methods for the Design of Computer, Communication and Software Systems, SFM-RT 2004, Bertinoro, Italy, September 13-18, 2004, Revised Lectures, *Lecture Notes in Computer Science*, vol. 3185, pp. 200–236. Springer (2004). DOI: 10.1007/978-3-540-30080-9_7
7. Benner, P., Grivet-Talocia, Quarteroni, A., Rozza, G., Schilders, W., Silveira, L.M.: Model-order reduction vol. 2: Snapshot-Based Methods and Algorithms. De Gruyter (2021). DOI: 10.1515/9783110671490
8. Benner, P., Grivet-Talocia, S., Quarteroni, A., Rozza, G., Schilders, W., Silveira, L.M. (eds.): Model-order reduction vol. 1: System- and Data-Driven Methods and Algorithms. De Gruyter (2021). DOI: 10.1515/9783110498967
9. Benner, P., Grivet-Talocia, S., Quarteroni, A., Rozza, G., Schilders, W., Silveira, L.M. (eds.): Model-order reduction vol. 3: Applications. De Gruyter (2021). DOI: 10.1515/9783110499001
10. Bibel, W.: Automated Theorem Proving. Springer Science & Business Media
11. Bjørner, D., Jones, C. (eds.): The Vienna Development Method: The Meta-Language, *Lecture Notes in Computer Science*, vol. 61. Springer-Verlag (1978). DOI: 10.1007/3-540-08766-4
12. Bjørner, D., Jones, C. (eds.): Formal Specification and Software Development. Prentice-Hall International (1982)
13. Blochwitz, T., Otter, M., Akesson, J., Arnold, M., Clauss, C., Elmqvist, H., Friedrich, M., Junghanns, A., Mauss, J., Neumerkel, D., Olsson, H., Viel, A.: The Functional Mockup Interface 2.0: The Standard for Tool independent Exchange of Simulation Models. In: Proceedings of the 9th International Modelica Conference. Munich, Germany (2012)
14. Blockwitz, T., Otter, M., Åkesson, J., Arnold, M., Clauss, C., Elmqvist, H., Friedrich, M., Junghanns, A., Mauss, J., Neumerkel, D., Olsson, H., Viel, A.: Functional Mockup Interface 2.0: The Standard for Tool independent Exchange of Simulation Models. In: Proc. 9th International Modelica Conference. Linköping University Electronic Press (2012)
15. Brück, D., Elmqvist, H., Mattsson, S., Olsson, H.: Dymola for multi-engineering modeling and simulation. Tech. rep., Proceedings of modelica (2002)
16. Butcher, J.C.: Numerical methods for ordinary differential equations, 2nd ed edn. Wiley, Chichester, England ; Hoboken, NJ (2008). OCLC: ocn191024153
17. CCITT: Recommendation Z.100. CCITT Specification and Description Language SDL (1988)
18. Cellier, F.E., Kofman, E.: Continuous System Simulation. Springer Science & Business Media
19. Chakraborty, S., Tomsett, R., Raghavendra, R., Harborne, D., Alzantot, M., Cerutti, F., Srivastava, M., Preece, A., Julier, S., Rao, R.M., Kelley, T.D., Braines, D., Sensoy, M., Willis, C.J., Gurram, P.: Interpretability of deep learning models: A survey of results. In: 2017 IEEE SmartWorld, Ubiquitous Intelligence and Computing, Advanced and Trusted Computed, Scalable Computing and Communications, Cloud and Big Data Computing, Internet of People

and Smart City Innovation (SmartWorld/SCALCOM/UIC/ATC/CBDCom/IOP/SCI), pp. 1–6 (2017). DOI: 10.1109/UIC-ATC.2017.8397411
20. Chen, R.T., Rubanova, Y., Bettencourt, J., Duvenaud, D.K.: Neural ordinary differential equations. Advances in neural information processing systems **31** (2018)
21. Clarke, E.M., Henzinger, T.A., Veith, H., Bloem, R. (eds.): Handbook of Model Checking. Springer (2018). DOI: 10.1007/978-3-319-10575-8
22. Constantin, P., Foias, C.: Navier-stokes equations. University of Chicago Press (2020)
23. Cuijpers, P.J.L., Reniers, M.A.: Hybrid process algebra. Journal of Logic and Algebraic Programming **62**(2), 191–245 (2005)
24. De Kooning, J.D., Stockman, K., De Maeyer, J., Jarquin-Laguna, A., Vandevelde, L.: Digital twins for wind energy conversion systems: a literature review of potential modelling techniques focused on model fidelity and computational load. Processes **9**(12), 2224 (2021)
25. Feng, H., Gomes, C., Sandberg, M., Thule, C., Lausdahl, K., Larsen, P.G.: Developing a Physical and Digital Twin: An Example Process Model. In: 2021 ACM/IEEE International Conference on Model Driven Engineering Languages and Systems Companion (MODELS-C), pp. 286–295. IEEE (2021)
26. Fitzgerald, J., Larsen, P.G.: Modelling Systems – Practical Tools and Techniques in Software Development. Cambridge University Press, The Edinburgh Building, Cambridge CB2 2RU, UK (1998). DOI: 10.1145/1668862.1668879. ISBN 0-521-62348-0
27. Fitzgerald, J., Larsen, P.G., Mukherjee, P., Plat, N., Verhoef, M.: Validated Designs for Object–oriented Systems. Springer, New York (2005). DOI: 10.1007/b138800
28. FMI: Functional mock-up interface tools (2014)
29. Frehse, G., Guernic, C.L., Donzé, A., Cotton, S., Ray, R., Lebeltel, O., Ripado, R., Girard, A., Dang, T., Maler, O.: SpaceEx: Scalable Verification of Hybrid Systems. In: G. Gopalakrishnan, S. Qadeer (eds.) Computer Aided Verification - 23rd International Conference, CAV 2011, Snowbird, UT, USA, July 14-20, 2011. Proceedings, *Lecture Notes in Computer Science*, vol. 6806, pp. 379–395. Springer (2011). DOI: 10.1007/978-3-642-22110-1_30
30. Frey, P., Ehrismann, B., Drück, H.: Development of artificial neural network models for sorption chillers. In: Proceedings of the ISES Solar World Congress (2011)
31. Fritzson, P.: Principles of Object-Oriented Modeling and Simulation with Modelica 3.3: A Cyber-Physical Approach, 2 edn. IEEE Press. Wiley (2015). DOI: 10.1002/9781118989166
32. Gomes, C., Thule, C., Broman, D., Larsen, P.G., Vangheluwe, H.: Co-simulation: a Survey. ACM Comput. Surv. **51**(3), 49:1–49:33 (2018)
33. Goodfellow, I., Bengio, Y., Courville, A.: Deep learning. MIT press (2016). http://www.deeplearningbook.org
34. Griffiths, D.F., Higham, D.J.: Numerical Methods for Ordinary Differential Equations. Springer Undergraduate Mathematics Series. Springer London, London (2010). DOI: 10.1007/978-0-85729-148-6
35. Gudivada, V., Apon, A., Ding, J.: Data quality considerations for big data and machine learning: Going beyond data cleaning and transformations. International Journal on Advances in Software **10**(1), 1–20 (2017)
36. Gurtin, M.E.: An Introduction to Continuum Mechanics. Academic press
37. Hahn, E.M., Hartmanns, A., Hermanns, H., Katoen, J.: A compositional modelling and analysis framework for stochastic hybrid systems. Formal Methods Syst. Des. **43**(2), 191–232 (2013). DOI: 10.1007/s10703-012-0167-z
38. Hansen, S.T., Thule, C., Gomes, C., Lausdahl, K.G., Madsen, F.P., Abbiati, G., Larsen, P.G.: Co-simulation at different levels of expertise with Maestro2. Journal of Systems and Software p. 111905 (2023). DOI: https://doi.org/10.1016/j.jss.2023.111905
39. Harel, D.: Statecharts: A visual formalism for complex systems **8**(3), 231–274. DOI: 10.1016/0167-6423(87)90035-9
40. Henzinger, T.A.: The Theory of Hybrid Automata. In: Verification of Digital and Hybrid Systems, pp. 265–292. Springer, Berlin, Germany (2000)

41. Henzinger, T.A., Ho, P., Wong-Toi, H.: HYTECH: A model checker for hybrid systems. Int. J. Softw. Tools Technol. Transf. **1**(1-2), 110–122 (1997). DOI: 10.1007/s100090050008
42. Jensen, K.: Coloured Petri Nets. In: W.R. W.B̃rauer, G. Rozenberg (eds.) Petri Nets: Central Models and Their Properties: Advances in Petri Nets 1986. Lecture Notes in Computer Science 254, Springer-Verlag, Berlin Heidelberg New York (1987)
43. Jia, W., Wang, W., Zhang, Z.: From simple digital twin to complex digital twin Part I: A novel modeling method for multi-scale and multi-scenario digital twin. Advanced Engineering Informatics **53**(July), 101706 (2022). DOI: 10.1016/j.aei.2022.101706
44. Jin, L., Nikiforuk, P., Gupta, M.: Approximation of discrete-time state-space trajectories using dynamic recurrent neural networks. IEEE Transactions on Automatic Control **40**(7), 1266–1270 (1995). DOI: 10.1109/9.400480
45. Junghanns, A., Blochwitz, T., Bertsch, C., Sommer, T., Wernersson, K., Pillekeit, A., Zacharias, I., Blesken, M., Mai, P., Schuch, K., Schulze, C., Gomes, C., Najafi, M.: The Functional Mock-up Interface 3.0 - New Features Enabling New Applications. In: Proceedings of the 14th International Modelica Conference. Linköping University Electronic Press, Linköpings Universitet, online (2021). DOI: 10.3384/ecp2118117
46. Kleijn, C.: Modelling and Simulation of Fluid Power Systems with 20-sim. Intl. Journal of Fluid Power **7**(3) (2006)
47. Kloeden, P.E., Platen, E., Kloeden, P.E., Platen, E.: Stochastic Differential Equations. Springer
48. Ladosz, P., Weng, L., Kim, M., Oh, H.: Exploration in deep reinforcement learning: A survey. Information Fusion **85**, 1–22 (2022). DOI: https://doi.org/10.1016/j.inffus.2022.03.003
49. Ladyzhenskaya, O.A.: Sixth problem of the millennium: Navier-Stokes equations, existence and smoothness **58**(2), 251
50. Larsen, K.G., Pettersson, P., Yi, W.: Uppaal in a nutshell. STTT **1**(1-2), 134–152 (1997)
51. Liu, Q., Liang, J., Ma, O.: A physics-based and data-driven hybrid modeling method for accurately simulating complex contact phenomenon. Multibody System Dynamics **50**, 97–117 (2020)
52. Lopez Pinaya, W.H., Vieira, S., Garcia-Dias, R., Mechelli, A.: Chapter 11 - autoencoders. In: A. Mechelli, S. Vieira (eds.) Machine Learning, pp. 193–208. Academic Press (2020). DOI: https://doi.org/10.1016/B978-0-12-815739-8.00011-0
53. Luo, W., Li, J., Yang, J., Xu, W., Zhang, J.: Convolutional sparse autoencoders for image classification. IEEE transactions on neural networks and learning systems **29**(7), 3289–3294 (2017)
54. Makhzani, A., Shlens, J., Jaitly, N., Goodfellow, I., Frey, B.: Adversarial autoencoders. arXiv preprint arXiv:1511.05644 (2015)
55. MathWorks: http://www.mathworks.com/ (2011). Simulink official website
56. McCulloch, W.S., Pitts, W.: A logical calculus of the ideas imminent in nervous activity. bulletin of mathematical biophysics, vol. 5, pp. 115-133. Journal of Symbolic Logic **9** (1943)
57. Moos, J., Hansel, K., Abdulsamad, H., Stark, S., Clever, D., Peters, J.: Robust reinforcement learning: A review of foundations and recent advances. Machine Learning and Knowledge Extraction **4**(1), 276–315 (2022)
58. Nelles, O.: Nonlinear System Identification: From Classical Approaches to Neural Networks and Fuzzy Models. Springer Science & Business Media
59. Platzer, A.: Differential dynamic logic for hybrid systems. J. Autom. Reas. **41**(2), 143–189 (2008). DOI: 10.1007/s10817-008-9103-8
60. Platzer, A.: Logical Foundations of Cyber-Physical Systems. Springer (2018). DOI: 10.1007/978-3-319-63588-0
61. Reddy, J.N.: Introduction to the Finite Element Method. McGraw-Hill Education
62. Sangiovanni-Vincentelli, A., Damm, W., Passerone, R.: Taming Dr. Frankenstein: Contract-Based Design for Cyber-Physical Systems* **18**(3), 217–238. DOI: 10.3166/ejc.18.217-238

63. Smirnov, G.V.: Introduction to the Theory of Differential Inclusions, vol. 41. American Mathematical Society
64. Usama, M., Qadir, J., Raza, A., Arif, H., Yau, K.L.A., Elkhatib, Y., Hussain, A., Al-Fuqaha, A.: Unsupervised machine learning for networking: Techniques, applications and research challenges. IEEE access **7**, 65579–65615 (2019)
65. Verhoef, M., Larsen, P.G., Hooman, J.: Modeling and Validating Distributed Embedded Real-Time Systems with VDM++. In: J. Misra, T. Nipkow, E. Sekerinski (eds.) FM 2006: Formal Methods, Lecture Notes in Computer Science 4085, pp. 147–162. Springer-Verlag (2006). DOI: 10.1007/11813040_11
66. Versteeg, H.K., Malalasekera, W.: An Introduction to Computational Fluid Dynamics: The Finite Volume Method. Pearson education
67. Vincent, P., Larochelle, H., Bengio, Y., Manzagol, P.A.: Extracting and composing robust features with denoising autoencoders. In: Proceedings of the 25th international conference on Machine learning, pp. 1096–1103 (2008)
68. Wiering, M.A., Van Otterlo, M.: Reinforcement learning. Adaptation, learning, and optimization **12**(3), 729 (2012)
69. Woodcock, J., Davies, J.: Using Z – Specification, Refinement, and Proof. Prentice Hall International Series in Computer Science (1996)

Chapter 6
Calibration of Models for Digital Twins

Cláudio Gomes, Hao Feng, Zahra Kazemi and Ken Pierce

Abstract We provide an overview of techniques that can be used to calibrate the models that form the basis of digital twins of cyber-physical systems. Calibration (or parameter estimation) involves finding values for the model parameters that optimise the model's capabilities, particularly the accuracy of its predictions about the future evolution of the modelled system. We consider techniques to mitigate noise in the data observed from the real system, beginning with a least squares method that is unable to calibrate nonlinear models. A gradient descent method that remedies this is described, as are alternative approaches such as Newton and Gauss-Newton methods. These methods bring the risk of only finding local minima. Genetic algorithms for design space exploration and other heuristic methods are briefly outlined.

6.1 Introduction

A Digital Twin (DT) is based on one or more models of the Physical Twin (PT) that it aims to complement. As described in Chap. 5, these models can take a variety of forms, but they all need to be a "good enough" representation of the PT to allow dependable services to be built around them. Models will typically have parameters that allow them to be tailored to individual systems. For example, a model of a generic robot manipulator will have parameters that can be tuned to describe not an abstract manipulator but a specific one, with its quirks and idiosyncrasies. This tailoring process is called *calibration* or *parameter estimation*, and the predictive power of the digital model depends on it being done well.

The PTs considered in this book are in fact *Cyber*-Physical Systems (CPSs) that can be deployed in a range of environments, not all of which can be known completely in advance. Calibration comes into its own in those situations in which a system is operated in conditions that were not previously expected.

J. Fitzgerald et al. (eds.), *The Engineering of Digital Twins*,
https://doi.org/10.1007/978-3-031-66719-0_6

Calibration relies on data produced from experiments with the PT. Different types of model require different calibration procedures. For example, a kinematic model of a robot manipulator requires careful experimental planning and extra measuring equipment. However, measurements can easily be contaminated by noise in the data, requiring the use of methods able to cope with this. Differential equations that are linear in the parameters (discussed later in this chapter) require the application of least squares techniques to mitigate noise. Neural network models, on the other hand, which are nonlinear in the parameters, require gradient descent-based techniques and special hardware to permit efficient calibration. Although it is beyond the scope of this book provide an exhaustive review of calibration, this chapter provides an introduction to this subject. A deeper treatment can be found in more general texts on system identification such as [7, 8, 9].

This chapter first briefly explains the nature of calibration Sect. 6.2 before exploring the process for linear systems (in Sect. 6.3) and non-linear systems (in Sect. 6.4). Calibration is rarely perfect, and so it is important to consider the practicalities of overfitting and underfitting (Sect. 6.5). The role of calibration in our running incubator example is discussed in Sect. 12.3.

6.2 What is Calibration?

If you come from a computing background, you might think of "constants" as unchanging values, but in this chapter we are thinking of these as values that can be different when a product is applied in different environments. Think of the friction between a vehicle and the road which changes in new conditions, or the force of gravity which is different in different locations. We may need to calibrate in order to get new and better values for these now and then.

The term *calibration* has different meanings in different disciplines. For our purposes, calibration means estimating the values for model parameters such that the model behaviour matches some behavioural data derived from the real world.

For example, consider an algebraic equation model of a line:

$$y = ax + b \tag{6.1}$$

This is an abstract model capable of representing many systems, where a and b are parameters that are related to system properties and x and y are quantities that we can observe in the system. Each system of this type corresponds to a pair of a and b which are the parameters that need to be calibrated. In addition to this model, systems can be abstracted by other types of models such as differential equations, finite state machines, etc. Each type of model requires different calibration techniques, from techniques that "instantly" compute the best parameter values for the data given if the model is linear or "simple", to techniques that try iteratively to calculate the best parameter values using gradients, in an optimisation process when the model is nonlinear or "complex". When dealing with nonlinear or complex models using gradients, algorithms have the potential to "get stuck" in a local minimum. To manage this risk, heuristic methods such as genetic algorithms can be applied. We go on to examine each of these kinds of technique in the remainder of the chapter.

6.3 Calibration of Linear Algebraic Models

Let us consider a system that has two numerical inputs and one numerical output. We can run experiments by stimulating the inputs and observing the output. The data gathered from such experiments gives us a basis for calibrating a model, trying to fit it to the observed data. We can then assess the predictive power of the model by checking its predictions against more experimental data. Let us work through this process.

Assume that we propose a simple system model in which $x_1 \in \mathbb{R}$ and $x_2 \in \mathbb{R}$ are the inputs, and $y \in \mathbb{R}$ is the output. We are assuming that all quantities are measurable[1]. Our proposed (linear) model takes the following form:

$$y = a \times x_1 + b \times x_2, \tag{6.2}$$

where $a \in \mathbb{R}$ and $b \in \mathbb{R}$ are the parameters in the model that can be adjusted in the light of the experimental data and $\times$ is multiplication. This model represents the relation between system inputs and outputs, which means that if we suggest values for a and b are known, then using Eq. (6.2) we can predict the outputs of the system that correspond to the given inputs. So, to use the model, the values for a and b need to be estimated, i.e., Eq. (6.2) needs to be calibrated for the real system that can be observed.

To look at the calibration of this linear model, let us first assume somewhat unrealistically, that there are no sources of noise in our measurements (we revisit this assumption in Sect. 6.3.1). We should collect two valid samples (input-output pairs) to construct a system of two equations that can be solved for the two parameters. It is not the case that just any samples will do: the calibration can fail if the collected samples are not valid. For example, if the samples are

$$x_1 = 1,\ x_2 = 2,\ y = 2$$
$$x_1 = 2,\ x_2 = 4,\ y = 4$$

[1] By "measurable" here we mean that the values of a quantity can be measured directly from sensors. Unmeasurable quantities cannot be measured by sensors directly, for example, the heater temperature in our incubator case. For more details, please refer to [2].

substituting them into Eq. (6.2) yields the following system of two equations:

$$\begin{aligned} 2 &= a \times 1 + b \times 2 \\ 4 &= a \times 2 + b \times 4. \end{aligned} \tag{6.3}$$

If we try to solve the above system, we end up with actually one equation, the other equation is only a scaled version of this equation, which adds no information as to what the values of the parameters might be (meaning, the parameters can take any value and still satisfy Eq. (6.3)). This is because the vectors $[1, 2]^T$ and $[2, 4]^T$ are not linearly independent. In fact, one is just a scaling of the other, so intuitively, it adds no information than we gain from the first vector.

To quickly and automatically determine whether the vectors are linearly independent, we can organise Eq. (6.3) into a matrix form:

$$\begin{bmatrix} 2 \\ 4 \end{bmatrix} = \underbrace{\begin{bmatrix} 1 & 2 \\ 2 & 4 \end{bmatrix}}_{A} \begin{bmatrix} a \\ b \end{bmatrix}, \tag{6.4}$$

and compute the rank of the matrix A above, denoted as $\text{rank}(A)$[2]. If A is not full rank (i.e., $\text{rank}(A)$ is equal to the number of columns), then the collected samples are not valid. We also called such samples *linearly dependent* or *not linearly independent*.

The consequence of not having enough linearly independent samples is that there are infinitely many solutions to a and b and it leads to invalid calibration. Instead, if the samples are for example

$$\begin{aligned} x_1 = 1, x_2 = 2, y = 2 \\ x_1 = 3, x_2 = 5, y = 6 \end{aligned}$$

we can construct a system of two independent equations:

$$\begin{aligned} 2 &= a \times 1 + b \times 2 \\ 6 &= a \times 3 + b \times 5, \end{aligned} \tag{6.5}$$

which the reader can verify corresponds to a unique solution for a and b. In such a case, one can also multiply the inverse of A (i.e., A^{-1}) to Eq. (6.4) and quickly get the parameters a and b for calibration.

[2] Rank is an operation that can be found with most linear algebra programming libraries.

6.3.1 Calibration Under Noisy Measurements

In practice, data from a real PT experiment will almost inevitably contain noise (as discussed in more depth in Chap. 7), and this can cause conflicts in calibration. For example, consider our linear model in Eq. (6.2). Noise may appear in any of our variables, y, x_1 and x_2, but suppose that we have collected three independent but noisy samples, which are:

$$x_1 = 1, x_2 = 2, y = 3 + 0.1$$
$$x_1 = 3, x_2 = 2, y = 5 - 0.1$$
$$x_1 = 2, x_2 = 2, y = 4 - 0$$

In this example, we only show the noise in y. There are three samples, so we can choose two, say the first two, to calibrate the system. Substituting the samples into Eq. (6.2), the equations become:

$$\begin{aligned} 3.1 &= a \times 1 + b \times 2 \\ 4.9 &= a \times 3 + b \times 2. \end{aligned} \tag{6.6}$$

Solving Eq. (6.6), we obtain $a = 0.9$ and $b = 1.1$. If we use the third sample to verify the obtained results by predicting the y value with the corresponding x_1, x_2, and using the calibrated a and b, we get $y_{predict} = 4$ while the measured y value is 3.9, and thus there is a conflict in solving Eq. (6.6). Similarly, we could use the last two samples to obtain the $a = 1$ and $b = 0.95$, leading to $y_{predict} = 2.9$ and there is a discrepancy of 0.2 while the discrepancy found using the first two samples was 0.1. If we had more, say n, samples, we could use two of them to calibrate the system and probably get $n - 2$ conflicting results. So the question is now how to find "the best" values for the model's parameters when there is noise.

Suppose that we have the data from a series of experiments. For the ith experiment, we have provided inputs $x_1^{(i)}$ and $x_2^{(i)}$, and observed a result $y^{(i)}$. The difference between $y^{(i)}$ and the value predicted by our model is

$$y^{(i)} - (a \times x_1^{(i)} + b \times x_2^{(i)})$$

We call this the *residual*, denoted by $\epsilon^{(i)}$, rather like the loss in Artificial Neural Networks (ANNs), as mentioned in Chap. 5. The least squares method, aims to help us reach model parameters that minimise the sum of the squares of all the residuals in the experimental data. We denote this as the loss L:

$$\begin{aligned} \text{L} &= \sum_i (\epsilon^{(i)})^2 \\ &= \sum_i (y^{(i)} - (a \times x_1^{(i)} + b \times x_2^{(i)}))^2, \end{aligned} \tag{6.7}$$

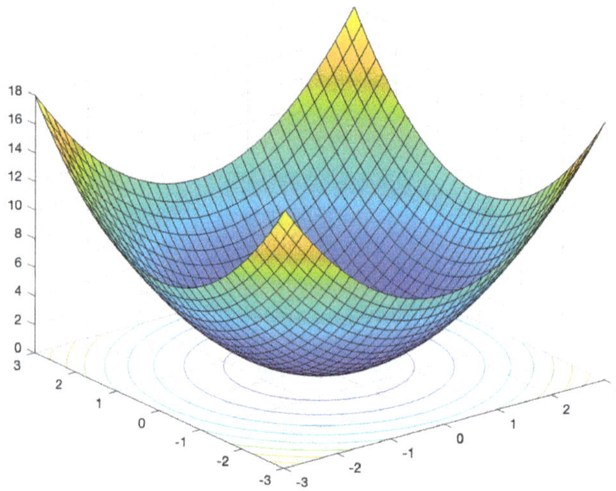

Fig. 6.1: An optimisation surface with one minimum valley, for two parameters on the horizontal axes. The vertical axis represents the loss function as in Eq. (6.10).

In calibration using this method, we aim to minimise L. There are many ways to solve Eq. (6.7), but one particularly efficient way is to form a matrix of all data samples, and then obtain a solution to the equation containing the matrices (the solution will be the parameters that minimise L in Eq. (6.7)). If we define

$$Y = \begin{bmatrix} y^{(1)} \\ y^{(2)} \\ \vdots \\ y^{(d)} \end{bmatrix}, X = \begin{bmatrix} x_1^{(1)} & x_2^{(1)} \\ x_1^{(2)} & x_2^{(2)} \\ \vdots & \vdots \\ x_1^{(d)} & x_2^{(d)} \end{bmatrix} \tag{6.8}$$

and

$$w = \begin{bmatrix} a \\ b \end{bmatrix}, \tag{6.9}$$

Eq. (6.7) can now be formulated in matrix form:

$$\mathrm{L} = (Y - Xw)^T (Y - Xw). \tag{6.10}$$

where the independent variables are the parameters a, b. Plotting Eq. (6.10) in a 3D plot would yield a surface plot, with one minimum resembling a bowl or valley bottom, as shown in Fig. 6.1. At the bottom of the valley, the derivative of L with respect to w will be zero, because the slope of the surface is 0. So to find the optimum w we just need to compute $\frac{\partial \mathrm{L}}{\partial w}$ and then solve the equation $\frac{\partial \mathrm{L}}{\partial w} = 0$:

$$
\begin{aligned}
\frac{\partial \mathrm{L}}{\partial w} &= \frac{\partial}{\partial w}\left((Y - Xw)^T (Y - Xw)\right) \\
&= \frac{\partial}{\partial w}\left(Y^T Y - 2w^T X^T Y + w^T X^T X w\right) \\
&= -2X^T Y + 2X^T X w = 0 \\
&\Leftrightarrow X^T Y = X^T X w
\end{aligned}
\tag{6.11}
$$

If the collected samples are linearly independent and the number of the samples is large than the number of the parameters, then the $X^T X$ can be inverted. Therefore we further arrange $X^T Y = X^T X w$ by multiplication of the inverse of $X^T X$, obtaining w:

$$
w = (X^T X)^{-1} X^T Y. \tag{6.12}
$$

As long as one can collect enough linearly independent samples, the least squares method always has a global optimal solution as Eq. (6.7) is convex[3]. In the next section, we will deal with non-linear models that are not necessary convex for which it cannot be guaranteed that there is one global optimal solution.

6.4 Calibration of Non-Linear Algebraic Models

6.4.1 Gradient descent

In Sect. 6.3.1, a residual error was constructed to stand for the closeness of the model prediction and the measurements in terms of a linear model. This section will introduce the calibration of non-linear models.

Regardless of the linearity of the algebraic models, the essence is to solve an optimisation problem that minimises the residual error between the model prediction $f(x_i; w)$, where x_i is an input sample, and w are the unknown model parameters, and a sampled output y^i of the model, which is

$$
\mathrm{L}(w) = \sum_i (\epsilon^i)^2 = \sum_i (y^i - f(x_i; w))^2. \tag{6.13}
$$

If $f(x_i; w)$ is not a linear function, then we cannot apply the least squares method introduced in Sect. 6.3.1 because the derivative of the residual errors described by Eq. (6.13) is non-trivial to solve and it is not likely to lead to the solution in the form of the Eq. (6.12). In such a case, the gradient descent [10] method is recommended.

Intuitively, the gradient descent method is like a person going down a mountain, which is illustrated in Fig. 6.2. Each time the person just looks around and finds the steepest descending direction nearby, then move forward a little toward that direction,

[3] If all points between any two points in a set are still in the set, then the set is called a convex set. More details can be found in [3].

Fig. 6.2: An intuitive understanding for gradient descent.

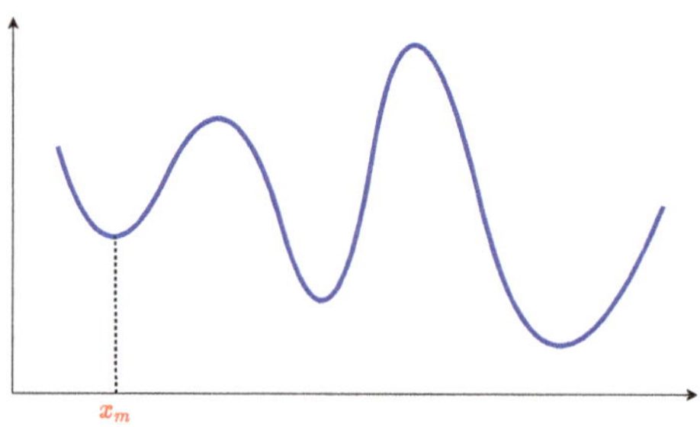

Fig. 6.3: An example for local minimums.

this is repeated until a flat area is reached. As the person only looks nearby, they might stop walking when a basin is reached, even though there is another basin that is lower than the current one. In summary, gradient descent is a first-order iterative optimisation algorithm, which is capable of finding a *local minimum*, i.e., a minimum within its neighbourhood. For example, in Fig. 6.3 the point x_m is a local minimum because it is a minimum within its neighbourhood, while a global minimum is the smallest local minimum. In some cases the shape or the surface formed by Eq. (6.13) is complicated and may contain many local minima, e.g., Fig. 6.4. In such cases, gradient descent could lead to a local minimum as it only looks nearby and only with good luck or prior knowledge can one reach the global minimum. All solution techniques for non-linear optimisation problems tend to suffer from this weakness.

The gradient descent method can be formulated as

$$w_{d+1} = w_d - \gamma \nabla \mathrm{L}(w_d). \tag{6.14}$$

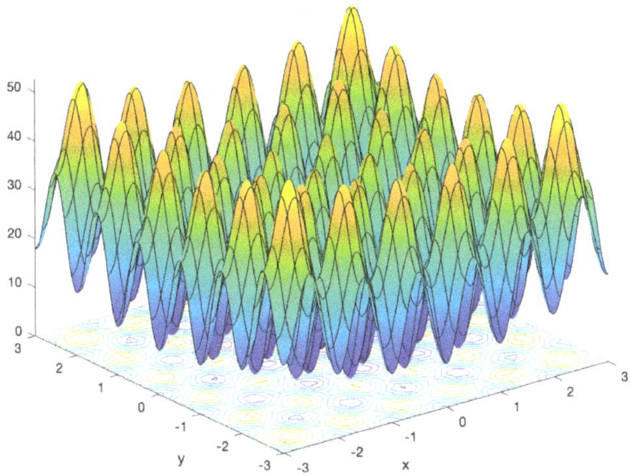

Fig. 6.4: An exemplar complicated optimisation surface.

where w_d is the parameter at the d-th step, γ is a learning rate (a positive value set by the user), and $\nabla \mathrm{L}(w_d)$ is the gradient of $\mathrm{L}(w)$ at w_d. The gradient at a point w_n represents the direction in which the function increases most quickly from this point w_n. So the minus between gradient in Eq. (6.14) means the direction decreasing most quickly. After a series of updates, a sequence of $(w_0, w_1, \ldots, w_n)$ is obtained and hopefully, after each update, the parameter w_n is closer to the values where the function $\mathrm{L}(w)$ arrives at a local minimum. This also means the element in the sequence of $(\mathrm{L}(w_0), \mathrm{L}(w_1), \ldots, \mathrm{L}(w_n))$ is smaller and smaller. Considering the proof, which proceeds by using Taylor's expansion yields some practical insights, and so is considered here.

If you have a software engineering background, think of the proof here as a proof of termination of the gradient descent algorithm in finding at least a local optimum (no guarantee for a global optimum). This is similar to a proof by induction. If you are not so interested in the proof, skip down to the discussion of the learning rate after Eq. (6.18).

We will take a one-dimensional function $\mathrm{L}(w)$ to demonstrate the proof. The function value at w_{n+1} can be expressed by

$$\mathrm{L}(w_{n+1}) = \mathrm{L}(w_n + \Delta w), \tag{6.15}$$

the Δw is the difference between w_{n+1} and w_n. The function of $\mathrm{L}(w_n + \Delta x)$ can be expanded by Taylor's expansion, so the first-order Taylor's expansion is

$$\mathrm{L}(w_n + \Delta x) \approx \mathrm{L}(w_n) + \mathrm{L}'(w_n)\Delta w, \tag{6.16}$$

where $\mathrm{L}'(w_n)$ is the derivative of $\mathrm{L}(w)$ at w_n. In addition, $\mathrm{L}(w + \Delta w)$ is the $\mathrm{L}(w_{n+1})$, substituting this into Eq. (6.16), we obtain

$$\mathrm{L}(w_{n+1}) - \mathrm{L}(w_n) \approx \mathrm{L}'(w_n)\Delta w. \tag{6.17}$$

According to Eq. (6.14), the $\Delta w = w_{n+1} - w_n$ is chosen to be $-\gamma \nabla \mathrm{L}(w_n)$ where $\nabla \mathrm{L}(w_n)$ is $\mathrm{L}'(w_n)$ in this case. By substituting $\Delta w = -\gamma \mathrm{L}'(w_n)$ into Eq. (6.17), equation of

$$\mathrm{L}(w_{n+1}) - \mathrm{L}(w_n) \approx -\gamma(\mathrm{L}'(w_n))^2 \leq 0 \tag{6.18}$$

is obtained. Therefore $\mathrm{L}(w_{n+1}) \leq \mathrm{L}(w_n)$.

In practice, there is a trade-off in choosing the learning rate. A large learning rate results in a faster convergence rate but at the risk of divergence; a smaller learning rate offers more chance of convergence but a slow convergence rate.

An example of calibration with the gradient descent method is shown in the following. Assuming the model we need to calibrate is wx^2, and we have obtained a large amount of data $(x^1, y^1), (x^2, y^2), \ldots, (x^n, y^n)$. Then the residual error is

$$\mathrm{L}(w) = \sum_{i=1}^{n} (y^i - w(x^i)^2)^2, \tag{6.19}$$

and the derivative of the residual error is

$$\nabla \mathrm{L}(w) = \sum_{i=1}^{n} -4wx^i(y^i - w(x^i)^2). \tag{6.20}$$

Therefore the updating rule is

$$w_{next} = w - \sum_{i=1}^{n} -4wx^i(y^i - w(x^i)^2). \tag{6.21}$$

The values of the residual error and the parameter w at each iteration are shown Fig. 6.5 and Fig. 6.6 (respectively), in which true w is 3, initial value is 70, the learning rate is 0.0001, and the maximum iterations is 1283.

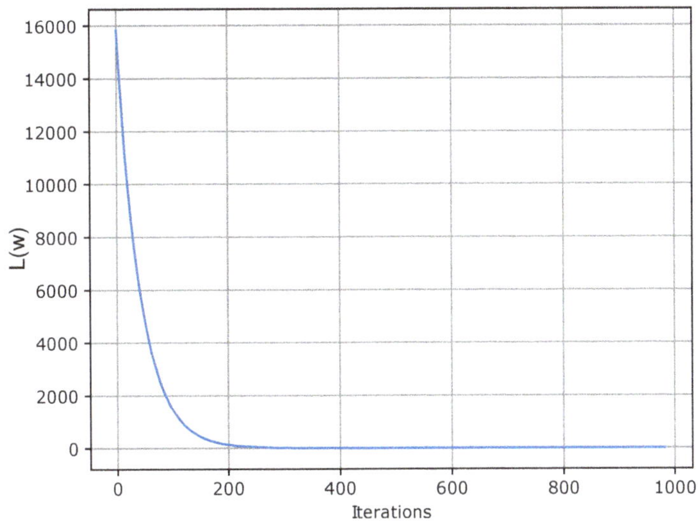

Fig. 6.5: Residual errors of $f(x) = wx^2$.

6.4.2 Alternative Non-Linear Optimisation Methods

6.4.2.1 Newton Method

An alternative to the gradient descent method is the Newton method. This method uses the first few terms of the Taylor series of a function $f(w)$ to approximate the function for solving the solution of $f(w) = 0$. For example, using the first two terms of $f(w) = f(w_0) + f'(w_0)(w - w_0) + 1/2f''(w_0)(w - w_0)^2 + \ldots$ means using a line to approximate the original function $f(w)$. We also say that the point satisfying $f(w) = 0$ is a root. If we are seeking the solution of $f(w) = n$ where n is a real number, we can set $g(w) = f(w) - n$ and if we get the root of $g(x)$ then we solve the original problem. Newton's method starts with an initial guess and uses the tangent line to approximate the roots of the desired real-value function, as illustrated in Fig. 6.7.

When the parameter closes to the root, the function value closes to 0 and the iteration stops. This can be formulated as

$$\mathrm{L}'(w_n) = \frac{\mathrm{L}(w_n) - 0}{w_n - w_{n+1}}. \tag{6.22}$$

If the tangent rate is not infinite (meaning, the surface is flat), then $w_n - w_{n+1} \neq 0$, then Eq. (6.22) can be re-organised as

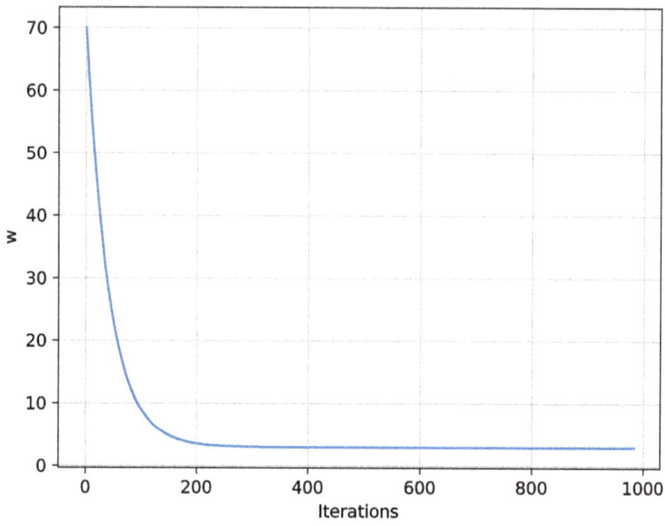

Fig. 6.6: The values of w when $f(x) = wx^2$.

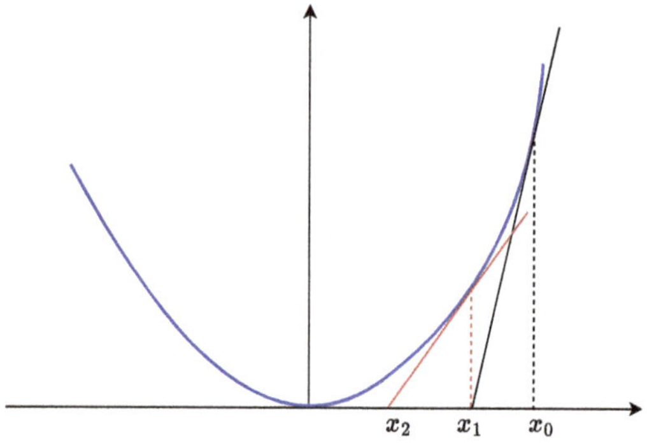

Fig. 6.7: An example of the process of Newton method iteration.

$$w_{n+1} = w_n - \frac{1}{\mathrm{L}'(w_n)} \mathrm{L}(w_n). \tag{6.23}$$

Equation (6.23) is the rule for updating parameters. Newton's method cannot guarantee convergence. Starting with different initial points might produce different convergence properties.

6.4.2.2 The Gauss-Newton Method

Gauss-Newton method is a variant of Newton's Method. In Newton's Method of Eq. (6.23), the derivative or the Hessian matrix for multivariable functions must be computed, while the Gauss-Newton method utilises the Jacobian matrix to approximate the Hessian matrix. The Gauss-Newton method also starts with initial guess parameters w_0 and updates similarly with Newton Method, which is

$$w_{n+1} = w_n - (J_{\mathrm{L}}^T J_{\mathrm{L}})^{-1} J_{\mathrm{L}} \, \mathrm{L}(w_n), \tag{6.24}$$

where J_{L} is the Jacobian matrix of $\mathrm{L}(w)$. As with the Newton method, the Gauss-Newton method doe not guarantee convergence. Prior knowledge about the parameters, for example, if one knows a range for the parameters, makes the Gauss-Newton method a better choice for convergence as the algorithm starts with parameters that might be close to the desired parameters.

In practice, it would be better to incorporate a step size α into Eq. (6.24), which is similar to the learning rate γ in Eq. (6.14). Then the new updating rule becomes

$$w_{n+1} = w_n - \alpha (J_{\mathrm{L}}^T J_{\mathrm{L}})^{-1} J_{\mathrm{L}} \, \mathrm{L}(w_n). \tag{6.25}$$

α can be assigned manually or it can be searched automatically using, e.g., the golden ratio search method. This parameter is a tuning mechanism to improve the performance of the method.

6.4.3 Calibration of Differential Equations

At first glance, the form of a differential equation is different from the functions above.

$$\frac{dx}{dt} = f(t, x, w) \tag{6.26a}$$

$$\widehat{y} = h(x), \tag{6.26b}$$

where t is the independent variable that usually refers to time, w represents the parameters, and $\widehat{y}$ is the measurement output from the model which is determined by the function $h(x)$. In general, there are some initial conditions of the form $x(t_0) = x_0$ for Eq. (6.26). The extra equation $\widehat{y} = h(x)$ represents the output of the differential equation that can be measured from the internal state. This is a more realistic formulation of this model, since in practice the observed behaviour may not exactly match the evolution of the state x.

As the model can output a value that can be measured, the residual similar to Eq. (6.13) can be constructed but now with an extra constraint:

$$\mathrm{L}(w) = \sum_i (\epsilon^{t_i})^2 = \sum_i (h(x^{t_i}) - y^{t_i})^2 \tag{6.27a}$$

$$s.t. \ \frac{dx}{dt} = f(t, x, w), \tag{6.27b}$$

$$s.t. \ \frac{dx^{t_i}}{dt} = f(t_i, x^{t_i}, w), \tag{6.28}$$

where y^{t_i} is the label which represents the observed data at discrete time t_i and x^{t_i} denotes the state at discrete time t_i. Before we can minimise the residual errors, the value x at time step t_i should be solved. As the x is constrained by the differential equation and the initial conditions are known, x is estimated by numerical simulation, as discussed in Sect. 5.4.1. For example $x^{t_{i+1}} = x^{t_i} + f(t_i, x^{t_i}, w)(t_{i+1} - t_i)$. After applying the numerical discretisation, we have many equations relating each of the values of $x^{t_{i+1}}$ with x^{t_i} all the way to the initial given value x^0, and most importantly, all these equations are parameterised by w. Hence we can collect them together in an equation of the form of Eq. (6.27) without the (now redundant) constraint, resulting in a non-linear calibration problem of the form already presented in Eq. (6.13) so that the methods introduced above can be used to obtain w.

Consider two simple examples, both of which assume that $h(x) = x$. If the dynamics are $\frac{dx}{dt} = wx$ and the initial condition is $f(0) = 0$, the analytical solution of the dynamic can be obtained, which is $\frac{1}{2}wx^2$, then this problem becomes the same as the example in Sect. 6.4.1 and the process for calibration is similar.

When it is non-trivial to acquire the analytical solution of the dynamics, a numerical solution can be deployed for calibration. We assume that the analytical solution to $\frac{dx}{dt} = wx$ is unknown, which should be $\frac{1}{2}wx^2$. Then the function values at next time, i.e. $h(x) = x$ can be obtained by

$$\begin{aligned} x^{t+\Delta t} &= x^t + \frac{dx}{dt}\Delta t \\ &= x^t + wx^t\Delta t, \end{aligned} \tag{6.29}$$

when we discrete the function (which is similar to Eq. (5.5)), it would be $x^{t_i} = x^{t_{i-1}} + wx^{t_{i-1}}\Delta t$. In addition to the above equations, the derivative of $\mathrm{L}(W)$ with respect to w, i.e. $\nabla \mathrm{L}(w_n)$ is

$$\begin{aligned} \nabla \mathrm{L}(w) &= \sum_i (h(x^{t_i}) - y^{t_i})^2 = \sum_i (x^{t_i} - y^{t_i})^2 \\ &= \sum_i 2(x^{t_i} - y^{t_i})\frac{d(x^{t_i})}{dw}. \end{aligned} \tag{6.30}$$

According to Eq. (6.30), we need $\frac{d(x^{t_i})}{dw}$, which is the derivative of the dynamics with respect to w, it is $x^{t_{i-1}}\Delta t$ in this example. Then the parameter can be calibrated by substituting all data and using the updating rules for gradient descent.

6.4.4 Genetic Algorithms and Design Space Exploration

Design Space Exploration (DSE) is a technique in which a set of designs can be (fully or semi) automatically explored in order to find those that best meet a set of desired characteristics. The term DSE originated in the integrated circuit manufacturing domain, where the goal is to allocate a set of elements to a chip. The placement of each element constrains, and is constrained by, the placement of the other elements. Each placement represents a *design choice*, and the set of all possible choices defines a *design space*. The placement of each element is controlled by a *design parameter*, and during DSE, design parameters are changed to explore a set of designs. One or more *objective functions* are defined that allow designs to be compared and ranked. Depending on the objective functions defined, the outcome might be a rank of the designs explored, or one or more Pareto fronts that show the design trade offs. The DSE idea has since been expanded to include exploration of design models through simulation or co-simulation [4].

A DSE analysis can be defined to run exhaustively, by analysing every possible combination from a set of design parameters. This approach can naturally suffer from the massive growth of the state space as the number of parameters and their ranges increase (a so-call *state-space explosion*). This can be managed to some extent through experiment design, by reducing the number of values for each possible design parameter or using orthogonal arrays to focus exploration on key parameter combinations [5]. An alternative way to use the same resources (e.g., number of simulations) adaptively through iteration, using the output of one round of analysis to inform subsequent rounds. In this sense DSE is closer to optimisation, and a variety of iterative techniques can be applied. Genetic Algorithms (GAs) are a classic technique inspired by biology, in which the best designs are selected and new generations of designs are produced by mutating and crossing over design parameters. Alternative approaches include metaheuristics based on probability, such as simulated annealing, and physics, such as Particle Swarm Optimisation (PSO).

In the context of calibration, DSE techniques can be used to help find a set of parameters for a model (or models) that best fits available calibration data [1]. In this approach, instead of trying to explore a design based on desired characteristics, objective functions are defined the measure how well a model aligns with calibration data. Given the constraints on calibration described above, for example, noise, it is necessary to expect that there is unlikely that DSE would find an exact fit to the data. It is more likely that the engineer will need to trade off in selecting a model which best fits the calibration data as part of multi-criteria decision making. In this case, it is important to consider which calibration data and characteristics of the model are most important. Additional techniques can be employed such as the Weighted Sum Model (WSM), where weights are assigned to the metrics in the objective function to reflect their importance.

6.5 Practical Considerations

We conclude with some practical advice based on experience of calibration for DTs of CPSs using the types of technique discussed in this chapter.

It is advisable to plan experiments with the PT to minimise the risk of collecting linearly dependent samples. The chances of doing so are low anyway provided sufficient data are collected. The almost inevitable noise that accompanies observations further reduces the risk here.

In most DT-related scenarios for CPSs, the model that needs to be calibrated is likely to be nonlinear and so the appropriate techniques have to be deployed. In practice, there is no guarantee that the nonlinear calibration methods will find a global minimum, which means that if we apply the calibrated models beyond the range of the training samples, they may fail.

When applying the gradient descent methods for calibration, keep an eye on the settings of the learning rate. A large learning rate means a faster convergence but a higher risk for divergence, which means fewer iterations are needed to obtain the results but there is a higher probability of the failure of the calibration. Conversely, a small learning rate means slower convergence but a lower risk of divergence. For example, if we set the learning rate in the example shown in Sect. 6.4.1 to be 0.02, then the absolute values of the residual errors soon go to infinity.

Two common issues that may arise during data-driven modelling are over-fitting and under-fitting, which is particularly probable in supervised learning scenarios [11]. Over-fitting and under-fitting refer to the inability of the model to generalise to new and unseen data. These problems are explained in more detail through a simple example [6]. Considering the given data point $\mathbf{u}(j)$ and supervision $S(\mathbf{u}(j))$, the ANN-based regression model can be formulated as:

$$\mathbf{P}(\mathbf{u}(j), \mathbf{W}) = \sum_{k=0}^{r} \mathbf{W}_k^\top \mathbf{u}^k(j) \tag{6.31}$$

where $\mathbf{u}^k(j)$ is the data point $\mathbf{u}(j)$ to the power k and $\mathbf{W}_k$ is its corresponding weights vector. The regression model Eq. (6.31) is linear in $\mathbf{W}_k$ and nonlinear in $\mathbf{u}(j)$. The objective is to calculate the weights $\mathbf{W}_k$ considering four different polynomial degrees, i.e. $r = 1, 2, 3$, and 9. For each polynomial degree, the Means Square Error (MSE) loss function is obtained as follows [6]:

$$\mathbf{L}(\mathbf{W}) = \frac{1}{2m} \sum_{j=1}^{m} ||\mathbf{P}(\mathbf{u}(j), \mathbf{W}) - S(\mathbf{u}(j))||^2. \tag{6.32}$$

To minimise loss function Eq. (6.32), for $r = 1$ the least square method from Sect. 6.3.1 and for $r = 2, 3$ and 9 the gradient descent from Sect. 6.4 can be used.

The results obtained considering polynomial degrees of 1, 2, 3, and 9 are shown in Fig. 6.8. The training data points and test data point are shown with blue dots and

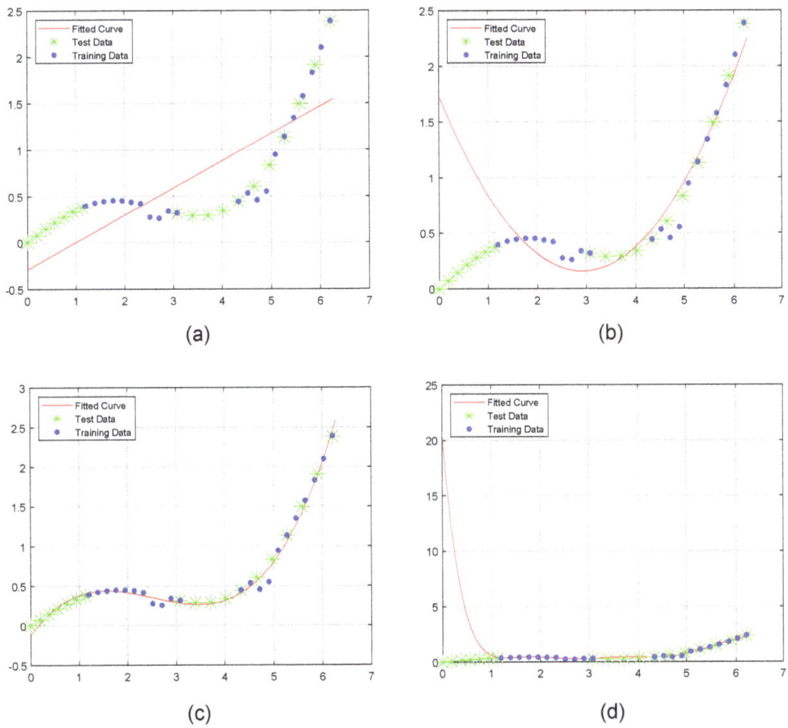

Fig. 6.8: Regression model trained considering polynomials with different degrees (a) $r = 1$, (b) $r = 2$, (c) $r = 3$, and (d) $r = 9$. Overfitting occurs for $r = 9$ and under-fitting is observed for $r = 1$ and 2.

yellow stars, respectively. The red graph shows the generated outputs of the trained regression model described in Eq. (6.31).

As Fig. 6.8 shows, the results for $r = 9$ have led to overfitting. This occurs when the model learns the training data's noise and random fluctuations rather than capturing the underlying patterns. In such circumstances, the model becomes very complex and performs well on the training data yet it fails to generalise to new data points not included in the training dataset. It essentially memorises the training data instead of learning the general patterns that would enable it to make accurate predictions on new data. Overfitting may caused by using over-complicated model structures with excessive numbers of parameters, and/or inadequate training data [11].

In Fig. 6.8, under-fitting is also observed in the top row for $r = 1$ and $r = 2$. The results indicate the model failure considering both the training data and validation data. In such cases the model cannot capture the complexity of the relationship between the features and the target variable, leading to poor predictions. Under-fitting can be caused by the use of over-simplified model structures or models with inadequate parameters to account for the complex data, insufficient feature engineering or excluding important features from the model [12]. A key challenge in data-driven modelling is to obtain a good trade-off and balance in the model structure to avoid under-fitting and over-fitting. The right model should be able to generalise well to new data while still capturing the underlying relationships in the training data. Regularisation techniques and cross-validation can help in this process.

References

1. Bogomolov, S., Fitzgerald, J., Foldager, F., Larsen, P.G., Pierce, K., Stankaitis, P., Wooding, B.: Tuning Robotti: the Machine-assisted Exploration of Parameter Spaces in Multi-Models of a Cyber-Physical System. In: J.S. Fitzgerald, T. Oda (eds.) Proceedings of the 18th International Overture Workshop, pp. 50–64. Overture (2020)
2. Boris Allan, G.J.: Variables – Observable, objective, measurable and constructed **11**(4), 347–350. DOI: 10.1007/BF00143959
3. Boyd, S., Xiao, L.: Least-squares covariance matrix adjustment. SIAM Journal on Matrix Analysis and Applications **27**(2), 532–546 (2005). DOI: 10.1137/040609902
4. Fitzgerald, J., Larsen, P.G., Verhoef, M. (eds.): Collaborative Design for Embedded Systems – Co-modelling and Co-simulation. Springer (2014). DOI: 10.1007/978-3-642-54118-6
5. Gamble, C., Pierce, K.: Design space exploration for embedded systems using co-simulation. In: J. Fitzgerald, P.G. Larsen, M. Verhoef (eds.) Collaborative Design for Embedded Systems, pp. 199–222. Springer Berlin Heidelberg (2014)
6. Geiger, A.: Deep learning lecture notes (2020). URL https://drive.google.com/file/d/16TaFr6d3eZXNkShgJJxaf6CN7xz1eOBs/edit
7. Isermann, R., Münchhof, M.: Identification of Dynamic Systems: An Introduction with Applications. Springer Science & Business Media
8. Keesman, K.J.: System Identification: An Introduction. Springer Science & Business Media
9. Nelles, O.: Nonlinear System Identification: From Classical Approaches to Neural Networks and Fuzzy Models. Springer Science & Business Media
10. Poliak, B.T.: Introduction to Optimization. Optimization Software, Publications Division
11. Ying, X.: An overview of overfitting and its solutions. In: Journal of physics: Conference series, vol. 1168, p. 022022. IOP Publishing (2019)
12. Zhang, H., Zhang, L., Jiang, Y.: Overfitting and underfitting analysis for deep learning based end-to-end communication systems. In: 2019 11th International Conference on Wireless Communications and Signal Processing (WCSP), pp. 1–6 (2019). DOI: 10.1109/WCSP.2019.8927876

Chapter 7
Sensing and Communication of Data from the Physical Twin

Cláudio Gomes, Daniel Enrique Lucani Rötter, Alexandros Iosifidis, Hao Feng, Henrik Ejersbo and Mirgita Frasheri

Abstract Much of the value of a digital twin depends on the extent to which the twin is kept up to date with the state of its real-world counterpart. This chapter surveys sensing and communication methods that play a key role in maintaining this relationship. The sensors making real-world observations, the characteristics of the communications infrastructure (which will often be wireless), communications delays, frequency of observations and the use of techniques such as data compression all therefore have an impact on the qualities of the digital twin and the services it offers. Data arriving at the digital twin will make use of time-series databases. When multiple sensors are present it may also be necessary to consider sensor fusion or use deep learning for deriving the optimal solution for the data that has been sensed.

7.1 Introduction

This chapter tackles the challenge of measuring quantities from a Physical Twin (PT) and communicating them to its Digital Twin (DT). The challenges introduced, as well as the solutions discussed, are well known and we cannot hope to be exhaustive in each topic. The chapter therefore reads more like a review, selecting the most important solutions, introducing them, and referring the reader to the state of the art for more details. When relevant, we introduce a simple example, to make the explanations clearer.

The chapter starts with Sect. 7.2 looking into sensors and the limits they have in extracting data that can be passed on subsequently. Afterwards, Sect. 7.3 deals with network communication of the sensor data to the DT side. Then Sect. 7.4 considers alternative communication technologies. Afterwards, Sect. 7.5 considers how data can be stored inside the DT in a time-series structure. Finally, Sect. 7.6 looks into the possibility of sensing without a physical sensor using either data fusion or deep learning.

J. Fitzgerald et al. (eds.), *The Engineering of Digital Twins*,
https://doi.org/10.1007/978-3-031-66719-0_7

For those readers from a cyber background, it is worth noting that sensors are subject to all the variations and uncertainties that come from working in the physical world. Some need to be calibrated regularly in order to avoid growing errors in sensed values. Errors are also introduced in the communications process as data is transferred from the PT to the DT. Consequently, on the digital side, effort is put into detecting and compensating for these weaknesses in detection and transmission.

7.2 Sensors and Their Limits

Sensing is the activity of detecting, measuring or capturing information from the environment using various devices. It involves converting physical phenomena, such as light, temperature, pressure, sound or chemical processes, into measurable signals that can be processed, analysed and interpreted. As an example, let is consider the sensing of temperature in our running incubator example.

Thermsistor Example

A thermistor is a temperature sensor that exploits the principle that the electrical resistance of certain materials changes predictably as their temperature changes. This change in resistance is not linear and is typically characterised by a specific resistance-temperature curve unique to each thermistor. In a practical application, a thermistor is often connected in a circuit as part of a voltage divider network. The thermistor and a fixed resistor are connected in series across a voltage source. The voltage across the thermistor is then measured and used to calculate the temperature based on the known resistance-temperature relationship. This relationship can be found empirically and is often given in the sensor data sheet.

In the remainder of this section we discuss some of the most common issues to bear in mind when working with sensors, using the thermistor as a running example.

7.2.1 Limited Sampling Frequency

Sampling is at the heart of many sensing applications, especially at the boundary between the physical and digital worlds. It involves reducing a *continuous-time* signal into a *discrete-time* signal. Basic sampling relies on taking samples at regular time intervals, with an equal time between each sample and the next. Thus, the system has a *sampling frequency*. What, then, is the correct sampling frequency for a given signal?

There are several approaches to selecting a sampling frequency. We begin here with the simplest and most widespread. A signal can be thought of as the composition of several different signals of various frequencies: we can plot a signal as a graph over time or we can plot it as a *spectrum* showing the amplitude of the various component frequencies. The Nyquist-Shannon sampling theorem [46] provides sufficient conditions for sampling continuous-time signals that contain no frequency components higher than B Hz. It states that, for such a signal, sampling at a frequency $f_s > 2 \cdot B$ is sufficient to completely determine (reconstruct) the original signal from the discrete-time signal. Thus, it is straightforward to select the appropriate sampling frequency for an individual signal if the signal characteristics are known a priori. If the goal of the system is to maintain as low a sampling frequency a possible, setting it marginally higher than $2 \cdot B$ will be sufficient. A wide range of practical sampling applications rely on this result due to its simplicity.

Perfect reconstruction may be possible under other assumptions or conditions. For example, there is an assumption of *uniformity* in the time spacing of the samples that is critical in the Nyquist-Shannon theorem. Uniform sampling below Nyquist's frequency can introduce *aliasing*, an effect in which frequency components higher than $f_s/2$ add their contribution to lower frequency components, resulting in distortion of the signal and preventing perfect recovery [37]. If a signal with high frequency components is *under-sampled*, those high frequency components will appear as low frequency components, changing the nature of the signal. In the context of control and DTs, this can limit the system's potential to interpret and react appropriately to measured signals. Sampling above Nyquist's frequency does not affect the signal recovery process as the signal is considered to be *over-sampled*.

Although sampling at the Nyquist frequency provides a guarantee for recovery, specific challenges arise when we consider DTs:

1. The DT affects the PT, which may change the characteristics of the signal being sampled from the PT. Having a description solely of the original system may not be sufficient.
2. The PT and communications environment may often not be precisely known a priori when the DT is initially set up and deployed, so the DT may have to estimate the environmental coefficients and constants for its models. A conservative solution is to use a worst case value for each constant, but this will usually result

in a higher sampling frequency than required, generating much more data traffic per sensor.
If no worst case estimation is done, and the environment changes substantially, then the sampling frequency may no longer be adequate, leading to the challenge of aliasing.
3. Sampling frequencies can be limited by the particular sensors and Analog-to-Digital Converters (ADCs) deployed. The selection of the sensors and their characteristics needs to occur at the early stages of design and accounting for worst-case conditions.
4. Although sampling may happen uniformly at the sampling frequency, other effects such as communication delays and jitter (discussed in Sect. 7.3.3) can affect the timeliness of the samples received by the DT. Appropriate protocols are required to incorporate samples in the DT estimators and models properly. If delays are large, the system may need to consider whether to re-run the simulation, introduce delays or simply remove problematic incoming samples. This can be particularly important if operating with heterogeneous sensors that have different sampling frequencies.

Non-uniform sampling has been widely studied. The Paley-Wiener-Levinson theorem [38, 30] extended the uniform sampling theorem to non-uniform sampling by considering the potential for a bounded uncertainty in the measurement time of each sample with respect to the uniform case. In other words, it considers that the sample could have been taken some time before or after the expected time for sampling if it were uniform. Recovery of the signal relies on more complex approaches based on interpolation, while the amount of drift between samples is fairly restricted. Recent studies have expanded the considerations for non-uniform sampling as well as exploring various applications e.g., [3, 2, 21]. The requirement for perfect recovery of the signal is changed to the average of the sampling frequency complying with Nyquists' frequency.

Sampling below Nyquist's frequency with perfect recovery is still possible if additional conditions are imposed on the signal. *Compressed sensing* is a relatively recent example that formalises the conditions for recovery of signals that are *sparse* in some domain, e.g., frequency domain [12]. This could be the case for signals that have a small number of frequency components, where their location in the spectrum is unknown. Recovery of such signals is generally more complex than that of uniform or non-uniform sampling. Fundamentally, instead of relying on a well-characterised equation for describing (recovery of) the signal, compressed sensing will rely on solving an optimisation problem.

Another common example for potential sampling below Nyquist occurs when considering signals that will not only be sampled, but also *quantised*. Since an error is already being introduced in the signal at the time of quantisation, there is additional flexibility in the recovery of the signal itself. For example, in a signal with high frequency components (e.g., due to noise) but of low amplitude, these may be reduced due to quantisation of the signal into finite levels. Further details on quantisation are given in Sect. 7.2.2.

Event-based sampling techniques only generate information when a significant change in the signal occurs, e.g., it exceeds a given threshold compared to the last sampled value. In a way, they provide an interesting combination of non-uniform sampling and signalling of a change of levels, where the latter can be directly mapped into a quantised signal. Of course, information of the signal behaviour in-between samples, i.e., while it does not exceed the set threshold, will be lost at the receiver. This makes event-based sampling technique that does not result in perfect reconstruction of the signal.

7.2.2 Quantisation

Quantisation is the process of mapping input signal values from a large (possibly infinite) set to a smaller (finite) set. For example, the incubator temperature sensors are digital and will therefore represent each sensor measurement as a sequence of binary numbers (0s and 1s). The choice of how many numbers to use in representing the measurement leads to quantisation. In general, a signal with a continuous and infinite set of values is converted to a finite set S of values representable in $\log_2 |S|$ bits, where $|S|$ is the cardinality of the set, i.e., the number of possible values. Typically, the original values are rounded to the closest representation within the new set. The difference between the original value of the signal and the *quantised* value, is defined as the *quantisation error*.

Quantisation occurs naturally when analog signals re sampled by ADCs in the sensors in order to generate digital signals. The speed of the ADCs is one of the key bottlenecks for a sensors's sampling frequency. Although ADCs run the gamut from low to very high speeds, higher speed comes at greater cost, for example in development, manufacturing and integration. By their design, ADCs also determine the number of bits that can be used to represent the signal value. In turn, this determines the number of levels of the quantisation as well as the quantisation error based on the range of input voltage values supported by the ADC.

7.2.2.1 Analog Signal Compression as Non-Uniform Quantisation

There are methods to modify the input signal prior to ADC, i.e., prior to quantisation. This can be done for signals that have large dynamic ranges, i.e., with both very large and very small amplitudes. For such signals, a uniform quantisation could place strong requirements on the number of bits used, particularly if the small amplitudes contain important information for the system such as voice, error or difference signals.

Different transformations or modifications can be introduced depending on the signal characteristics. One example is Dynamic Range Compression (DRC) in which a signal is either attenuated or amplified depending on its amplitude. This effectively *compresses* a signal that may have a large dynamic range into a much smaller range. This can be useful if low or high amplitudes of the signal contain more information/value. DRC can therefore help reduce the number of bits used during quantisation without unduly compromising signal quality.

With adequate designs, it is possible to reverse the process in order to recover the original values. For example, in the case of audio communications which involve ADC, digital transmission followed by Digital To Analog (DTA) conversion for rendering the audio, this process was called *companding*, a compound of compression and expansion. Compression prior to the ADC process can involve a transformation with a logarithmic transformation, at least for part of the signal range. This provides the potential to have a higher precision for lower amplitudes during quantisation. The quantisation error need not be uniform and can be tailored to the more important ranges of the signal, so that fewer bits can be used for digital representation.

7.2.2.2 Dynamic Quantisation

After the signal is quantised by the ADC, there are still different opportunities for further choosing the quantisation in software, for example by removing some of the least significant bits at the cost of a higher quantisation error to reduce the data transmission cost. There are interesting opportunities in the field of DTs, as the quantisation requested could be related to the importance or relevance of specific sensors during different states of the system. In effect, the DT could decide on the quantisation error dynamically for each sensor.

Incubator Dynamic Quantisation

In our incubator example, changes in temperature tend to be slow. If the heater has just been started, the change in temperature over a certain period will not require as much precision as may be needed when the incubator is closer to the target temperature. In the first stage, a coarser quantisation can be used to generate fewer bits in the data stream. As the temperature becomes closer to the target, more precision may be needed.

7.2.3 Immeasurable Quantities

Physical properties or phenomena that are challenging or currently impossible to quantify accurately using existing measurement techniques or technologies are termed *immeasurable*. These are attributes or factors that cannot be directly observed or detected because of limitations in sensor capabilities, measurement methods or our understanding of the underlying science. In the thermistor example, an immeasurable quantity is the temperature far from the heat conductive material. If we cannot rely on the assumption that the system is in thermal equilibrium, such temperature cannot be measured directly. Another example is when, due to the structure of the system, sensors cannot be positioned appropriately. For example, a thermal sensor cannot be placed under the tail of a rocket as the flame would melt the sensor immediately. Another example is the incubator, where it would be impractical to place a temperature sensor inside the fermenting fungus, to track its temperature, since this would potentially lead to contamination. Finally, there are quantities for which no sensor exists. For example velocity cannot be measured directly but is always calculated from other measurable quantities.

We discuss potential solutions to handle immeasurable quantities in Sect. 7.6. As the reader may already suspect, they involve the experimentation and creation of models that allow the transformation of directly measurable quantities into reasonable approximations.

7.2.4 Noise

Various sensing systems can be deployed on PTs for perception. A sensing system may consist of different sensors such as pressure sensor, a temperature sensor, an NOx sensor and an NH3 sensor. Each sensor can be considered as a small "chain" system with the general form shown in Fig. 7.1. The chain has four parts: a quantity under measurement (the measurand), a preprocessing system, an ADC and a digital processor. Noise can appear in anywhere of the four parts. For example, while in the incubator the intention of the temperature sensors is to measure the air temperature, they also measure the radiation reflected by the box walls, which means the quantity under measurement is already contaminated by noise. In this section, any mention of noise refers to that introduced in the scope of the sensors itself in Fig. 7.1. This contrasts with later chapters, in which we consider noise arising due to communication and modelling uncertainties.

There are various types of noise and we here introduce three: thermal, shot and flicker.

Thermal Noise [4]: Also known as Johnson or Johnson-Nyquist noise, this is a type of electromagnetic noise due to the random motion of charged particles in conducting materials. Always present in electronic circuits, it is caused by the

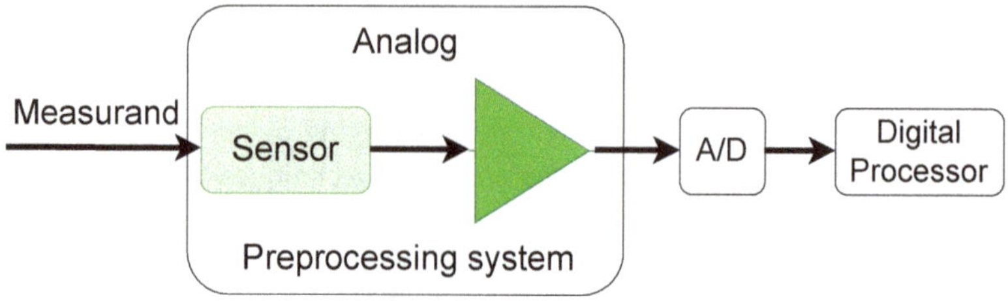

Fig. 7.1: Schematic of a sensor measuring chain, modified from [8].

thermal agitation of electrons in resistors, and its magnitude is proportional to the temperature of the conductor. Thermal noise is considered to be white noise, as it has a flat power spectral density versus frequency and its power increases with temperature.

Shot Noise [5]: Shot noise arises due to the quantisation of charge in an electrical system. Unlike thermal noise, shot noise specifically emerges from the discrete nature of electrons and their stochastic behaviour during transport through a conductor.

Flicker Noise [52]: Flicker noise or "1/f noise" arises from fluctuations caused by different mechanisms, including impurities in conductive channels and the complex interactions of charges within transistors. It is commonly observed in various electronic devices and systems, often associated with the flow of direct current. The name comes from the fact that as the frequency or offset increases, the power density of the noise decreases.

7.2.5 Clock Drift

In a large distributed system, such as a manufacturing plant or a power grid, sensors will often be physically distant from the supervisor systems that aggregate the data that they gather into meaningful insights. In order to deal with the delays that can be introduced into digital networks (discussed in Sect. 7.3.3), these sensors, which are often deployed in computational units that are not the same as the ones running the supervisor, are equipped with a clock that is used to timestamp the data readings.

The term *clock drift* refers to the gradual deviation of a clock from another (reference) clock. It can be caused by a variety of phenomena such as manufacturing flaws, ageing, and external factors such as variations in power sources, temperature and electromagnetic disturbances.

Addressing clock drift involves employing various strategies. Using precision clock sources, such as atomic clocks, can significantly reduce drift. Temperature compensation techniques adjust clock frequencies based on temperature changes. Clock synchronisation protocols like the Network Time Protocol (NTP) [34] ensure alignment across distributed systems, while periodic calibration identifies and rectifies deviations. We refer the reader to the following surveys on the topic [47, 57]. These techniques are not themselves perfect, and consequently some error in the timestamp of sensor measurements is a fact of life.

7.3 Network Communication

To an engineer with a background in largely physical systems, the area of computer communications and networking may be rather alien. It is a large and complex topic and it is not possible provide an introduction within the confines of this chapter. Instead, this section introduces one aspect that is very significant for digital twinning, namely the sources of delays.

Computers normally communicate over a digital network by following a set of *protocols*, each of which is a set of rules that describe how processes transmit and receive information. They will cover the form of messages, their meanings, how they are ordered or synchronised and often how errors are identified and resolved. Protocols are organised conceptually into layers that collectively form a *network stack*. For example, the Open Systems Interconnection (OSI) reference architecture consists of the layers and protocols shown in Table 7.1. There are several excellent references that provide full details [28, 40].

The previous sections have introduced Noise and Quantisation, challenges that are related to the Physical layer in Table 7.1. In the remainder of this section, we focus on the Network and Transport layers, to exemplify how the protocols used in these layers can lead to network delays and degradation, assuming the network is well designed for the number of computers joining it. The remaining layers suffer from similar issues, so we do not cover them further in the book. Finally, we shift our focus from timing to space, and in Sect. 7.3.4 we discuss data compression, so that more data can be transmitted using fewer bits.

7.3.1 Data Link – Medium Access Control Protocols

Medium Access Control (MAC) protocols make it possible for multiple sources to share the capacity offered by a communication channel. A range of such protocols are described comprehensively in [33]. The first such protocols were quite simple, as they were devised for wired networks which are reliable in terms of data loss, and which provide abundant bandwidth. However, moving toward wireless networks, the

Table 7.1: OSI Network Model.

Layer	Function	Example Protocols
7. Application	Network services for end-user applications	Hypertext Transfer Protocol (HTTP), File Transfer Protocol (FTP), Simple Mail Transfer Protocol (SMTP), Advanced Message Queuing Protocol (AMQP)
6. Presentation	Data format translation and encryption	Secure Sockets Layer (SSL)
5. Session	Establishes/maintains/terminates sessions	Remote Procedure Call (RPC)
4. Transport	End-to-end communication, error recovery	Transmission Control Protocol (TCP), User Datagram Protocol (UDP)
3. Network	Routing and logical addressing	Internet Protocol, Internet Control Message Protocol (ICMP)
2. Data Link	Physical addressing, error detection	Ethernet, Point-to-Point (PPP) Protocol, Medium Access Control (MAC) protocols
1. Physical	Physical medium and signalling	Universal Serial Bus (USB), Ethernet

design of MACs is not straightforward any longer. Certain issues arise that involve the deterioration of the radio signal, e.g., due to reflection, as well as potential interference that comes as a result of increasing the power transmission of the sources to deal with signal fading.

In schemes like these, we think of messages being transmitted and received as sets of discrete packets. There are two main MAC schemes, namely *contention-free* and *contention-based*. In the former, different sources are assigned specific network resources (frequency bands, time-slots etc.) by a central scheduler, ensuring that packets from different devices to not access the same resource at the same time (such transmissions are termed *conflict-free*). The use of a scheduler helps to manage a heavy loaded network. Typical examples include: token-ring, poll, static Time Division Multiple Access (TDMA)/ Frequency Division Multiple Access (FDMA).

In contention-based schemes, sources send packets when they become available. With no central coordination, responsibility for ensuring no collision of packets falls to the MAC protocol itself. These protocols work quite well when there is relatively light traffic in the network. Typical examples include ALOHA and Carrier Sense Multiple Access (CSMA) with Collision Avoidance (CA) and Collision Detection (CD), etc. In any case, for both groups, more signalling is introduced with the newer protocols in order to allow for more dynamic allocation of network resources in the contention-free schemes, and to minimise the number of packet collisions in the contention-based cases.

It is easy to see how delays (also known as latency) can arise when too many sources attempt to send packets at the same time or need to wait a long time for their turn.

7.3.2 Protocols and their Tradeoffs

There are many reasons why data might not be transmitted perfectly across a network. A network protocol is intended to ensure that devices can communicate regardless of their individual characteristics, and that the information being transmitted reaches its destination despite network failures. In addition, they ensure efficient data transmission by determining optimal packaging, transmission, and routing methods, which reduces network congestion and enhances overall performance. Security is also bolstered because many protocols incorporate encryption, authentication, and authorisation features, safeguarding data from unauthorised access. Additionally, protocols facilitate scalability, enabling effective management of data addressing and routing within expansive networks. By establishing guidelines for proper sequencing of data packets, communication protocols are crucial for applications relying on sequential or synchronised data processing. Robust error-handling mechanisms detailed by protocols ensure effective resolution of communication issues and prevention of data loss. Moreover, certain protocols offer Quality of Service (QoS) features, empowering network administrators to prioritise critical data for enhanced service levels.

The most common network failures (and the layer that addresses them) are:

Corruption of data: This often happens while the data are being transmitted in the physical medium, and protocols in this layer are designed to prevent them.

Data Omission: This happens when a message fails to arrive at the destination, due to poor physical medium connectivity or contention. It usually is handled at the Transport layer.

Data duplication: This happens when the same message arrives twice at the same destination, usually due to a failure in receiving an acknowledgement (designed in turn to prevent corruption or loss of of data) from the Data Link protocols, and are usually handled in the Transport layer protocols.

Data reordering: Happens when messages emitted at a particular sequence, arrive at a different sequence. These problems are usually handled at the Network layer.

Network Drop: Happens when the physical connectivity is broken.

Depending on the services offered by a DT, different protocols may be chosen, allowing the engineer to determine when and where should network delays and data loss be handled. There are several tradeoffs in the design and selection of network protocols, the main ones being:

Reliability vs. Efficiency: A reliable protocol ensures data integrity and accurate delivery but can introduce additional overhead, leading to slower communication.

Latency vs. Throughput: Minimising latency (delay) in data transmission can impact overall throughput (data rate). Protocols optimised for low latency might sacrifice throughput and vice versa.

Security vs. Performance: Incorporating robust security measures, such as encryption and authentication, can introduce processing overhead that affects performance.

Energy Efficiency vs. Communication Quality: In resource-constrained environments like Internet of Things (IoT) devices, optimising for energy efficiency can impact communication quality, which may affect data accuracy and completeness.

As an example, it is worth highlighting the tradeoffs in two widely used protocols, namely TCP (Transmission Control Protocol) and UDP (User Datagram Protocol). TCP provides reliable, connection-oriented communication with error checking, flow control, and guaranteed delivery of data. It establishes a connection between sender and receiver, ensures data integrity, and retransmits lost packets. UDP, on the other hand, offers connectionless, lightweight communication without error checking or guaranteed delivery. It is often used for applications where speed and low overhead are prioritised, such as real-time streaming. The main differences between TCP and UDP are that TCP provides reliable and ordered data transmission with acknowledgment and retransmission mechanisms, while UDP offers faster and simpler communication without acknowledgments or retransmissions.

In summary, it is challenging in practice to eliminate network delays (latency) and connection drops, especially when throughput is important such as in DT applications. As such, the following section discusses how to mitigate such delays at the application layer level (recall Table 7.1).

7.3.3 Mitigating the Effects of Network Delays and Drops

This section introduces a protocol that can be used when the network is unstable (i.e., delays and drop can occur). In essence, the protocol just signals when the network delays are too high, so that DT services using that data can avoid taking an action on the PT under those circumstances.

A DT can in principle be deployed alongside its PT in different ways, e.g., by running natively on the PT, in a different machine in the same local network or even on the cloud. In most cases where the elements of the DT-enabled system are deployed on separate machines, the connection between them depends on a computer network. Communication delays cannot be neglected, as the DT is expected to operate alongside its PT and be able respond in a timely manner to potential changes in circumstances, by reconfiguring the PT as needed. On one hand, the expected network delay is an important parameter that has to be taken into account when setting up the DT to ensure proper operation. On the other hand, measures must be taken to deal with network disturbances, which are bound to occur at some point in time, when a WiFi connection drops temporarily or the cellular network degrades, e.g., from 4G to 2G. In these cases, it is crucial for the operation of the DT/PT system to continue safely. Here we consider two examples of network disturbances, namely *network degradation* and *network drops*, and how these could potentially be mitigated at the DT side.

7.3.3.1 Network Degradation

Network degradation is here defined as an abnormal increase in delay. Consider a DT-enabled system in which there is a network delay in normal conditions of d_{exp}. We assume the DT and PT share a synchronized clock. We will refer to the data timestamps as the "simulated time", since it can deviate from the wall clock time (when delays occur). Any DT Service that relies on that data is invariably keeping track of simulated time as it represents the last known state of the PT.

In order for the simulated time to progress, the DT needs a regular inflow of data from the PT. If no messages arrive, the simulated time is stalled. As shown in Fig. 7.2a, a message that is sent during this time, e.g., at t_1, will arrive at the DT between h_0 and h_1, which is when the DT is attempting to reach h_1. The time difference between the wall-clock and simulated time is given by some measure t_{diff}, and would be below a predetermined threshold θ_{safe} to ensure the safe operation of the system. Assume that after some time, e.g., after t_2, an increased delay is being experienced in the network given by $d(t) = d_{exp} + \gamma(t)$, where $\gamma(t)$ is assumed constant for the purposes of this discussion. As the messages come at a slower rate than before, the simulated time will progress at a slower pace as well, with t_{diff} possibly going above the accepted time difference θ_{safe}. In this case, the simulated time and wall-clock time can be considered out-of-sync, and messages sent by a DT service might have become obsolete by the time they reach the PT. The PT may have been equipped with mechanisms that enable such messages to be discarded. However, if a CPS has been retrofitted with a DT, this might not be the case. The DT service on the other hand would be oblivious to the out-of-sync situation that has risen, and would be burdening a potentially bandwidth-limited connection with meaningless messages.

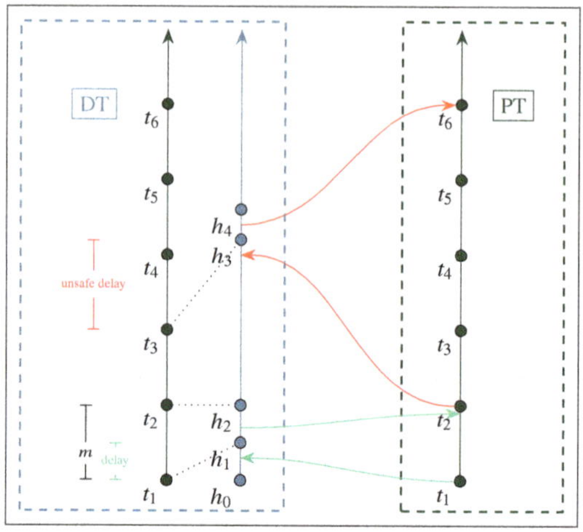

(a)

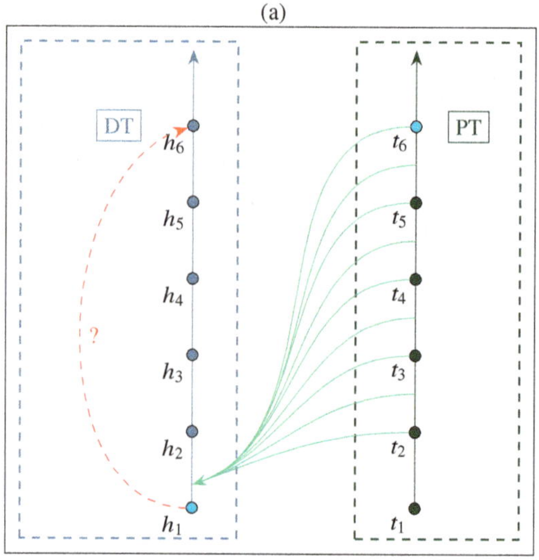

(b)

Fig. 7.2: a) Depiction of network degradation leading to higher delays. Green arrows represent message transmission within the expected delays; red arrows represent transmission during a period with network disturbances. t_i is the wall-clock time, (assumed) shared by both DT and PT; m is the age constraint. Dotted lines show the relation between DT and PT time. b) Message burst after connection is re-established. The current time-step (simulation and wall-clock time) is represented as a circle filled in cyan. The dashed red arrow shows the option to go directly from h_1 to h_6, rather than taking steps one-by-one, i.e. h_1, h_2, ..., h_6. Figure based on [16].

7.3.3.2 Network Drop

We define network drop as the loss of connectivity. Transport protocols such as TCP will attempt to retransmit messages when this happens, leading to a potential influx of delayed messages connectivity is regained.

In case of a network drop, the PT may continue sending data to the DT, with the wall-clock advancing as it would, while the DT time would be stalled at h_1 as seen in Fig. 7.2b. In this situation, the t_{diff} would be continuously increasing, and the DT/PT system would get more and more out-of-sync. Should the connection be re-established after some time, a burst of messages will be seen in the DT. These messages could be dealt with one-by-one. Assuming that a simulated step can be performed faster than real-time, it could be possible that after some time the DT/PT becomes synced on its own. Another option would be to jump ahead to the latest message, skipping the bulk in between, in order to reach the synced state in shortest real time possible. The choice of strategy depends on how the data is used, e.g., does the DT service need only the latest state or the whole history?

7.3.3.3 Requirements on the DT

Whatever the nature of delays, the DT service should at least be able to identify if the simulated time is out of sync with the wall-clock time, and handle this occurrence as soon as possible. Thereafter, the DT service should be enhanced with mechanisms to mitigate the effects of such a scenario when possible. While there is not much to do if no data is flowing into the DT, in the case of message bursts different strategies could be implemented with which the DT can attempt to get back in sync as soon as possible. Note that the wall clock time it takes to process a message on the DT service is a crucial factor. Preferably, a DT service processing time should be faster than the difference between the average difference of simulated time between continuous data messages, the faster the better. However, speed may come at the price of accuracy and performance of the DT service. This is a trade-off that has to be considered when setting up the DT-enabled system.

7.3.3.4 Simulation of Network Degradation and Drop

Setting up a DT service requires, among other things, configuring parameters like θ_{safe}. Different applications can place different requirements on such parameters, e.g., for human-robot interaction an appropriate range may be $100 - 300$ms, whereas for other systems they can be as strict as 10ms [36].

One method for evaluating θ_{safe} involves estimating the maximum delay for which the DT-enabled system can still operate safely (which should be defined in the DT enabled system requirements, recall Chap. 4), and using that as an upper bound. As an example, consider a DT service that detects that the lid is open in the incubator, and adjusts the controller accordingly to save energy, later detailed in Sect. 10.5. If there is no mechanism to detect such an event, the controller will make the heater work more, to try and maintain the desired temperature. As a result, heat will be dissipated in the environment outside of the box. Further, should the lid then be put back on incubator with the heater working harder, the box may become overheated, resulting in a fire hazard. A detection mechanism can be used by the controller to determine that, if the lid is open, the heater should be turned off. The network delay of the message from the controller to the heater is thus related to the safe operation of the system as a whole. A series of simulations could be performed to evaluate the upper bound of such delay that can lead to dangerous situations. It would also be possible to use tools like NS3FMU [35] (which encapsulates the ns-3 network simulator in an FMU) to simulate these delays in the DT itself, without additional changes in the rest of the system. Additionally, more realistic scenarios can be achieved by placing delay nodes in the running system, which simulate scenarios of network disturbances like degradation and drop.

7.3.4 Data Compression

Given the importance of the communications infrastructure to the successful operation of a DT-enabled system, it is worth considering techniques that help to manage network load and costs by reducing the size of the data transmitted. These *data compression* techniques effectively encode data into a more compact format, reducing transmission and storage costs. Since this encoding happens at or near the data source prior to transmission, data compression is also known as *source coding*. Of course, this comes at a price in terms of the computational power required to encode and decode the data, and in terms of the risk of losing information content in the encoding. There are thus important choices to be made about where encoding and decoding take place, and how much loss may be acceptable for a given application.

A compression algorithm considers a stream of symbols which could represent any data of interest such as sensor readings or images. The result of compression is a smaller representation which can subsequently be decoded to a usable form. Compression algorithms are typically classified as either *lossless* or *lossy*. Lossless compression maps the symbols in the sequence from a given source to a more compact representation (e.g., fewer bits in average for the data stream), while ensuring the later recovery of the original sequence. This is usually done by eliminating statistical redundancy. For example, a frequently occurring sequence of symbols might be represented as a single symbol. Similarly, repetitions of strings can be encoded by the string and the number of repetitions.

Lossy compression is used for data that can suffer distortions without compromising the purpose for which it is intended. For example, audio data intended for human listening may be compressed in this way if the resulting distortion cannot be perceived by the listener. In other cases, some distortion is acceptable given that the data source is naturally noisy. Note that some techniques for compression of analog data such as quantisation may have preceded the digital compression of the bit sequences discussed here.

Data compression can help reduce the load and costs for the overall digital infrastructure in IoT and CPS applications, including DTs. It can be beneficial to perform compression in energy-constrained sensors because the compression operations tend to consume less energy than wireless transmissions. Thus, there is a large potential for energy savings. Although a wide variety of compression schemes exist, until recently there was no scheme that could adequately satisfy the key requirements of emerging CPS applications, namely:

- Have good compression ratios (i.e., the size of the compressed sequence compared to the original sequence size) for time-series sensor data.
- Deliver computationally efficient and simple (de)compression.
- Allow for compression of small data chunks to allow for timely sensor updates (compression typically requires accumulating large chunks to be efficient [49]).
- Support fine-grained random access to the compressed data for retrieval and queries (compression typically requires decompression of the entire data for recovery).

There are also potential trade-offs between these requirements. For example, random access and operation on small chunks can reduce the compression ratio. Schemes that address these challenges are now appearing, e.g., [50]. These have prompted contributions describing systems that operate in online mode [56], compressing after a small number of samples (potentially, one sample) have been received or that reduce the need for data transmissions by providing approximate data analytics that can be run close to the end-device (e.g., at an edge computing device) and only transmit in the event of uncertainty [22].

7.4 Message-Based Communication

The section identifies application layer protocols and tools (recall Table 7.1) that surveys suggest are commonly applied in DT contexts [18, 9, 26] to enable the transfer and storage of data across applications running in heterogeneous computing environments.

ZeroMQ[1] is a high-performance messaging library that provides patterns including publish-subscribe, request-reply and more. It allows fast and scalable communication between applications and processes. ZeroMQ is lightweight and designed for simplicity, making it suitable for various distributed systems and messaging needs.

[1] `https://zeromq.org/`

Robot Operating System (ROS[2]) is an open-source middleware framework designed for building robotic systems. It provides a set of tools and libraries for communication, hardware abstraction, device drivers and more. ROS facilitates the development of complex robotic applications by offering a modular and distributed architecture. Communication in ROS is often based on a publish-subscribe model, allowing different components to communicate seamlessly.

Apache Kafka[3] is a distributed streaming platform that is often used for building real-time data pipelines and streaming applications. It is designed for high-throughput, fault tolerance and horizontal scalability. Kafka uses a publish-subscribe model, where producers publish messages to topics and consumers subscribe to those topics to receive the messages. It is widely used for event-driven architectures and data streaming.

Finally, RabbitMQ[4] is a message broker that facilitates communication between distributed systems. The architecture implements the Advanced Message Queuing Protocol (AMQP) as its core protocol and has additional protocols such as MQTT, STOMP, HTTP and WebSockets supported via a plugin mechanism. The AMQP model defines *publishers* and *consumers* as its basic elements. Publishers send messages to brokers which then forward them to consumers. Published messages are routed via *exchanges*, with the routing behaviour being contained in the message properties or header. An exchange routes messages to a specific *message queue* for storing based on specific binding rules and a queue may be configured to keep messages in Random Access Memory (RAM) or in persistent storage. The AMQP supports message acknowledgements from consumers to the broker so that messages will only be removed from a broker message queue once acknowledged by the consumer. Consumers retrieve messages from a queue by subscribing to have messages pushed from the queue by the broker. The underlying protocol uses TCP for reliable communication and most of the AMQP API is based on an RPC-like structure to set up connections and subscribe to broker queues.

7.5 Storing Data in Time-Series Databases

Time-series data generated from sensors in the PT, as well as predictions, monitoring tasks and abnormality events from the DT are typically generated on-the-fly and require rapid storage and processing. These data streams have unique properties and are expected to be accessed and processed in real-time by some critical tasks from the DT. However, medium-term or even long-term storage can provide valuable data for training and learning tasks that may help optimise different services, including

[2] `https://www.ros.org/`

[3] `https://kafka.apache.org/`

[4] `https://rabbitmq.com/`

monitoring and abnormality detection mechanisms and can also help develop data-driven models of the PT itself.

Time-series databases are designed and optimised for storing and serving time-series data, usually structured or associated as pairs of time and value(s). The values stored by the time-series database can be, in part, measured data from sensors, e.g., deployed in the PT. However, they could also be used for storing predictions from models in the DT or other events. Time-Series DataBases (TSDBs) usually rely on a *column-oriented* design (i.e., a columnar database), e.g., Apache IoTDB[5], Apache Druid[6], Apache Pinot[7], Apache Kudu[8], although other approaches have been used.

Databases in general, but TSDBs in particular, use various compression strategies to reduce memory usage, increase storage capacity and overcome speed limitations of (slow) hard drives (less data to read). Compression also provides benefits for long-term storage of the data.

Traditionally, compression in databases was seen as a "nice-to-have" feature. However, TSDB and in-memory databases aiming to accelerate processing time have changed the role of compression to an essential (must-have) technique, required to store and analyse large datasets. This is particularly important when considering that the memory size is usually at least an order of magnitude smaller than hard-drive capacity in a server. Simply adding memory would be a costly and, ultimately, unscalable solution. Many of the leading TSDBs implement several compressors to cope with the requirements of different data streams.

Column-oriented databases are more suitable to compression than traditional databases. This is explained due to the fact that the data in the same column originates from the same data source (e.g., a sensor). Modern in-memory column-stores further split columns into segments which are independently compressed, e.g., [13]. This helps to achieve high-performance execution of transactional and analytical workloads, particularly if independent segments are compressed using the techniques best suited to the data type and data characteristics. For example, CodecDB [23] trains a model to determine which compressor is likely to have the best compression for a segment. The encoding selection module of Hyrise [13] determines the optimal encoder based on a given memory budget [7]. Recent techniques have resulted in better compression-access trade-offs by optimising time-series oriented, highly configurable compressors for each data segment [14].

[5] https://iotdb.apache.org/

[6] https://druid.apache.org/

[7] https://pinot.apache.org/

[8] https://kudu.apache.org/

7.6 Software Sensing

Compared to hardware sensors, which are designed based on physical laws, soft sensors (software sensors) have more flexility and may be designed by means of first principle models or use data-driven techniques. Soft sensors are virtual, making the use of the underlying relations between the accessible process variables and providing readouts of the required physical quantities and/or the performance indices [24]. Specifically, they are algorithms which are able to reconstruct state variables from the outputs of common hardware sensors such as gyroscopes and accelerometers. For example, few temperature sensors inside the insulated container of our example incubator can be used to derive the temperature distribution of the air inside the insulated container. Another example is that a system might track readings from a device's gravity sensor to infer complex user gestures and motions, such as tilt, shake, rotation or swing.

The main reason we introduce software sensing here is that the techniques can be applied in the PT before the data is transmitted to the DT. However, the same techniques can be deployed as DT services for the purposes of monitoring (discussed in Chap. 9). The main difference is in whether the data that forms the input to the software sensing algorithm should be transmitted to the DT or whether only the outputs of the algorithm should be transmitted. In the following, we will give more details of soft sensing based on first principle models and data-driven techniques.

7.6.1 Sensor Fusion

In situations where one sensor cannot provide complete or accurate states of PTs, data from multiple sensors may be fused to obtain a more reliable and/or precise representation.

One simple method is to use an average of sensor outputs, but weighting each of them based on the sensor's accuracy or reliability. This approach is effective for sensors that are delivering similar types of data.

The Kalman Filter (KF) [55, 15] can combine data from multiple sensors to produce a more accurate and reliable estimation of the system's state. It has two main components: the prediction step and the update step. In the prediction step, the filter estimates the system's state at the next time step based on the current state which is the input from various sensors and the system's dynamics (usually linear models). In the update step, the KF combines the predicted state with new measurements from the sensors to produce a more accurate estimate of the system's state. An advantage of the KF is that it can be implemented recursively. When facing nonlinear dynamics or non-Gaussian distribution, an Extended Kalman Filter (EKF) [45] can be deployed and this has become commonplace in many applications, including inertial navigation, robotics and computer vision. This uses a linear approximation to the nonlinear system model and a Gaussian distribution for the state estimate.

An alternative is the Particle Filter (PF) [11], which is a probabilistic approach for sensor fusion and widely used in sensor fusion for multi-sensor systems. It uses a set of particles to represent the state of a dynamic system, each with a probability distribution. The particles are updated based on sensor measurements and the state estimate is obtained by averaging the particles' probability distributions.

In addition to those outlined above, there are other methods such as Bayesian inference technique and Dempster-Shafer theory of evidence. Interested readers are referred to the overviews provided by Fung et al. [17] and Krishnamurthi et al., [25].

7.6.2 Deep Learning Perception

Deep Learning refers to the set of Machine Learning models which receive as input raw (or pre-processed) sensor data and a target defined on their output and learn multiple data transformations for mapping their inputs to the target. Even though such models can be used for a variety of problems belonging to the supervised, semi-supervised, unsupervised learning and reinforcement learning paradigms, their most successful applications to date address supervised learning problems. In these cases, learning occurs during the a training phase by tuning the parameters of the model using a set of data accompanied by targets given by humans.

Depending on the type of input data and the corresponding targets, different types of Deep Learning model can be used. The most common approach is to express the model as an Artificial Neural Network formed by multiple layers. The modular structure of such models, i.e., the fact that one can combine neural layers of different types in a variety of ways, in combination with the existence of easy-to-use public software, like PyTorch [39] and TensorFlow [1], greatly facilitates the development of specialised Deep Learning models targeting a variety of problems where the task is to determine a mapping from an input to an output through detecting patterns in the input data. A highly successful implementation of Deep Learning models and the one which first showcased their performance against other Machine Learning paradigms, is that of image classification [27]. Here, an image is introduced to a Deep Learning model and the task is to classify it to one of the classes belonging to a set of predefined classes, the classes of the images used to train the model.

Image classification can be used in combination with a sliding window process to detect objects in images [20, 10]. In this case, a window of fixed size scans the image and crops out image patches, each of which is introduced to the model and classified. Using windows of multiple sizes, differently sized objects can be detected. This approach has been successful when combined with efficient classification schemes. For example, one of the earliest successful object detection methods using this approach is the Viola Jones detector [51] developed in 2001.

Deep Learning models are computationally demanding, so simple adoption of this two-stage process is prohibitive in many applications. An alternative is to perform window selection based on the contents of the input image, through so-called object proposal methods. Such methods perform a fast pre-processing of the input image in order to determine which locations (and at which scales) of the image are likely to contain objects through image over-segmentation and region grouping [48], by calculating an "objectness" score of each image window based on either the input image [58] and by re-ranking regions based on deep features [54]. Another approach combines the region selection and classification within the model and performing Region of Interest (RoI) pooling at an intermediate layer of the model [19] or by incorporating a dedicated part of the model called Region Proposal Network (RPN) [44], which is jointly trained with the overall model.

The object detectors discussed above implement two processes, namely region selection and classification. Combining these leads to *one-stage detectors*. Two early and successful examples are the Single Shot Detector (SSD) [32] and You Only Look Once (YOLO) [41], which have been followed by many improvements [42, 43, 6, 31, 53]. These detectors typically replace the fully-connected layers of pre-trained Convolutional Neural Networks with convolutional layers leading to outputs which have a tensor form providing object detection results at different locations of the input image. For example, YOLO splits the input image to an $S \times S$ grid and, for each cell in the grid, it predicts B bounding boxes, their confidence scores and the object class. The confidence score of each bounding box is the corresponding Intersection over Union score which measures the level of agreement between a detected object location and its actual location in the image. Thus, the output of the model for an input image is a tensor of $S \times S \times (B * 5 + C)$. These one-stage detectors encode the object detections in the form of anchor boxes represented by a central point and the corresponding width-height sizes. Anchor-free detectors, e.g., CornerNet [29], parameterise the detected bounding boxes using top-left and bottom-right corner points, i.e., it predicts the corners directly and eliminates the need for anchor box creation as part of the model.

References

1. Abadi, M., Agarwal, A., Barham, P., Brevdo, E., Chen, Z., Citro, C., Corrado, G.s., Davis, A., Dean, J., Devin, M., Ghemawat, S., Goodfellow, I., Harp, A., Irving, G., Isard, M., Jozefowicz, R., Jia, Y., Kaiser, L., Kudlur, M., Levenberg, J., Mané, D., Schuster, M., Monga, R., Moore, S., Murray, D., Olah, C., Shlens, J., Steiner, B., Sutskever, I., Talwar, K., Tucker, P., Vanhoucke, V., Vasudevan, V., Viégas, F., Vinyals, O., Warden, P., Wattenberg, M., Wicke, M., Y., Y., Zheng, X.: Tensorflow: Large-scale machine learning on heterogeneous systems. In: 10.5281/zenodo.4724125 (2015)
2. Al-Hammali, H., Faridani, A.: Uniform and non-uniform sampling of bandlimited functions at minimal density with a few additional samples. Sampling Theory, Signal Processing, and Data Analysis **21**(1), 2 (2022). DOI: 10.1007/s43670-022-00041-7

3. Aldroubi, A., Gröchenig, K.: Nonuniform sampling and reconstruction in shift-invariant spaces. SIAM Review **43**(4), 585–620 (2001). URL http://www.jstor.org/stable/3649716
4. AlHinai, N.: Introduction to biomedical signal processing and artificial intelligence. In: W. Zgallai (ed.) Biomedical Signal Processing and Artificial Intelligence in Healthcare, Developments in Biomedical Engineering and Bioelectronics, pp. 1–28. Academic Press (2020). DOI: https://doi.org/10.1016/B978-0-12-818946-7.00001-9
5. Blanter, Y., Büttiker, M.: Shot noise in mesoscopic conductors. Physics Reports **336**(1), 1–166 (2000). DOI: https://doi.org/10.1016/S0370-1573(99)00123-4
6. Bochkovskiy, A., Wang, C.Y., Mark Liao, H.Y.: YOLOv4: Optimal speed and accuracy of object detection. In: arXiv:2004.10934 (2020)
7. Boissier, M.: Robust and budget-constrained encoding configurations for in-memory database systems. Proc. VLDB Endow. **15**(4), 780–793 (2021). DOI: 10.14778/3503585.3503588
8. Bordoni, F., D'Amico, A.: Noise in sensors. Sensors and Actuators A: Physical **21**(1), 17–24 (1990). DOI: https://doi.org/10.1016/0924-4247(90)85003-M. Proceedings of the 5th International Conference on Solid-State Sensors and Actuators and Eurosensors III
9. Böttjer, T., Tola, D., Kakavandi, F., Wewer, C.R., Ramanujan, D., Gomes, C., Larsen, P.G., Iosifidis, A.: A review of unit level digital twin applications in the manufacturing industry. CIRP Journal of Manufacturing Science and Technology pp. 162–189 (2023). DOI: https://doi.org/10.1016/j.cirpj.2023.06.011
10. Chumachenko, K., Gabbouj, M., Iosifidis, A.: Object detection and tracking. In: A. Iosifidis, A. Tefas (eds.) Deep Learning for Robot Perception and Cognition, pp. 243–278. Academic Press (2022)
11. Djuric, P.M., Kotecha, J.H., Zhang, J., Huang, Y., Ghirmai, T., Bugallo, M.F., Miguez, J.: Particle filtering. IEEE signal processing magazine **20**(5), 19–38 (2003)
12. Donoho, D.: Compressed sensing. IEEE Transactions on Information Theory **52**(4), 1289–1306 (2006). DOI: 10.1109/TIT.2006.871582
13. Dreseler, M., Kossmann, J., Boissier, M., Klauck, S., Uflacker, M., Plattner, H.: Hyrise re-engineered: An extensible database system for research in relational in-memory data management. In: EDBT, pp. 313–324 (2019)
14. Fehér, M., Lucani, D.E., Chatzigeorgiou, I.: An adaptive column compression family for self-driving databases. In: Thirteenth International Workshop on Accelerating Analytics and Data Management Systems Using Modern Processor and Storage Architectures (ADMS) – A Workshop of VLDB (2022)
15. Feng, H., Gomes, C., Larsen, P.G.: Model-based monitoring and state estimation for digital twins: The kalman filter (2023). DOI: https://doi.org/10.48550/arXiv.2305.00252
16. Frasheri, M., Ejersbo, H., Thule, C., Gomes, C., Kvistgaard, J.L., Larsen, P.G., Esterle, L.: Addressing time discrepancy between digital and physical twins. Robotics and Autonomous Systems **161**, 104347 (2023). DOI: https://doi.org/10.1016/j.robot.2022.104347
17. Fung, M.L., Chen, M.Z.Q., Chen, Y.H.: Sensor fusion: A review of methods and applications. In: 2017 29th Chinese Control And Decision Conference (CCDC), pp. 3853–3860 (2017). DOI: 10.1109/CCDC.2017.7979175
18. Gil, S., Mikkelsen, P.H., Gomes, C., Larsen, P.G.: Survey on open-source digital twin frameworks–A case study approach. Software: Practice and Experience (December 2023), 1–32 (2024). DOI: 10.1002/spe.3305
19. Girshick, R.: Fast R-CNN. In: International Conference on Computer Vision (2015)
20. Girshick, R., Donahue, J., Darrell, T., Malik, J.: Rich feature hierarchies for accurate object detection and semantic segmentation. In: IEEE Conference on Computer Vision and Pattern Recognition (2014)
21. Hall, P., Penev, S.: Wavelet-based estimation with multiple sampling rates. The Annals of Statistics **32**(5), 1933–1956 (2004). URL http://www.jstor.org/stable/3448560
22. Hurst, A., Lucani, D.E., Assent, I., Zhang, Q.: Glean: Generalized-deduplication-enabled approximate edge analytics. IEEE Internet of Things Journal **10**(5), 4006–4020 (2023). DOI: 10.1109/JIOT.2022.3166455

23. Jiang, H., Liu, C., Paparrizos, J., Chien, A.A., Ma, J., Elmore, A.J.: Good to the Last Bit: Data-Driven Encoding with CodecDB. In: Proceedings of the 2021 International Conference on Management of Data, SIGMOD '21, p. 843–856. Association for Computing Machinery, New York, NY, USA (2021). DOI: 10.1145/3448016.3457283
24. Jiang, Y., Yin, S., Dong, J., Kaynak, O.: A Review on Soft Sensors for Monitoring, Control, and Optimization of Industrial Processes. IEEE Sensors Journal **21**(11), 12868–12881 (2021). DOI: 10.1109/JSEN.2020.3033153
25. Krishnamurthi, R., Kumar, A., Gopinathan, D., Nayyar, A., Qureshi, B.: An overview of IoT sensor data processing, fusion, and analysis techniques. Sensors **20**(21), 6076 (2020)
26. Kritzinger, W., Karner, M., Traar, G., Henjes, J., Sihn, W.: Digital twin in manufacturing: A categorical literature review and classification. IFAC-PapersOnLine **51**, 1016–1022 (2018). DOI: 10.1016/j.ifacol.2018.08.474
27. Krizhevsky, A., Sutskever, I., Hinton, G.E.: Imagenet classification with deep convolutional neural networks. In: Conference on Information Processing Systems (2012)
28. Kurose, J., Ross, K.: Computer Networking: A Top-down Approach, Global Edition. Pearson Education Limited
29. Law, H., Deng, J.: CornerNet: Detecting objects as paired keypoints. In: European Conference on Computer Vision (2018)
30. Levinson, N.: Gap and Density Theorems, vol. 26 (1940)
31. Li, C., Li, L., Jiang, H., Weng, K., Geng, Y., Li, L., Ke, Z., Li, Q., Cheng, M., Nie, W., Li, Y., Zhang, B., Liang, Y., Zhou, L., Xu, X., Chu, X., Wei, X., Wei, X.: YOLOv6: A single-stage object detection framework for industrial applications. In: arXiv:2209.02976 (2022)
32. Liu, W., Anguelov, D., Erhan, D., Szegedy, C., Reed, S., Fu C.Y.and Berg, A.: SSD: single shot multibox detector. In: European Conference on Computer Vision (2015)
33. Miao, G., Zander, J., Sung, K.W., Slimane, S.B.: Fundamentals of mobile data networks. Cambridge University Press (2016)
34. Mills, D.: Internet time synchronization: The network time protocol **39**(10), 1482–1493. DOI: 10.1109/26.103043
35. Nguyen, T.L., Tran, Q.T., Caire, R., Besanger, Y., Hoang, T.T., et al.: FMI compliant approach to investigate the impact of communication to islanded microgrid secondary control pp. 1–6 (2017)
36. Obermaier, C., Riebl, R., Al-Bayatti, A.H., Khan, S., Facchi, C.: Measuring the realtime capability of parallel-discrete-event-simulations. Electronics **10**(6), 636 (2021)
37. Oppenheim, A., Willsky, A., Nawab, S.: Signals & Systems. Prentice-Hall signal processing series. Prentice Hall (1997). URL https://books.google.dk/books?id=LwQqAQAAMAAJ
38. Paley, R.C., Wiener, N.: Fourier transforms in the complex domain, vol. 19 (1934)
39. Paszke, A., Gross, S., Massa, F., Lerer, A., Bradbury, J., Chanan, G., Killeen, T., Lin, Z., Gimelshein, N., Antiga, L., Desmaison, A., Kopf, A., Yang, E., DeVito, Z., Raison, M., Tejani, A., Chilamkurthy, S., Steiner, B., Fang, L., J., B., Chintala, S.: Pytorch: An imperative style, high-performance deep learning library. In: Conference on Neural Information Processing Systems (2019)
40. Peterson, L.L., Davie, B.S.: Computer Networks: A Systems Approach. Elsevier
41. Redmon, J., Divvala, S., Girshick, R., Farhadi, A.: You only look once: unified, real-time object detection. In: IEEE Conference on Computer Vision and Pattern Recognition (2016)
42. Redmon, J., Farhadi, A.: YOLO9000: better, faster, stronger. In: IEEE Conference on Computer Vision and Pattern Recognition (2017)
43. Redmon, J., Farhadi, A.: YOLOv3: an incremental improvement. In: arXiv:1804.02767 (2018)
44. Ren, S., He, K., Girshick, R., Sun, J.: Faster R-CNN. In: Advances in Neural Information Processing Systems (2015)
45. Ribeiro, M.I.: Kalman and extended kalman filters: Concept, derivation and properties. Institute for Systems and Robotics **43**(46), 3736–3741 (2004)

46. Shannon, C.: Communication in the presence of noise. Proceedings of the IRE **37**(1), 10–21 (1949). DOI: 10.1109/jrproc.1949.232969
47. Sundararaman, B., Buy, U., Kshemkalyani, A.D.: Clock synchronization for wireless sensor networks: A survey **3**(3), 281–323. DOI: 10.1016/j.adhoc.2005.01.002
48. Uijlings, J., Van De Sande, K., Gevers, T., Smeulders, A.: Selective search for object recognition. International Journal of Computer Vision **104**(2), 154–171 (2012)
49. Vestergaard, R., Zhang, Q., Lucani, D.E.: Enabling random access in universal compressors. In: IEEE INFOCOM 2021 - IEEE Conference on Computer Communications Workshops (INFOCOM WKSHPS), pp. 1–6 (2021). DOI: 10.1109/INFOCOMWKSHPS51825.2021.9484460
50. Vestergaard, R., Zhang, Q., Sipos, M., Lucani, D.E.: Titchy: Online time-series compression with random access for the Internet of Things. IEEE Internet of Things Journal **8**(24), 17568–17583 (2021). DOI: 10.1109/JIOT.2021.3081868
51. Viola, P., Jones, M.: Rapid object detection using a boosted cascade of simple features. In: IEEE Conference on Computer Vision and Pattern Recognition (2001)
52. Voss, R.F., Clarke, J.: Flicker (1 f) noise: Equilibrium temperature and resistance fluctuations. Physical Review B **13**(2), 556 (1976)
53. Wang, C., Bochkovskiy, A., Mark Liao, H.Y.: YOLOv7: Trainable bag-of-freebies sets new state-of-the-art for real-time object detectors. In: arXiv:2207.02696 (2022)
54. Waris, M., Iosifidis, A., Gabbouj, M.: CNN-based edge filtering for object proposals. Neurocomputing **266**, 631–640 (2017)
55. Welch, G., Bishop, G., Hill, C.: An Introduction to the Kalman Filter. Tech. rep. (2006)
56. Yazdani, N., Lucani, D.E.: Online compression of multiple iot sources reduces the age of information. IEEE Internet of Things Journal **8**(19), 14514–14530 (2021). DOI: 10.1109/JIOT.2021.3064511
57. Yiğitler, H., Badihi, B., Jäntti, R.: Overview of Time Synchronization for IoT Deployments: Clock Discipline Algorithms and Protocols **20**(20), 5928. DOI: 10.3390/s20205928
58. Zitnick, C., Dollár, P.: Edge boxes: locating object proposals from edges. In: European Conference on Computer Vision (2014)

Part III
Services for Digital Twins

Part II introduced the types of models data that can form the core assets of Digital Twins (DTs). In this part, we explore the services that can be built on these assets to add value to the Physical Twin (PT) in the whole DT-enabled system. Chap. 8 describes techniques that support visualisation of system state and trends over time. Chap. 9 describes how it is possible to monitor data from a PT as a basis for detecting potential faults. Chap. 10 considers capabilities that can be included in a DT to support decision-making through what-if simulations, fault injection, fault diagnosis and response, predictive maintenance and autonomous or semi-autonomous reconfiguration of the PT.

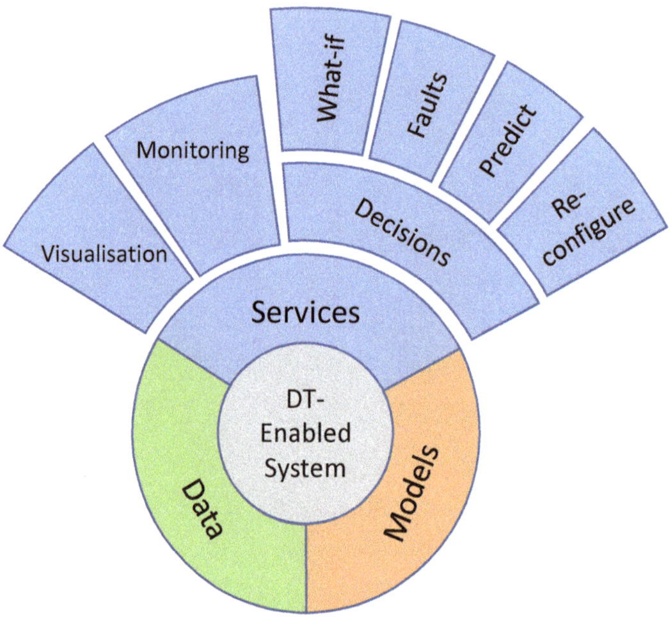

Topics related to DT services.

Chapter 8
Visualisation in a Digital Twin Context

Christian H. Bohlbro, Hugo Daniel Macedo, Daniella Tola, Lukas Esterle, and Peter Gorm Larsen

Abstract This chapter describes the potential of visualisation as a service offered by the digital twin of a cyber-physical system. Much of the value of visualisation technology comes from the ability to present stakeholders with a representation of the state and dynamics of the scyber-physical system, augmented with the results of simulation, analysis and prediction services offered by the digital twin. Forms of visualisation are briefly enumerated and compared.

8.1 Introduction

The overall objective of visualisation is to make it easier for humans to extract the desired information from data [13]. Consequently, visualisation services have a key role to play in enabling the users of a Digital Twin (DT) to gain the most value from the models, data, enabling and analytic services that go to make up the twin. They allow the state and history of the Physical Twin (PT) to be presented alongside the outcomes of the rich set of analyses that can be performed using the models in the DT and the data that has been obtained from the PT, and thereby support insight and decision-making [10]. In this short chapter, we will not deal with the human cognition aspects of visualisation, but simply offer a guide to some techniques that the DT engineer may wish to consider using [11].

In order to select a specific visualisation technique one first needs to consider the purpose of the visualisation and the dimension of the data to be shown. The *purpose* is related to the information the user is interested in deriving, while the dimension influences the optimal way of displaying the data. The dimension can range from 1D to 4D[1]. 1D is essentially just one piece of data such as a number. 2D can either be

[1] In principle one could have 10 variables that one would wish to consider (maybe over time) but it is difficult to visualise in more than 3D + time.

J. Fitzgerald et al. (eds.), *The Engineering of Digital Twins*,
https://doi.org/10.1007/978-3-031-66719-0_8

data that can be represented in an ordinary coordinate system or data that changes over time, e.g., a graph where the x-axis corresponds to the time progressing. 3D can either be 2D data that changes over time or a three dimensional vision of the data. Finally, 4D will be 3D data which changes over time, e.g., the vision a person will have moving around in space. visualisation is frequently used in connection with the modelling, simulation and calibration activities described in Chap. 5 and Chap. 6. Thus, it is not surprising that visualisation also can be a valuable service in a DT context.

In this chapter, we first consider the purpose and importance of visualisation in the DT context (Sect. 8.2), as well as a selection of visualisation services and techniques (Sect. 8.3). Section 8.4 presents several frameworks that can be used for assisting with establishing a visualisation inside a DT. Finally, Sect. 8.5 provides examples illustrating visualisation in the DT setting.

8.2 Visualisation

Visualisation maps complex information into figures, 2D graphs (numbers updated over time), 3D charts (2D visualisation updated over time), or even 4D images, which are likely to match a user's understanding of the current state of a Physical Twin (PT). Even if only one dimension is required a figure showing a value accompanied by the range of nominal values coloured green may be better at conveying a message of normality than simply displaying a number in the standard colour. Digital visualisations can take a wide variety of forms, depending on the application and the user's needs. Of the many different approaches, using software running on a desktop computer to create the visualisation through graphical visuals is common. However, visualisations are not limited to only being displayed on a traditional screen or a display alone. The visualisation can be made using any digital interface to convey the functionality and interaction with the system, depending on the system.

Deciding how to implement the interface that visualises the DT is commonly conducted as a study in Human Computer Interaction (HCI), which focuses on understanding how humans experience and interact with digital technology, successfully or otherwise. HCI-related studies in the context of DT are often crucial to ensure the DT is easily understood. Dingli and Haddod point out that the interactions in virtual environments, such as Virtual Reality (VR) and Augmented Reality (AR), are very different to more traditional computer-generated interfaces, such as graphics displayed on a screen [3].

There are several benefits that flow from building a visualisation capability into a DT. Most obviously, visualisations can make it easier for DT stakeholders to understand and analyse the data extracted from a PT. From 2D upwards, they may provide a graphical representation of system state, history or possible future progress that a wide range of users can understand. An appropriately designed visualisation helps users to focus on the features in the data that are most important in a given scenario, including situations in which rapid decision-making is called for. in case of an emergency. As Vekiri states, "*... graphical representations are computationally efficient when they minimise the processing required for their interpretation.*"[14]. Based on studies presented in that paper, using graphs and charts as graphical representations can reduce demand on the cognitive processing of information for learners. The same also applies to the end-user of digital visualisation, making it less taxing to comprehend the information presented if graphical visualisations are applied.

Visualisations can also help the DT user to identify features, patterns, trends, and relationships that may not be apparent in raw data. By extension, the visualisation can also help the users identify potential issues or dangers that might be obscured in a virtual model alone. For example, the DT of an automotive assembly plant could include detailed information about the layout of the plant, encompassing the locations of machines, materials, and workers. By visualising potential fire hazards in the plant, such as areas with high concentrations of flammable materials or inadequate fire suppression systems, managers and workers could identify and address potential risks before they result in a real-world incident. Additionally, visualising the spread of a fire within the DT could help emergency responders plan and coordinate their response in the event of a real-world fire.

In addition to ease-of-use and comprehension, visualisations can clearly and concisely communicate information about the DT to others, such as stakeholders or decision-makers. They also allow DT users with a modest system-level understanding and technical skills to use and act on the information presented. This consequently widens the target group for the visualisation, from the data engineer to workers in the field with a non-technical background.

8.3 Visualisation Services in a Digital Twin

Fig. 8.1 illustrates conventional forms of visualisation techniques that are often used with DTs, using the example of the tempeh incubator first introduced in Chap. 3. The left-hand-side of Fig. 8.1 shows a dashboard view in which the evolution of selected signals are shown over time as curves (in this case also with indications of interesting events). The right-hand-side of Fig. 8.1 provides an overview of how the current state of the PT is displayed so the user can follow the development. These are different ways for visualising the current state. However, the visualisation can also contain system controls (if the user wishes to make decisions) and interactive elements (if the user wishes to examine a part of the data in further detail). The visualisation represents the various aspects of the PT in a graphical or visual format, including the PT's components and characteristics, its environment and external factors, and any real-time data associated with it.

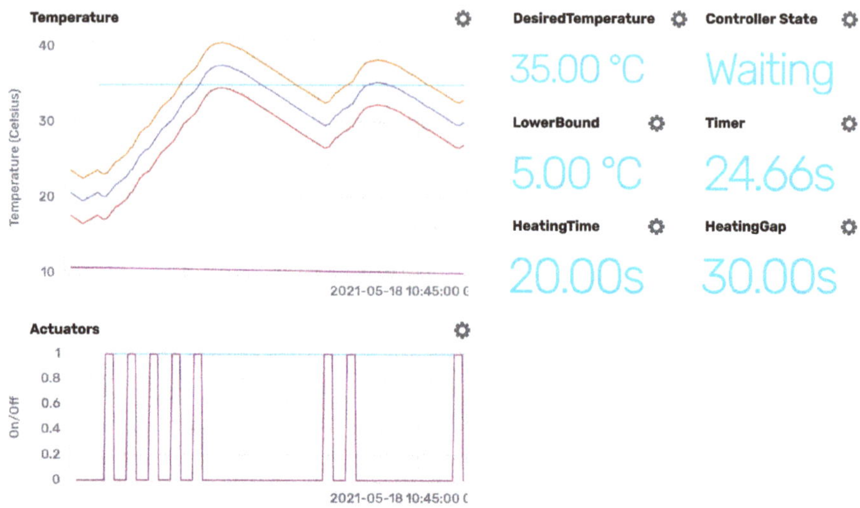

Fig. 8.1: A dashboard overview for the Incubator.

The visualisation may also have interactive elements and system controls that could range from simple buttons and fields to clickable components on a 3D Computer-Aided Design (CAD) model. The physical attribute data coming from the PT may include not only the PT's components and characteristics but also external environmental factors and components that could impact the behaviors being analysed or simulated in the DT.

8.3.1 Visualisation Techniques

The choice of visualisation techniques offered by the DT is determined by the decisions to be supported, the needs of the relevant human users, the characteristics of the the specific object or system being represented and the information that needs to be conveyed. There are many techniques from which to choose. As a starting point for the reader, and based on our own experience, we consider five types that can prove useful a DT. In order of increasing complexity:

1. **Data graphs and charts:** Real-time data associated with the PT can be visualised using graphs and charts, allowing users to see how the data changes over time and identify patterns or trends. This can be particularly useful in DTs of systems that generate much data, such as transportation networks or production lines. Fig. 8.1 falls into this category.

2. **Schematic diagrams:** A schematic diagram or 2D diagram evolving over time (so a 3D visualisation) is a simplified visual representation of a system's structure, showing the relationships between its components and how they work together. This type of visualisation is often used in DTs of complex systems, such as industrial machinery, transport networks, or power plants, to help users understand the system's functioning and identify points of concern. Such schematic diagrams are often presented as a 2D floor plan or a map where geographical information is significant.
3. **3D models:** A 3D model of the PT can be used to show physical characteristics, such as size, shape, and internal components. This type of visualisation can be rotated and zoomed, allowing the user to see the object from different perspectives and in greater detail in a 4D simulation.
4. **Augmented Reality:** AR can be used to overlay digital information on top of the physical world, allowing users to see information from the DT in the context of the real-world object or system it represents. This can be particularly helpful for users who need to interact with the PT, as it allows them to see both the digital and physical information simultaneously.
5. **Mixed Reality:** Mixed Reality (MR) utilises AR/VR and interactive controls, such as haptic feedback, to create an encompassing 4D experience. MR can be essential to provide an experience that mirrors a physical system that relies heavily on user precision and controls.

It is certainly not always necessary to implement an advanced 4D environment to visualise a simple system. For example, Wang et al. present a case study on creating a DT for a die-cutting machine in which the visualisation is a dashboard [15]. The dashboard serves several purposes, such as monitoring, planning, and communication of the various elements of the die-cutting machine. An advanced 3D representation of the system was not necessary as the dashboard alone provided access to the information and functions required.

Some cases do demand more advanced visualisation for accurate simulation. For example, advanced robotic systems involving a range of physical movement, such as robotic arms, often require visualisation with a 3D model to represent and simulate the robot's movement accurately. For example, Matulis et al. [9] present a case study on creating a DT for a robotic arm as an approach for training using Artificial Intelligence (AI) in a virtual space. A 4D visualisation was required to correctly simulate the robot's movement in the surrounding environment.

Data graphs and charts of the types shown in Fig. 8.1 are almost always useful in DT development. It is important that the signals of interest for the user can be configured/selected somehow. In both cases looking at such overviews of data are far easier than looking at the raw data from the PT. In general requirements for visualisation services would be documented in the specific requirements process as briefly described in Sect. 4.3.2.

The use of 2D or 3D graphical diagrams is very relevant when physical structure of the PT or its elements is significant, or if the part of the PT moves through a physical volume and the user needs the current position/dimensions and/or the path that led to the current state. As an example, Fig. 8.2 shows the path of a small autonomous vehicle predicted by a kinematics model within a DT. The significant deviation from the observed data can help to guide model calibration. It is worth including functionality to scale and zoom visualisations like these so that the user can appreciate small-scale features of the state or trajectory.

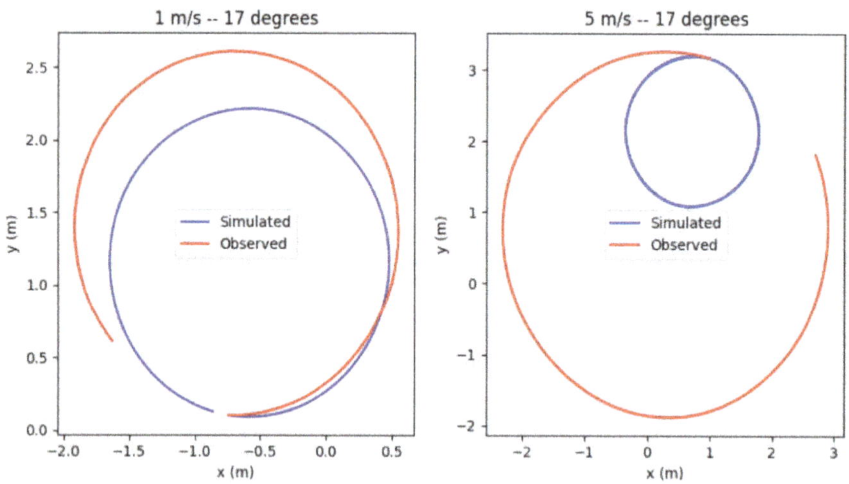

Fig. 8.2: 3D visualisation illustrated for an F1tenth vehicle moving autonomously at different speeds.

When visualising motion in a PT or elements of its environment over time, 4D models are effective but present greater development and operational costs. For example, this is illustrated for the incubator in Fig. 8.3 where it has been incorporated into Unity. Since this is a rather static PT it will not give much value to the user. In contrast, for robotic arms, it is easy to imagine how a 4D visualisation can add value. It is also possible to show the difference between the PT as observed and the status predicted by the DT. Fig. 8.4 shows how that can look in a visualisation constructed using Unity (see Sect. 8.4.2) where the important parameters for each joint are adjusted dynamically and additional information about each joint (displayed in 1D form) augments the visualisation.

We do not go into depth on VR, AR and MR in this book. However, we anticipate that the technologies developed for the industrial metaverse (see Chap. 15) will drive DT capabilities in relevant areas such as training [5].

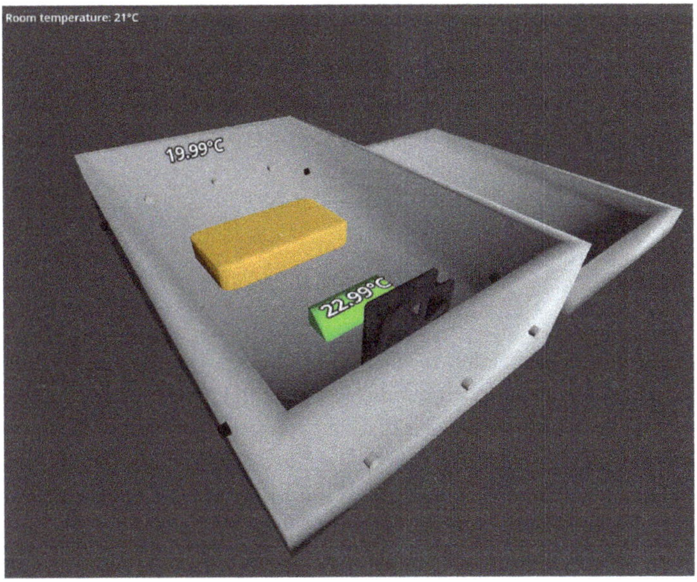

Fig. 8.3: Snapshot of a 4D visualisation of the incubator.

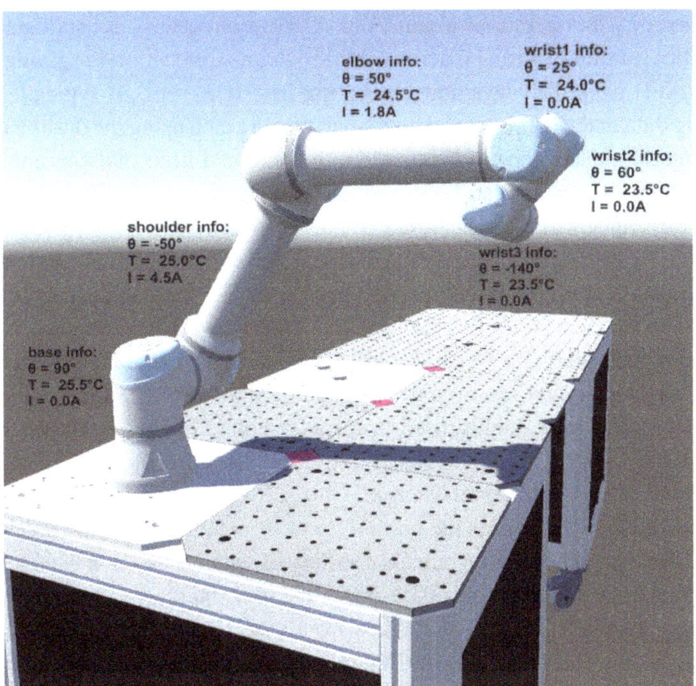

Fig. 8.4: Snapshot of a 4D visualisation of a Universal Robot cobot.

8.4 Frameworks used for DT Visualisation

In this section, we review several popular frameworks that can be used to deliver visualisations in DTs. We distinguish between two types of visualisation frameworks: dashboards and 4D visualisations.

8.4.1 Dashboards

Dashboards bring data together to give a broad overview of system state and performance against business-specific indicators at a high level. They will typically use graphs, schematic diagrams and charts, often alongside individual data points. They can be used to analyse data, show trends, visualise thresholds and requirements, and display the system state and loads. There are many visualisation frameworks that support the design of dashboards, and many of them have been created with IoT devices with cloud connections in mind. As examples, we look into a little more detail on three such frameworks: InfluxDB, JMobile and Grafana.

InfluxDB: The cloud-based time-series database InfluxDB provides a dashboard functionality for operations monitoring, application metrics, sensor data and data analytics[2]. InfluxDB can be accessed through an Application Programmer Interface (API) using C#, Java and JavaScript. The framework is typically used for storing data in the database and for visualising the data using the dashboard which can be customised, and has been used in IoT and real-time analytics applications. Data from PT sensors can be stored and visualised using InfluxDB, which can help users understand and analyse parts of the system, determine whether there is a fault or if there is a potential improvement that can be made. As system metrics are crucial in DTs, InfluxDB can be a beneficial tool for visualising these data.

JMobile: The industrial Human-Machine Interface (HMI) framework JMobile provides tools that can be used with IoT systems for device and process management, data visualisation and exchanging data from the edge to the cloud[3]. It is integrated with highly used third-party communication protocols such as PROFINET, EtherCAT, EtherNet/IP and Modbus, allowing for ease of support and development. Structured Query Language (SQL) databases are also supported with JMobile, giving users the possibility to store data using such databases and visualise using the HMI. It contains a gallery of widgets based on Scalable Vector Graphics (SVG) for a graphical design of the user interface.

Grafana: Grafana is a multi-platform web-based visualiser that allows users to visualise their data through interactive web-applications that can be hosted on

[2] https://www.influxdata.com/

[3] https://www.exorint.com/products/software/jmobile

the cloud or locally[4]. It provides the ability to visualise metrics, logs and traces through dashboards, reports and similar interfaces. The interfaces can be accessed through various data and application plugins, such as InfluxDB, Amazon Web Services, MySQL and MongoDB. Grafana is a flexible framework that can be used for different purposes, such as monitoring a Windows-based computer and setting up alerts depending on the usage or monitoring the signals of a CPS and visualising its traces. The visualisations are web-based with no installation requirements, allowing for rapid setup and visualisation of a system. It has open-source and enterprise licenses, depending on the user's needs.

8.4.2 4D Visualisation

A 4D visualisation encompasses 3D models that can be evolved over time, simulations, AR or MR. A brief analysis shows that the current 3D visualisation frameworks for DTs are typically based on 3D graphics engines. Such engines can be used for multiple purposes in the lifecycle of a DT. They can be used in the development stage to create 3D CAD models, e.g., by training algorithms on the simulation data or by testing the developed models in a realistic environment. 4D graphics engines can also be used in the deployment stage where the visualisation is a part of the DT [2]. We present four software tools for 4D visualisation of DTs: Unreal Engine, Unity, Godot and iTwin.

Unreal Engine: Unreal Engine is a 4D computer graphics engine that was initially developed for the video games industry[5]. With the increasing use of 4D visualisations in other industries, such as film and television, architecture, simulation, and automotive and transport, it has been extended to meet these industries' needs. Development using Unreal Engine is performed using an editor with drag and drop capabilities and programming using the programming language C++, which is good for resource-intensive applications. Unreal Engine offers a free license for students, educators and creators who do not require support. In addition, custom and enterprise licences are available for professional development.

Unity: Unity is another 4D computer graphics engine that was developed for the video games industry, but has since been extended to other industries including mobile applications[6]. Unity has a desktop application with a drag-and-drop interface, ability to set colours and materials, simulate physics and program the behaviour of objects using the programming language C#. DTs have been developed using the Unity game engine, for example, for automated vehicles [16]

[4] https://grafana.com/

[5] https://www.unrealengine.com/

[6] https://unity.com/

and robotic arms [12, 4]. Some also used Unity and Virtual Reality to assess the ergonomics and safety of an industrial work environment [6].

Godot: Godot is a cross-platform game engine that can also be used for visualisations in DTs[7]. It supports applications that can be run on desktop platforms, web browsers, mobile platforms and selected VR and AR devices. Applications can be programmed using the native GDScript language that is similar to Python, C++ and C#. It is an open-source platform released under the MIT license. Godot is a good option for lightweight visualisations, while Unity and Unreal Engine are more suitable for applications with advanced 3D graphics.

iTwin: iTwin is an open platform developed for infrastructure DTs by Bentley Solutions. It allows using Software-as-a-Service (SaaS) solutions for the development of infrastructure assets and the iTwin platform has integrated support for data integration and visualisation[8]. The platform is generally developed to reduce overhead when developing and maintaining infrastructure DTs. Bentley Solutions also provides an open-source web-based DT platform called iTwin.js which allows for initial development of DTs using NodeJS. It can visualise 3D models and data over time (so achieving a 4D effect), geographical information and IoT datastreams, which allows the combination of both 4D visualisations of the PT and dashboard-like visualisations using graphs or charts. The iTwin platform offers different licensing schemes, from subscriptions to a one-time purchase.

Of the frameworks considered here, InfluxDB, JMobile and Grafana are well-placed for dashboard-based visualisations, whilst Unreal Engine, Unity, Godot and iTwin have features enabling 3D and 4D visualisations. For Virtual or Mixed Reality applications, Unreal Engine and Unity may be the best starting points. Table 8.1 presents the frameworks, their visualisation type and how they can be run, i.e., on a web-browser or through a desktop application. Naturally all of these frameworks are under constant development so the comparisons presented here only reflect status at the time of writing.

Framework	**Visualisation type**	**Platform**
InfluxDB	Dashboard	Web
JMobile	Dashboard	Web, Desktop
Grafana	Dashboard	Web, Desktop
Unreal Engine	3D and 4D	Desktop, VR, AR, MR
Unity	3D and 4D	Desktop, VR, AR, MR
Godot	3D and 4D	Desktop, VR, AR, MR
iTwin	4D	Web

Table 8.1: Overview of frameworks.

[7] `https://godotengine.org/`

[8] `https://www.bentley.com/software/itwin-platform/`

Comparing the dashboard frameworks identified above, InfluxDB is used by developers for analysis of time-series databases, JMobile is used by mobile app developers for interfacing with industrial automation machines, and Grafana is used by data analysts for monitoring and visualising time-series data.

The selection of a 4D visualisation framework may depend on a range of factors including the DT requirements, as well as pragmatic aspects such as the previous experience of the developers and the licensing requirements of the project as well as features of the frameworks themselves. An overview of properties related to the 4D frameworks discussed above are shown in Table 8.2. These are based on the most

Tool	Programming languages	CAD formats	Community
Unreal Engine	C++, Blueprint	FBX, OBJ, USD, Alembic	Large
Unity	C#	FBX, OBJ, USD, DXF, DAE	Large
Godot	GDScript, C++, C#	OBJ, DAE, glTF	Growing
iTwin	JavaScript	FBX, OBJ, DXF, DAE	New and growing

Table 8.2: Comparison of relevant properties between 4D visualisation frameworks. The primary programming language of each framework is listed first.

recent versions of the frameworks. To determine the simplicity of the frameworks, we can compare the programming languages, where C# is programmed at a higher abstraction level than other languages such as C++, and therefore it may be anticipated that Unity and Godot are more abstract frameworks for programming parts of the visualisations. Comparing the supported CAD formats, all of the frameworks support OBJ, which can be used to visualise both colours and textures, supporting the exchange of models across these frameworks if needed. For beginners working with a new framework, community support is essential for a smooth learning process. Of the frameworks discussed, Unreal Engine and Unity are the two frameworks with most mature community support at the time of writing.

8.5 Visualisation Examples

In this section we introduce two types of visualisation using the incubator system as an example. An InfluxDB dashboard is used to visualise information and an AR application is used to visualise the objects inside the incubator.

8.5.1 Dashboards: Incubator Prototype

This case study provides an example of visualisation using dashboards. We delve into the details of the incubator presented in Sect. 3.2, which uses InfluxDB dashboards as a framework. Our imaginary developers Twinpeh chose InfluxDB to provide data storage for the DT, and so it was also used to create the dashboard shown in Fig. 8.1, providing simple visualisations including temperature data from the sensors, the state of the fan and heater, and parameters of the controller and the control states.

The dashboard in the figure uses a visual hierarchy. It uses the viewing pattern to guide the user's focus order to the most significant elements. For example, following the F-pattern [1], the first element that comes into focus is the top-left graph displaying the various temperature measurements and calculations, which is also the most significant aspect of the incubator, making its positioning optimal. The text size on the right side displaying the controller state and the timer is also more prominent than the neighbouring text. The size also naturally attracts the user's focus, making it a good use of visual hierarchy, suggesting that the controller state and timer are of higher importance.

Colours could be incorporated into the controller state text to accentuate the importance of different states, such as red or yellow, if the controller reaches a critical state. The dashboard already uses the light-blue colour, associated with calm and cold, regarding the "Cooling down" controller state. Therefore, additional colours might already be present in the dashboard, though it is not established from the information acquired on the case.

The dashboard's visual hierarchy makes it a suitable tool for conveying the system information of the DT. For the incubator a dashboard overview can be seen in Fig. 8.1. However, using graphs and text-based metrics might still affect the understanding of system states and monitoring for non-expert users. Furthermore, if we consider a future extension to, for example, include humidity factors and controlled airflow, the number of metrics needed might increase drastically, further affecting the comprehension of the system.

8.5.2 Augmented Reality: Incubator Prototype

There are many applications of AR for improving the quality and experience of CPSs in general, making them valuable through the DT life cycle and even before the PT is operational. There are many examples of its use in guidance and information visualisation, such as in providing instructions for training and manual assembly [8]. AR applications are mediated throguh a range of devices. Head-mounted devices that are hands-free, such as the Microsoft HoloLens[9], allow users to manipulate

[9] `https://www.microsoft.com/en-us/hololens`

objects while viewing the augmented visual overlay, while hand-held devices, such as smart phones, allow users to observe additional information about a system [7].

In our incubator example, AR can be used for both guidance and information visualisation. When engineering such a DT, it is important to consider how the user may interact with the system and if potential interactions may affect operation. For example, humans are curious creatures, so Twinpeh will find that many of their users choose to open the lid of the incubator from time to time to check on their tempeh. Doing this will affect the temperature within the incubator, potentially prolonging fermentation and reducing tempeh quality. AR can be used to reduce this risk by providing users with the possibility to visualise the tempeh and its temperature using a mobile device, as in Fig. 8.5 which merely shows an example visualisation developed using Unity. A better approach might be to create a visual model of the tempeh's growth to provide a more realistic image. Additional examples where AR could be used in the incubator are:

Before purchase: Visualising the dimensions of the incubator using AR allows users to determine how the incubator can be positioned before purchasing and therefore also determine if there is enough physical space for the system. In this case using a smart phone is appropriate as they are widely accessible.

Before operation: Guiding the user with assembly instructions of how to set up and connect the different parts of the incubator, for instance the wiring connections, can both reduce the need for providing a printed manual and reduce the risk of incorrect assembly of the system. In this case using a head-mounted device would be easiest to allow the user to set up the system using their hands.

During operation: If a fault occurs in the system and a part needs to be fixed or replaced, AR can be used to guide the user with this process.

Fig. 8.5: A potential AR application for the incubator.

References

1. Bach, B., Freeman, E., Abdul-Rahman, A., Turkay, C., Khan, S., Fan, Y., Chen, M.: Dashboard design patterns. IEEE Transactions on Visualization and Computer Graphics **29**(01), 342–352 (2023). DOI: 10.1109/TVCG.2022.3209448
2. Böttjer, T., Tola, D., Kakavandi, F., Wewer, C.R., Ramanujan, D., Gomes, C., Larsen, P.G., Iosifidis, A.: A review of unit level digital twin applications in the manufacturing industry. CIRP Journal of Manufacturing Science and Technology pp. 162–189 (2023). DOI: https://doi.org/10.1016/j.cirpj.2023.06.011
3. Dingli, A., Haddod, F.: Interacting with Intelligent Digital Twins. In: A. Marcus, W. Wang (eds.) Design, User Experience, and Usability. User Experience in Advanced Technological Environments, Lecture Notes in Computer Science, pp. 3–15. Springer International Publishing, Cham (2019). DOI: 10.1007/978-3-030-23541-3_1
4. Garg, G.: Digital Twin for Industrial Robotics,. Master's thesis, University of Tartu (2021)
5. Harichandran, A., Johansen, K.W., Jacobsen, E.L., Teizer, J.: A conceptual framework for construction safety training using dynamic virtual reality games and digital twins. In: C.e.a. Feng (ed.) Proceedings of the 38th International Symposium on Automation and Robotics in Construction (ISARC), pp. 621–628. International Association for Automation and Robotics in Construction (IAARC), Dubai, UAE (2021). DOI: 10.22260/ISARC2021/0084
6. Havard, V., Jeanne, B., Lacomblez, M., Baudry, D.: Digital twin and virtual reality: a co-simulation environment for design and assessment of industrial workstations. Production & Manufacturing Research **7**(1), 472–489 (2019). DOI: 10.1080/21693277.2019.1660283
7. Kolla, S., Sanchez, A., Plapper, P.: Comparing effectiveness of paper based and augmented reality instructions for manual assembly and training tasks. SSRN Electronic Journal (2021). DOI: 10.2139/ssrn.3859970
8. Künz, A., Rosmann, S., Loria, E., Pirker, J.: The potential of augmented reality for digital twins: A literature review. In: 2022 IEEE Conference on Virtual Reality and 3D User Interfaces (VR), pp. 389–398 (2022). DOI: 10.1109/VR51125.2022.00058
9. Matulis, M., Harvey, C.: A robot arm digital twin utilising reinforcement learning. Computers & Graphics **95**, 106–114 (2021). DOI: 10.1016/j.cag.2021.01.011
10. da Silva, G.S., Kruger, K., Basson, A.H.: Opportunities for visualising complex data by integrating virtual reality and digital twins. In: K. von Leipzig, N. Sacks, M. Mc Clelland (eds.) Smart, Sustainable Manufacturing in an Ever-Changing World, pp. 553–563. Springer International Publishing, Cham (2023)
11. Sweller, J., et al.: Visualisation and instructional design. In: Proceedings of the International Workshop on Dynamic Visualizations and Learning, vol. 18, pp. 1501–1510. Citeseer (2002)
12. Tola, D., Böttjer, T., Larsen, P.G., Esterle, L.: Towards modular digital twins of robot systems. In: 2022 IEEE International Conference on Autonomic Computing and Self-Organizing Systems Companion (ACSOS-C), pp. 95–100 (2022). DOI: 10.1109/ACSOSC56246.2022.00040
13. Tufte, E.R.: Envisioning Information. Graphics Press (1990)
14. Vekiri, I.: What is the value of graphical displays in learning? Educational psychology review **14**(3), 261–312 (2002)
15. Wang, K.J., Lee, Y.H., Angelica, S.: Digital twin design for real-time monitoring–a case study of die cutting machine. International Journal of Production Research **59**(21), 6471–6485 (2021)
16. Wang, Z., Han, K., Tiwari, P.: Digital twin simulation of connected and automated vehicles with the unity game engine. In: 2021 IEEE 1st International Conference on Digital Twins and Parallel Intelligence (DTPI), pp. 1–4 (2021). DOI: 10.1109/DTPI52967.2021.9540074

Chapter 9
System Monitoring through a Digital Twin

Mirgita Frasheri, Panagiotis Katsaros, Alexandros Iosifidis, Simon Thrane Hansen, Cláudio Gomes, Valdemar Tang, and Peter Gorm Larsen

Abstract A Digital Twin (DT) has great potential as a tool for monitoring, detecting and reporting anomalies. Such advance warning allows stakeholders to sidestep failures, or to be better prepared to recover from them. This chapter introduces *model-based* and *data-driven* approaches to monitoring in DTs. Model-based monitoring uses definitions of properties (expressed in *temporal logics*) that describe the desired behaviour of the Physical Twin (PT). Data-driven approaches build monitors from historical data collected from the PT using *machine learning* models. We discuss how these types of techniques can be used to provide anomaly detection, at either the component or system levels.

9.1 Introduction

One of the ways in which a DT adds value to a PT is by bringing together models of the PT's intended or expected behaviour alongside data gathered from observations of the PT itself and showing what is "really going on". The DT can act as a monitor or a critical friend to the PT, identifying behaviour that is "out of character", sub-standard, or that may betray an impending or actual failure [5, 27, 21, 19]. This in turn means that the DT and its users can trigger prevention or recovery. This role is called *monitoring* and is key to supporting DT stakeholders in their decision-making (described in Chap. 10) and is the subject of this chapter [5, 38].

To enable monitoring, the DT could implement a monitor that checks data derived from the PT against a set of properties that form a specification of desired behaviour, or that highlight unwanted behaviour, producing a verdict to indicate whether the behaviour satisfies the property or not, as depicted in Fig. 9.1. The PT provides data to the DT, where the data is transformed into a behavioural trace (e.g., a sequence of states), given as an input to the monitor. The monitor then analyses the behaviour

J. Fitzgerald et al. (eds.), *The Engineering of Digital Twins*,
https://doi.org/10.1007/978-3-031-66719-0_9

and produces a verdict indicating whether the behaviour satisfies the given property. This verdict could be fed back to the PT or a human interacting with the system.

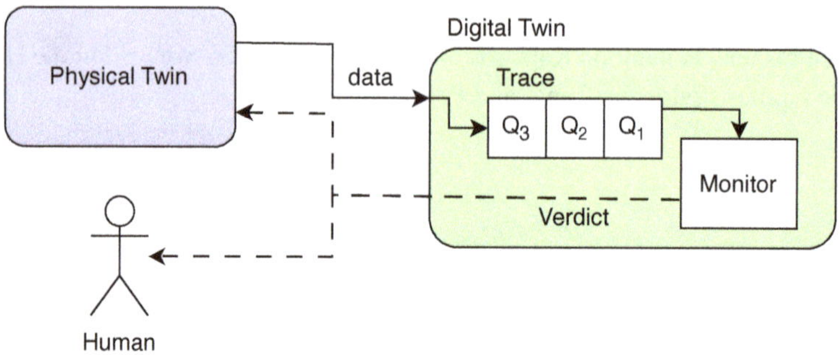

Fig. 9.1: The basic idea of monitoring.

Monitoring can be performed offline or online. *Offline monitoring* examines stored/historical data and is typically used for debugging purposes, as particular situations can be replayed to investigate the cause of a fault or other observed issue. *Online monitoring* operates on data as it becomes available from the PT and can be used to track properties during operation [33, 35]. Online monitoring can feed directly into the decision-making components in the DT, as well as allowing a human operator to intervene if necessary. There has, for example, been work on the role of such online monitoring in heightening the security of CPSs [67, 58].

Incubator example: Monitoring

Suppose we wish to monitor whether anything has changed regarding the incubator plant (box, volume of air, power supply and fan). The simplest monitor that has been deployed to this effect consists of a Kalman Filter (see Sect. 7.6.1) that uses a model of the plant (see Eq. (5.1)). Such a filter will also provide an estimate the temperature of the heater T_h which cannot be measured directly. Implementing this monitor requires a calibrated model (see Chap. 6). Results are visualised in Fig. 3.10.

Monitors can be chained, and in the incubator, as we show next, we can use other monitors for anomaly detection that exploit results from the Kalman Filter.

DT platforms supporting monitoring have become significant in manufacturing. For example, Chhetri et al. describe the monitoring of a process through indirect observation, thinking about situations in which legacy equipment cannot readily be equipped with sensors [17]. Our focus on foodstuffs is not unique – Karadeniz et al. have applied monitoring in the manufacture of ice cream [48]!

Broadly speaking, we can distinguish between *model-based* and *data-driven* monitoring. Model-based monitors rely on a model of the system under observation [5], for example the list of properties that specify the desired behaviour. Data-driven monitors rely on data already collected from the PT. The approach taken to monitoring a PT depends on many factors, such as the goals of the monitoring process and the criticality of its verdicts (risks of false positives and negatives), the type of properties to be monitored and the type and availability of data. For example, if the PT is a system that is critical to the safety of a process, and has therefore been kept relatively simple, one might want to use a monitor based on a precise specification of the conditions to be checked, which may form part of the safety case for the DT-enabled system. However, for more complex or poorly understood PTs, a data-driven monitor may be appropriate.

In the remainder of this chapter, we first focus on the ways in which the monitored behaviour of a PT can be specified, using *temporal logics* (Sect. 9.2). This leads to Sect. 9.3 which describes how such properties can be monitored using model-based monitors constructed from the properties. Data-driven monitors are discussed in Sect. 9.4.

9.2 Describing Desirable Properties

> This section focuses on the ways in which we can specify the conditions that the monitoring DT should be checking against the data gathered from the PT. These conditions are not often about a single observation of the PT state, but concern the history of observations, for example reflecting the fact that an observed event (such as overheating) should always be followed by another (such as the heater being switched off). They are therefore described using notations called *temporal logics* which are likely to be unfamiliar to you if your background is more in engineering of physical systems. Do not be afraid if you do not understand the details of these.

The desired behaviour of the PT is specified as a set of properties that describes how the PT should behave over the course of its operation. Monitoring the PT's behaviour then amounts to checking these properties for violations in the observed data. The properties are specified using a specific language called a *Temporal Logic (TL)* [5, 53]. Using a TL makes the property specification independent of the particular implementation of the monitor in the DT.

Before introducing TLs, we need to understand how the PT's behaviour, and that of its environment, are seen by the monitor. The PT behaviour is described as a sequence of states, each state denoting a snapshot of the system at a given point in time and referring to the values of the observables. A sequence of states is called a (behavioural) trace, typically denoted by π. A trace can be finite or infinite (although in practice a DT is concerned only with finite traces) and describes the system's evolution over time. If the data is for a physical entity which evolves continuously this trace will be a discretisation produced using techniques such as those described in Chap. 7 (time for the data also matters).

Using our running example of a tempeh incubator, some fictional finite traces of the temperature in a tempeh incubator are shown below:

$$\pi_1 = \langle 20°\text{C}, 30°\text{C}, 21°\text{C}, 42°\text{C}, 50°\text{C}, 80°\text{C} \rangle$$
$$\pi_2 = \langle 20°\text{C}, 20°\text{C}, 20°\text{C}, 20°\text{C}, 20°\text{C}, 21°\text{C} \rangle$$
$$\pi_3 = \langle 30°\text{C}, 31°\text{C}, 39°\text{C}, 38°\text{C}, 35°\text{C}, 32°\text{C} \rangle$$

Properties can be simply specified using a set of *propositions*. A proposition is a statement about a specific object. For example, "the sky is blue", or "the incubator temperature is over 40°C". A proposition is normally regarded as being either true or false. Propositions can be combined using logic connectives or operators such as "not", "or", etc. to make more sophisticated assertions. There are many full treatments of logics for reasoning using propositions, e.g., that by Büning and Lettmann [13].

To enable us to specify the evolution of a system over time, we need to be able to express properties over traces of many states, rather than a single state. This is done by expressing properties as *predicates* which may be true or not, depending on the object to which they are applied. In a predicate, variables stand in for the specific objects being discussed, thus we might state that "$x > 40°$C" using the variable x to stand for any temperature reading.

A predicate will be true for some values of the variables, and false for others. For example the current temperature reading might exceed 40°C, but the last one was in range. We can combine predicates as with propositions, but we can also use quantifiers such as "for all" ($\forall$) and "there exists" ($\exists$) to make assertions about whole ranges of values for the variables. For example, for a whole set of temperature readings, we might state that they are all over 40°C, written $\forall x \cdot x > 40°$C, or there exists at least one reading that satisfies our constraint: $\exists x \cdot x \leq 40°$C.

In order to allow conditions to be expressed over traces, we use a *first-order* TL which supports the use of predicates [36] parameterised by variables ranging over diverse domains (e.g., natural numbers or strings). We call predicates over constants *events*. For example, we might use a predicate $heat(y)$ to indicate the condition of the heater, switched on ($y = 1$) or off ($y = 0$). In this way $heat(1)$ is an event indicating that the heater is on. Events in this sense carry data. Similarly, we might use $inc(x)$ is an event indicating that the temperature is x and is increasing, and $dec(x)$ indicates that the temperature value is x and decreasing. Whether the temperature is increasing or decreasing may be either observed in the sensor measurements of the PT or derived from the sign of $\frac{dT_b}{dt}$ in Eq. (3.1). We may consider the following trace of events with data for the temperature in the incubator:

$$\pi_4 = \langle heat(1), inc(20°\text{C}), dec(30°\text{C}), inc(27°\text{C}), heat(0), dec(42°\text{C}), heat(1), inc(39°\text{C}), inc(42°\text{C})\rangle$$

An interesting property in first-order logic may be that it is not the case that the temperature is greater than 40°C and it is still increasing, i.e. $\neg(inc(t) \wedge (t > 40°\text{C}))$ with t denoting the observed temperature value. However, if we want to express a constraint that relates the current state with the occurrence of an event $heat(y)$ in trace π_4, we need to use a first-order TL.

Given a set of properties, one can express a system's desired behaviour as a set of properties in the language. These properties can be divided into two categories: safety and liveness properties.

- *Safety properties* state that something undesirable never happens, e.g., the temperature inside the incubator never exceeds a certain threshold.
- *Liveness properties* state that something desirable eventually happens, e.g., the temperature eventually reaches a certain threshold in the incubator.

A system doing nothing might be considered safe because nothing bad happens, but not live as nothing good happens either. System descriptions are therefore often combinations of safety and liveness properties.

Safety properties are *monitorable* on all finite traces, since their violation can be detected within finite time. On the contrary, while liveness may have not been violated, it might be delayed indefinitely. Therefore, liveness can be *non-monitorable* [9].

In this chapter we examine the use of TLs as a way of specifying the properties that we wish a DT to monitor. However, there is a limit to this. Some relatively sophisticated properties, for example to do with information security and service level agreements, cannot be expressed as predicates over single traces, but may relate multiple traces. Techniques to monitor these *hyperproperties* are the subject of current research [30].

9.2.1 Temporal Logic in General

TL extends predicate logic with operators that allow us to describe properties over multiple states of the same system. Many TLs have been developed over the years to address different needs and differing in expressiveness, complexity and capabilities., for example covering properties of reactive and concurrent systems [53] and supporting direct execution [31]. Two well-known and distinctive TLs are Linear Temporal Logic (LTL) [59] and Signal Temporal Logic (STL) [51] which will be introduced in this section. Different TL variants can be classified by how they view time (as discrete or continuous) and what verdict they provide (quantitative or binary). In the sub-sections below, each method is classified and the concrete meaning of TL classifications is explained. The classification is summarised in Table 9.1 together with references to some available tools for the given method. The tools listed in Table 9.1 are targeted at offline or design-time analysis. A list with examples of tools for runtime monitoring is provided in Table 9.2.

Table 9.1: Types of temporal logic, how they view time and their verdict. Relevant tools are suggested.

TL	Discrete/continuous	Verdict	Tools
LTL	Discrete	Binary	SPIN [44], PAT3 [50], NuSMV2 [18]
STL	Continuous	Binary or Quantitative	AMT 2.0 [55], Breach [24], TALIRO [3]

Tutorial introductions to this topic are available [6, 27], and in-depth survey ofLs and their use in runtime verification can be found in [11, 62].

9.2.2 Linear Temporal Logic

LTL is a "point-based" TL in that considers discrete time instants [11]. Although it does not allow for the expression of time interval constraints, it can express a wide range of safety and liveness properties. LTL uses *temporal operators* to specify properties over multiple states of the system. The *next* ($\bigcirc$) operator with a single LTL formula as argument requires that the formula holds true in the next state of the one, in which the specification is evaluated. A binary temporal operator in the basic syntax is *until* ($\mathcal{U}$), whereas two important derived operators are *always* ($\square$) and *eventually* ($\diamond$) [4]. By combining these operators with the standard boolean operators, one can express a wide range of properties such as the following: $\phi_1 \wedge \phi_2 \rightarrow \bigcirc\phi_3$ which reads: "*when* ϕ_1 *and* ϕ_2 *are true,* ϕ_3 *must be true in the next state*". Another example could be $\square(S.send \rightarrow \diamond R.receive)$ which reads: "*if S sends a message, then R must eventually receive it*".

For those readers interested in the detail, the semantics of the LTL operators is explained in the box below. An example of its use in specifying properties to monitor in the tempeh incubator follows.

For readers from a computing background, the semantics of LTL operators is defined as follows. Let $\pi = \langle s_0, s_1, \ldots \rangle$ be a trace and ϕ be an LTL formula that is evaluated over π. The satisfaction relation $\pi \models \phi$ (read"π *satisfies* ϕ") is defined by:

- All traces satisfy $\top$ and no traces satisfy $\bot$.
- The truth value of an atomic proposition p is determined by the current state in the trace, i.e., $\pi \models p$ iff p is true in the current state of the trace.
- The truth value of a negation $\neg\phi$ is the opposite of the truth value of ϕ, i.e., $\pi \models \neg\phi$ iff $\pi \not\models \phi$.
- The truth value of a conjunction $\phi_1 \wedge \phi_2$ is true iff both ϕ_1 and ϕ_2 are true, i.e., $\pi \models \phi_1 \wedge \phi_2$ iff $\pi \models \phi_1$ and $\pi \models \phi_2$.
- The truth value of a next operator $\bigcirc\phi$ is true iff the next state in the trace satisfies ϕ, i.e., $\pi \models \bigcirc\phi$ iff $\pi[1:] \models \phi$.
- The truth value of an until operator $\phi_1 \,\mathcal{U}\, \phi_2$ is true iff there exists a state s_i denoted by the index i such that all states after s_i satisfy ϕ_2 and all states before s_i satisfy ϕ_1, i.e., $\pi \models \phi_1 \,\mathcal{U}\, \phi_2$ iff $\exists i \geq 0 \cdot \pi[i:] \models \phi_2$ and $\pi[j:] \models \phi_1$ for all $0 \leq j < i$.
- The meanings of the always and eventually operators are defined in terms of the until operator, i.e., $\square\phi \triangleq \neg\diamond\neg\phi$ and $\diamond\phi \triangleq \top\,\mathcal{U}\,\phi$.

Further detail on LTL semantics is given by Baier and Katoen in their introduction to the principles of model checking [4].

Incubator example: Describing properties in LTL

To ensure that the incubator does not overheat even if the lid is opened, one can express the following safety property: The temperature of the heater T_h inside the incubator must never exceed 40°C. This can be expressed in LTL as:

$$\square(T_h < 40°\mathrm{C})$$

This property can be read as: "*for all states in the trace, the temperature is less than* 40°C." Looking at the traces π_1, π_2 and π_3 defined in Section 9.2, one can see that the property holds for π_2 and π_3, but not for π_1 as the fourth state in the trace has a temperature of 42°C.

Nevertheless, the above property does not describe that the incubator air T_b should reach a certain temperature, which is also desirable. To express this, one can use the following liveness property: The temperature of the heater inside the incubator must eventually reach 37°C. This is captured by the following LTL formula:

$$\diamond(T_b \geq 37°\text{C})$$

This property can be read as: "*there exists a state in the trace where the temperature T_b is eventually greater than or equal to* 37°C."

The above property holds for π_1 and π_3, but not for π_2 as the temperature never reaches 37°C in the trace. We may want to further verify a more specific behaviour. For example, the incubator may be programmed to lower the temperature below a certain threshold after exceeding another threshold. Such a property could be expressed as follows:

$$\square(\textit{temperature} \geq 38 \rightarrow \diamond(\textit{temperature} \leq 34))$$

The above property can be read as: "*for the entire trace, if the temperature is ever greater than or equal to* 38°C, *then the temperature eventually drops below* 34°C". The property holds for traces π_3 and π_2 but not for trace π_1. The above property is satisfied for π_2 even though the temperature does not exceed 38°C.

The same approach can be taken to specifying conditions that are not based on quantitative values such as temperature. For example, if we wished to indicate that, following an alarm being raised, a notification is sent to the incubator operator, we could approach that in the same way.

There is a broad consensus on the formal definition of LTL semantics over infinite length traces. However, in when checking whether a system's behavior trace satisfies a given set of properties (termed runtime verification and introduced below), we can only have a finite view of the system's trace and for this case there are multiple semantics making the problem of runtime verification itself more complex [9]. An alternative approach is to specify properties using operators only from the *past-time portion of LTL* [40]. This restricts the expressive power of the language to specifying only safety properties, but at the same time circumvents the problem of multiple semantics of the future-time portion of LTL for finite length traces.

In *past-LTL*, the $\ominus$ operator is the past dual of *next* and when preceding a formula ϕ it is true iff ϕ is true in the trace obtained from the current trace by omitting the last state. The past dual of *until* is the *since* ($\mathcal{S}$) operator. Formula $\phi_1\ \mathcal{S}\ \phi_2$ is true iff ϕ_2 was seen in the past (including last state) and since then ϕ_1 has been true. Beyond the boolean operators, there are two more, very useful, temporal operators that are derived from the basic syntax: *previously* ($\diamondminus$) and *historically* ($\boxminus$), which is also mentioned as *always in the past*. The semantics of these two operators is defined as follows: $\diamondminus\ \phi \triangleq \top\ \mathcal{S}\ \phi$ and $\boxminus\phi \triangleq \neg \diamondminus \neg\phi$.

QTL [36] is a first-order variant of past LTL for expressing properties for traces with events that include data (see Sect. 9.2). Each event is represented by a predicate symbol with parameters e.g. $p(3), r(x)$ that can take values from diverse domains (e.g., integers, reals, strings etc.). Boolean variables are represented as degenerate predicates with 0 parameters. A predicate $p(\alpha)$ for some constant α is true, if the last event in the trace is $p(\alpha)$. For some variable x, $r(x)$ is true if x is bound to a constant α such that $r(\alpha)$ is true. The formula $\exists x \phi$ is true, if there is a value α for which ϕ is true when x is bound to α. QTL inherits all other temporal operators of past-LTL and in addition includes the universal quantifier: $\forall x \phi \triangleq \neg \exists x \neg \phi$. Finally, formulas may also include terms with arithmetic expressions and comparisons between variables and constants. QTL has been used in the specification of CPSs, since it allows capturing the quantitative progress of time using time predicates, as opposed to other LTL-based formalisms [12]. The DejaVu tool [36] for the automated synthesis of runtime monitors also supports the interpretation of QTL formulas over all prefixes of the analysed trace, which allows violations of safety properties to be detected. The analysis of liveness properties is not supported, but it is possible to check if some progress has been achieved by a given deadline, in which case liveness takes the form of a safety property. The semantics of QTL is illustrated on our incubator example below.

Incubator example: Describing properties in QTL

To ensure that the incubator does not overheat, we require that, if the temperature of the heater has reached 40°C, it is not still increasing and the heater has been already switched off. This can be expressed in QTL as:

$$\forall t \; (t \geq 40°\text{C}) \rightarrow dec(t) \wedge \ominus(\neg heat(1) \; \mathcal{S} \; heat(0))$$

This property can be read as: "*if the temperature is 40°*C *or higher, it is decreasing and the heater has been switched off and has not been switched on since then.*" The property can be evaluated over the trace π_4 defined in Sect. 9.2.

The next property shows how to formalise quantitative progress of time using time predicates. Consider a property mentioned in Sect. 3.3: the incubator system takes at most 10 seconds to warm the air from 20°C to 30°C. In this case, we need to define predicates of the form $time(t_1)$ which can be part of the events in a trace. These predicates will refer to a timer that reports a value of variable t_1. Then, the aforementioned property can be expressed in QTL as:

$$\forall T_b \, \forall t_1 (time(t_1) \wedge (T_b < 30°\text{C})) \rightarrow \neg \exists t_2 \diamondminus \, (T_b < 30°\text{C} \wedge time(t_2) \wedge t_1 - t_2 > 10)$$

This can be read as: "*if the temperature of the air is less than 30°*C *at time* t_1*, there is no time* t_2 *preceding for more than 10 sec, in which the temperature was also less than 30°*C."

Both properties of this example can be also interpreted using the DejaVu tool over all prefixes of the analysed trace, in which case they take the form:

$$\Box \phi$$

with ϕ denoting the property specification. This enables the detection of safety violations, as opposed to the default QTL formalisation, which allows monitoring how the property alternates between true and false verdicts.

9.2.3 Signal Temporal Logic

The lack of ability to impose timing constraints makes classic LTL unsuitable for verifying real-time systems that include such constraints. To address this, TLs with timing constraints have been developed and one of the most notable is Signal Temporal Logic (STL) [51]. STL extends LTL by adding an interval to the $\mathcal{U}$ operator (written $\mathcal{U}_I$). This can then be used to construct the forms *eventually within I* ($\diamond_I$) and *always within I* ($\square_I$) where I is the time interval. This lets us express complex time-constrained properties very concisely such as

$$\square_{[0,t_e]}(S.send \Rightarrow \diamond_{[t_1,t_2]}(R.receive))$$

which reads: "*if S sends a message to R in the time interval* $[0,t_e]$*, then R must receive it within the time interval* $[t_1,t_2]$". Being able to specify such conditions is essential for monitoring the temporal properties of real-time systems. In contrast to LTL, STL treats time as continuous, meaning that the time instants are infinitely precise and thus STL does not have the $\bigcirc$ operator as it cannot be defined in a continuous time setting.

STL is a quantitative logic, meaning that it can produce a quantitative output that determines how close a property is to being satisfied or not, which is not possible in LTL as it is a binary logic only capable of producing a binary output (true or false). The quantitative output of STL is called the *robustness* of a property and is defined as a function that maps a trace to a real number, more specifically a function $\rho : \pi \longrightarrow \mathbb{R}$.

A positive robustness value indicates that the property is satisfied, whereas a negative value indicates that the property is violated and the magnitude of the value indicates how well the property is satisfied or violated. The robustness measure has been shown to be useful for runtime verification and control purposes, as it can be used to determine how close a system is to violating a property and thus it can be used to trigger a control action to prevent the violation from occurring [25].

The robustness of a property can be defined both *spatially* and *temporally* [25]. Temporal robustness enables distinguishing between cases where a property was barely satisfied within the time interval constraint or where it was surely satisfied. Moreover, Donse and Maler [25] distinguish between the left and right temporal robustness, which determines how far in time forwards (right) and backward (left) the time interval can be extended such that the property is still satisfied. Spatial robustness provides a quantitative output of how close a property is to being satisfied in a spatial sense, e.g., how close the temperature is to exceeding a certain threshold.

To see how STL allows for a more flexible monitoring approach in a real-time context compared to LTL, an example with the incubator is given below.

Incubator example: Describing properties in STL

The previous example using LTL is only able to verify that the temperature does not exceed 40°C. This is a hard limit and may not accurately represent the real-world limitations of the incubator. For example, a situation may arise where it is tolerable that the incubator heater temperature T_h rises above 40 degrees for a short period and then drops again. An example of an STL property that monitors that this period is not exceeded, can be given by:

$$\square(\neg \diamond \square_{[0,2]} T_h \geq 40°\text{C})$$

which can be read as "*For the entire trace, there will never be an interval of length* $[0, 2]$, *where the temperature is consistently above* 40°C". For the sake of the example, each measurement in the traces is one time unit apart. The above property will hold for trace π_2 and π_3 but not for trace π_1. The spatial robustness for the $x \geq c$ operator is defined as $x - c$, (and vice versa for $x \leq c$). The negation corresponds to inverting the sign, from which we can calculate the spatial robustness trace of the three traces π_1, π_2 and π_3:

$$\begin{aligned}
\rho_{\pi_1} &= \langle 20, 10, 19, -2, -10, -40 \rangle \\
\rho_{\pi_2} &= \langle 20, 20, 20, 20, 20, 19 \rangle \\
\rho_{\pi_3} &= \langle 10, 9, 1, 2, 5, 8 \rangle
\end{aligned}$$

Now let us consider the last LTL example, where we specified that the temperature of the air T_b will eventually decrease to $\leq$ 34°C after rising to or above 38°C. We may want to impose a performance constraint that the temperature should reach 34°C or below within two time units. Such a property could be specified as follows:

$$\square(T_b \geq 38 \rightarrow \diamond_{[0,2]}(T_b \leq 34))$$

The above expression holds for trace π_2 but not for π_1 and π_3. For more information on how to calculate the spatial and temporal robustness of the remaining STL operators, please refer to the work of Donzé and Maler [25].

9.3 Monitoring using Runtime Verification

As outlined in Sect. 9.1, model-based monitors rely on having a description of the system under observation. Such a description could take many forms, as we discussed in Chap. 5. In this chapter we have focused on TL formulae for specifying the behaviour being monitored. Many approaches for generating monitors from such formulae have been proposed [10, 37, 34, 60]. These transform the TL formula into a program that accepts or rejects a trace based on whether the trace satisfies the formula [32, 34]. Once the monitor is generated, it can be woven into the system using aspect-oriented programming [15] or used as a separate component to monitor the system's behaviour [68, 66].

Many frameworks have been developed to support runtime verification. Examples are provided in Table 9.2. For an in-depth overview, the reader is referred to the taxonomy of Falcone et al. [28].

Name	Programming Language	Supported Logics
Java PathExplorer [39]	Java	LTL, CTL [20], TCTL [2], Custom
Java-MOP [16]	Java	LTL, ERE [14]
MonPoly [8]	OCaml	MFOTL [7]
RTAMT [56]	Python	STL, IA-STL [29]
DejaVu [36]	Scala	QTL [36]
NuRV [19]	C, C++, Lisp, Java	LTL
Storm [42]	Python	PCTL, CSL

Table 9.2: Examples of Runtime Verification Tools

Runtime verification in a DT can be used in simple or more complex tasks such as monitoring the behaviour of the PT for detecting violations of safety specifications (when the verdict is stable) or for diagnosis purposes (when the property alternates between true and false verdicts) [9]. It can also be used in enforcing compensating actions to counter a (predicted) violation of a monitored property.

Runtime verification may pose specific requirements for communication between the PT and the DT (Fig. 9.1), since the input trace originates in the PT and monitoring the PT should be as non-intrusive as possible. One option is to rely on the RabbitMQ server or other solutions that implement message queue protocols (see Sect. 7.4).

The output of runtime verification (the verdict) may be a Boolean in the case of an LTL-based monitor or a float from an STL-based monitor. If the property is violated (or violated intolerably often), some compensating or safe action (e.g., a reconfiguration) can be enforced to restore the satisfaction of the property. If the monitoring approach adopts a predictive semantics [47, 57], it is possible to enforce safe actions even before the occurrence of a property violation so that the violation is avoided.

9.4 Data-driven Anomaly Detection

Sensors placed in different parts of the PT provide real-time information about the state of the PT. This information is used by the DT for reflecting the accurate state of the PT and can be valuable for real-time monitoring and decision-making by operators who infer relationships between the different parts of the PT through the DT interface. Data collected by sensors can also be used for automating this process, either by providing early warnings to operators or by performing automatic anomaly detection. This is achieved by estimating the parameters of data-driven anomaly detection models using historical data corresponding to different types of operations. It should be noted here that, even though anomaly detection can be formulated as a multi-class or binary (i.e., two-class) classification problem, e.g., as in [46], it is commonly approached as a one-class classification problem in which the objective is to *learn* what constitutes normal operation and detect deviations from that normal operation sufficient to constitute anomalies. This is due to the fact that anomalous data is usually scarce compared to normal data. Moreover, creation of datasets that can include a sufficiently high number of samples from all possible anomalies is too time-consuming and costly. For this reason, one-class classification is considered in the following.

Incubator example: Anomaly Detection

For the purpose of this example, assume that the incubator contains multiple independently opening doors. In order to precisely measure the temperature at different locations, multiple ($S = 9$) temperature sensors are installed as shown in Fig. 9.2. Such fictional incubator could be used finer regulation of temperature uniformity.

By opening and closing the doors one can influence the temperature at different locations of the incubator in different ways. For instance, opening door d_{45} will influence the temperature at locations close to its neighbouring sensors s_4 and s_5 more compared to locations closer to sensor s_1. To influence the temperature close to s_1, one needs to open either d_{12} or d_{18}. Such a capability can be important in case some anomaly takes place, which for instance leads to sharply increased temperature, for avoiding damaging the products being located close to the place where this anomaly was detected.

Additionally, assume that historical data corresponding normal operation of the incubator has been collected. Such historical data needs to be pre-processed into a form suitable for introduction to the ML models. For simplicity, consider that all sensors $s_i, i = 1, \ldots, S$ collected measurements with the same frequency and that these measurements were synchronised. Let us denote by $\mathcal{X}_i^{(1:t)} = \{x_i^{(1)}, x_i^{(2)}, \ldots, x_i^{(t-1)}, x_i^{(t)}\}$ the measurements collected by the i-th sensor starting from the point in time when it was switched on and up to a time instance t. Each measurement $x_i^{(\tau)}$ corresponds to the temperature at the location of sensor s_i at a specific time instance τ. Since

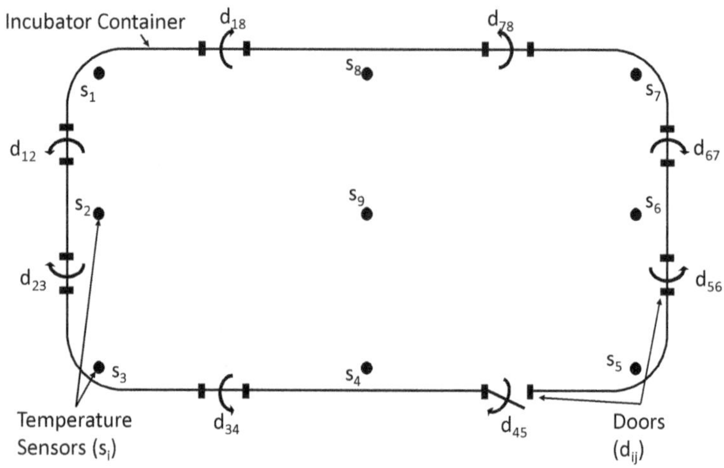

Fig. 9.2: An incubator with multiple temperature sensors (noted as s_i, $i = 1, \ldots, 9$) placed at a regular grid and multiple doors (noted as d_{ij}, $i, j = 1, \ldots, 8$) used to regulate its temperature.

temperatures at successive time instances are not independent, the measurements collected by each sensor form a time series.

Most ML models for anomaly detection which can be used in this setting receive as input a real-valued vector or a (multi-dimensional) time-series which corresponds to the status of the incubator in a temporal window. By using such an input, both static measures (e.g., temperature measurements at each sensor location) and dynamic measures (e.g., rate of temperature changes at each sensor location) can be used to define normal operation and detect anomalies. Thus, the time-series $\mathcal{X}_i^{(1:t)}$ containing the temperature measurements for each sensor s_i, $i = 1, \ldots, S$ need to be temporally segmented to create time-series samples of fixed length T.[1] Let us denote by $\mathcal{X}_i^n = \{x_i^{(n,1)}, x_i^{(n,2)}, \ldots, x_i^{(n,T)}\}$ the n-th historical time-series sample for sensor s_i obtained by temporally segmenting $\mathcal{X}_i^{(1:t)}$ using a step size of m. Step sizes $m < T$ lead to a higher number of time-series data with overlapping measurements, while step sizes $m \geq T$ lead to a lower number of time-series data which do not contain overlapping measurements.

Using the above definitions, one can define multiple anomaly detection tasks which can be solved by ML models at different levels:

Anomaly detection at the individual sensor level: In this case, historical time-series samples from each sensor are used to train an anomaly detection model.

[1] Even though there are Machine Learning (ML) models which can operate on time-series data of varying length, practical implications related to their training almost always leads to choosing to use time-series data corresponding to fixed-length time windows.

One can choose to train a different model for each sensor or combine the historical data of all sensors to train one anomaly detection model which is applied to data from all the sensors. If the amount of available historical data for each sensor is sufficiently high, using an anomaly detection model for each sensor can lead to better performance as these can better capture properties at a local level. However, if the amount of available historical data is not high, following the combined approach can lead to better training of the anomaly detection model and avoid over-fitting.

In the example, each time-series sample is formed of univariate time-instances of fixed length T, so each time-series sample $\mathcal{X}_i^n = \{x_i^{(n,1)}, x_i^{(n,2)}, \ldots, x_i^{(n,T)}\}$ can be reshaped to a T-dimensional vector $\mathbf{x}_i^{(n)}$, allowing us to represent the univariate time-series anomaly detection problem as a standard one-class classification problem. Numerous one-class classification models exist, e.g., models based on support vectors [65, 41, 1], models learning data transformations [43, 45, 61, 23] and models jointly performing data transformation and one-class classification [54, 64, 63].

Another way of performing anomaly detection is to adopt a data reconstruction approach. This can be achieved by using regression models which are trained to reconstruct their input, like Autoencoder networks (see Sect. 5.5.2). Such networks are trained by using only normal data and are expected to provide low-error reconstruction of their input when it corresponds to normal operation, while the reconstruction error is expected to be high for anomalous inputs [22, 69, 49]. A similar approach exploits a recurrent neural network which receives as input the time-series $\mathcal{X}_i^n = \{x_i^{(n,1)}, x_i^{(n,2)}, \ldots, x_i^{(n,T-1)}\}$ and regresses to the value $x_i^{(n,T)}$ [52]. When trained on normal time-series data, such an approach will be able to predict the value at the time-instance T while it will lead to high-error predictions for anomalous time-series. Approaches combining different models also exist, e.g., combination of a recurrent neural network to produce time-series data representations introduced to one-class support vector machine [26].

Anomaly detection at the incubator system level: In this case, historical data from all sensors can be combined, e.g., by concatenating the univariate time-series data $\mathcal{X}_i^{(1:t)}$, $i = 1, \ldots, 9$ to create multi-dimensional time-series $\mathcal{X}^{(1:t)} = \{\mathbf{x}^{(1)}, \mathbf{x}^{(2)}, \ldots, \mathbf{x}^{(t)}\}$, where $\mathbf{x}^{(\tau)} = [x_1^{(\tau)}, x_2^{(\tau)}, \ldots x_9^{(\tau)}]^T \in \mathbb{R}^9$ is a vector formed by the measurements of the all sensors at time-instance τ. Such multi-dimensional time-series data can then be introduced to recurrent neural networks which perform reconstruction or prediction in a similar manner to that described above.

References

1. Alam, S., Sonbhadra, S.K., Agarwal, S., Nagabhushan, P.: One-class support vector classifiers: A survey. Knowledge-Based Systems p. 105754 (2020)
2. Alur, R., Courcoubetis, C., Dill, D.: Model-checking for real-time systems. In: [1990] Proceedings. Fifth Annual IEEE Symposium on Logic in Computer Science, pp. 414–425 (1990). DOI: 10.1109/LICS.1990.113766
3. Annpureddy, Y., Liu, C., Fainekos, G., Sankaranarayanan, S.: S-TaLiRo: A Tool for Temporal Logic Falsification for Hybrid Systems. In: P.A. Abdulla, K.R.M. Leino (eds.) Tools and Algorithms for the Construction and Analysis of Systems, Lecture Notes in Computer Science, pp. 254–257. Springer, Berlin, Heidelberg (2011). DOI: 10.1007/978-3-642-19835-9_21
4. Baier, C., Katoen, J.P.: Principles of Model Checking. MIT press, Cambridge, MA, USA (2008)
5. Bartocci, E., Deshmukh, J., Donzé, A., Fainekos, G., Maler, O., Ničković, D., Sankaranarayanan, S.: Specification-Based Monitoring of Cyber-Physical Systems: A Survey on Theory, Tools and Applications. In: E. Bartocci, Y. Falcone (eds.) Lectures on Runtime Verification, *Lecture Notes in Computer Science*, vol. 10457, pp. 135–175. Springer International Publishing, Cham (2018). DOI: 10.1007/978-3-319-75632-5_5
6. Bartocci, E., Falcone, Y., Francalanza, A., Reger, G.: Introduction to runtime verification. In: E. Bartocci, Y. Falcone (eds.) Lectures on Runtime Verification - Introductory and Advanced Topics, *Lecture Notes in Computer Science*, vol. 10457, pp. 1–33. Springer (2018). DOI: 10.1007/978-3-319-75632-5_1
7. Basin, D., Klaedtke, F., Müller, S., Zălinescu, E.: Monitoring Metric First-Order Temporal Properties. Journal of the ACM **62**(2), 15:1–15:45 (2015). DOI: 10.1145/2699444
8. Basin, D., Klaedtke, F., Zalinescu, E.: The MonPoly Monitoring Tool. In: RV-CuBES 2017. An International Workshop on Competitions, Usability, Benchmarks, Evaluation, and Standardisation for Runtime Verification Tools, pp. 19–8 (2017). DOI: 10.29007/89hs
9. Bauer, A., Leucker, M., Schallhart, C.: The good, the bad, and the ugly, but how ugly is ugly? In: O. Sokolsky, S. Taşıran (eds.) Runtime Verification, pp. 126–138. Springer Berlin Heidelberg, Berlin, Heidelberg (2007)
10. Bauer, A., Leucker, M., Schallhart, C.: Runtime Verification for LTL and TLTL. ACM Trans. Softw. Eng. Methodol. **20**(4) (2011). DOI: 10.1145/2000799.2000800. URL https://doi.org/10.1145/2000799.2000800
11. Bellini, P., Mattolini, R., Nesi, P.: Temporal logics for real-time system specification. ACM Computing Surveys **32**(1), 12–42 (2000). DOI: 10.1145/349194.349197
12. Bensalem, S., Cheng, C.H., Huang, X., Katsaros, P., Molin, A., Nickovic, D., Peled, D.: Formal specification for learning-enabled autonomous systems. In: O. Isac, R. Ivanov, G. Katz, N. Narodytska, L. Nenzi (eds.) Software Verification and Formal Methods for ML-Enabled Autonomous Systems, pp. 131–143. Springer International Publishing, Cham (2022)
13. Büning, H.K., Lettmann, T.: Propositional Logic: Deduction and Algorithms. Cambridge University Press (1999)
14. Câmpeanu, C., Salomaa, K., Yu, S.: A formal study of practical regular expressions. International Journal of Foundations of Computer Science **14**(06), 1007–1018 (2003). DOI: 10.1142/S012905410300214X
15. Chen, F., Roşu, G.: MOP: An Efficient and Generic Runtime Verification Framework. SIGPLAN Not. **42**(10), 569–588 (2007). DOI: 10.1145/1297105.1297069
16. Chen, F., Roşu, G.: Java-MOP: A Monitoring Oriented Programming Environment for Java. In: N. Halbwachs, L.D. Zuck (eds.) Tools and Algorithms for the Construction and Analysis of Systems, Lecture Notes in Computer Science, pp. 546–550. Springer, Berlin, Heidelberg (2005). DOI: 10.1007/978-3-540-31980-1_36

17. Chhetri, S.R., Faezi, S., Canedo, A., Faruque, M.A.A.: QUILT: Quality inference from living digital twins in IoT-enabled manufacturing systems. In: Proceedings of the International Conference on Internet of Things Design and Implementation, IoTDI '19, pp. 237–248. Association for Computing Machinery (2019). DOI: 10.1145/3302505.3310085
18. Cimatti, A., Clarke, E., Giunchiglia, E., Giunchiglia, F., Pistore, M., Roveri, M., Sebastiani, R., Tacchella, A.: NuSMV 2: An OpenSource Tool for Symbolic Model Checking. In: E. Brinksma, K.G. Larsen (eds.) Computer Aided Verification, Lecture Notes in Computer Science, pp. 359–364. Springer, Berlin, Heidelberg (2002). DOI: 10.1007/3-540-45657-0_29
19. Cimatti, A., Tian, C., Tonetta, S.: NuRV: A nuXmv Extension for Runtime Verification. In: B. Finkbeiner, L. Mariani (eds.) Runtime Verification, pp. 382–392. Springer International Publishing, Cham (2019)
20. Clarke, E.M., Emerson, E.A.: Design and synthesis of synchronization skeletons using branching time temporal logic. In: D. Kozen (ed.) Logics of Programs, Lecture Notes in Computer Science, pp. 52–71. Springer, Berlin, Heidelberg (1982). DOI: 10.1007/BFb0025774
21. Convent, L., Hungerecker, S., Leucker, M., Scheffel, T., Schmitz, M., Thoma, D.: TeSSLa: Temporal Stream-based Specification Language. In: T. Massoni, M.R. Mousavi (eds.) Formal Methods: Foundations and Applications 21st Brazilian Symposium, SBMF 2018, Salvador, Brazil, November 26-30, 2018, Proceedings, *Lecture Notes in Computer Science*, vol. 11254, pp. 144–162. Springer (2018)
22. Cut, X., Cat, J., Wang, T., Lai, X.: Robust randomised autoencoder and correntropy criterion-based one-class classification. IEEE Transactions on Circuits and Systems II: Express Briefs **68**(4), 1517—-1521 (2021)
23. Day, H., Cat, J., Wang, T., Deng, M., Yang, Z.: Multilayer one-class extreme learning machine. Neural Networks **115**(7), 11—-22 (2019)
24. Donzé, A.: Breach, A Toolbox for Verification and Parameter Synthesis of Hybrid Systems. In: T. Touili, B. Cook, P. Jackson (eds.) Computer Aided Verification, Lecture Notes in Computer Science, pp. 167–170. Springer, Berlin, Heidelberg (2010). DOI: 10.1007/978-3-642-14295-6_17
25. Donzé, A., Maler, O.: Robust Satisfaction of Temporal Logic over Real-Valued Signals. In: K. Chatterjee, T.A. Henzinger (eds.) Formal Modeling and Analysis of Timed Systems, Lecture Notes in Computer Science, pp. 92–106. Springer, Berlin, Heidelberg (2010). DOI: 10.1007/978-3-642-15297-9_9
26. Ergen, T., Kozat, S.S.: Unsupervised Anomaly Detection with LSTM Neural Networks. IEEE Transactions on Neural Networks and Learning Systems **31**(8), 3127—-3141 (2020)
27. Falcone, Y., Havelund, K., Reger, G.: A Tutorial on Runtime Verification. Engineering Dependable Software Systems **34**, 141–175
28. Falcone, Y., Krstić, S., Reger, G., Traytel, D.: A taxonomy for classifying runtime verification tools. International Journal on Software Tools for Technology Transfer **23**(2), 255–284 (2021). DOI: 10.1007/s10009-021-00609-z
29. Ferrère, T., Nickovic, D., Donzé, A., Ito, H., Kapinski, J.: Interface-aware signal temporal logic. In: Proceedings of the 22nd ACM International Conference on Hybrid Systems: Computation and Control, pp. 57–66. ACM, Montreal Quebec Canada (2019). DOI: 10.1145/3302504.3311800
30. Finkbeiner, B., Hahn, C., Stenger, M., Tentrup, L.: Monitoring hyperproperties. Formal Methods Syst. Des. **54**(3), 336–363 (2019)
31. Fisher, M.: An introduction to executable temporal logics. The Knowledge Engineering Review **11**(1), 43–56 (1996). DOI: 10.1017/S0269888900007670
32. Gastin, P., Oddoux, D.: Fast LTL to Büchi Automata Translation. In: G. Berry, H. Comon, A. Finkel (eds.) Computer Aided Verification, 13th International Conference, CAV 2001, Paris, France, July 18-22, 2001, Proceedings, *Lecture Notes in Computer Science*, vol. 2102, pp. 53–65. Springer (2001)

33. Gawand, H.L., Bhattacharjee, A.K., Roy, K.: Online monitoring of a cyber physical system against control aware cyber attacks. In: Procedia Computer Science, vol. 70, pp. 238–244. Elsevier B.V. (2015)
34. Gerth, R., Peled, D., Vardi, M.Y., Wolper, P.: Simple On-the-fly Automatic Verification of Linear Temporal Logic. In: Protocol Specification, Testing and Verification XV, pp. 3–18. Springer, Boston, MA, Boston, MA, USA (1996). DOI: 10.1007/978-0-387-34892-6_1
35. Hallé, S., Soueidi, C., Falcone, Y.: Leveraging runtime verification for the monitoring of digital twins. In: S. Hallerstede, E. Kamburjan (eds.) Proceedings of the Workshop on Applications of Formal Methods and Digital Twins co-located with 25th International Symposium on Formal Methods (FM 2023), Lübeck, Germany, March 06, 2023, *CEUR Workshop Proceedings*, vol. 3507. CEUR-WS.org (2023)
36. Havelund, K., Peled, D., Ulus, D.: First-order temporal logic monitoring with BDDs. Form. Methods Syst. Des. **56**(1–3), 1–21 (2020)
37. Havelund, K., Roşu, G.: Synthesizing Monitors for Safety Properties. In: Tools and Algorithms for the Construction and Analysis of Systems, pp. 342–356. Springer, Berlin, Germany (2002). DOI: 10.1007/3-540-46002-0_24
38. Havelund, K., Roşu, G.: Runtime Verification - 17 Years Later. In: Runtime Verification, pp. 3–17. Springer, Cham, Switzerland (2018). DOI: 10.1007/978-3-030-03769-7_1
39. Havelund, K., Roşu, G.: Monitoring Java Programs with Java PathExplorer. Electronic Notes in Theoretical Computer Science **55**(2), 200–217 (2001). DOI: 10.1016/S1571-0661(04)00253-1
40. Havelund, K., Roşu, G.: Synthesizing Monitors for Safety Properties. In: J.P. Katoen, P. Stevens (eds.) Tools and Algorithms for the Construction and Analysis of Systems, pp. 342–356. Springer Berlin Heidelberg, Berlin, Heidelberg (2002)
41. Hejazi, M., Singh, Y.P.: One-class support vector machines approach to anomaly detection. Applied Artificial Intelligence **27**(5), 351–366 (2013)
42. Hensel, C., Junges, S., Katoen, J.P., Quatmann, T., Volk, M.: The probabilistic model checker Storm. International Journal on Software Tools for Technology Transfer pp. 1–22 (2021)
43. Hoffmann, H.: Kernel PCA for novelty detection. Pattern Recognition **40**(3), 863–874 (2007)
44. Holzmann, G.: The model checker SPIN. IEEE Transactions on Software Engineering **23**(5), 279–295 (1997). DOI: 10.1109/32.588521
45. Iosifidis, A., Mygdalis, V., Tefas, A., Pitas, I.: One-class classification based on extreme learning and geometric class information. Neural Processing Letters **45**(2), 577–592 (2017)
46. Kakavandi, F., De Reus, R., Heidari, N., Iosifidis, A., Larsen, P.G.: Product quality control in assembly machine under data restricted settings. In: IEEE International Conference on Industrial Informatics, pp. 735–741 (2022)
47. Kallwies, H., Leucker, M., Sánchez, C.: General anticipatory monitoring for temporal logics on finite traces. In: P. Katsaros, L. Nenzi (eds.) Runtime Verification, pp. 106–125. Springer Nature Switzerland, Cham (2023)
48. Karadeniz, A.M., Arif, I., Kanak, A., Erguen, S.: Digital Twin of eGastronomic Things: A Case Study for Ice Cream Machines. In: 2019 IEEE International Symposium on Circuits and Systems (ISCAS), pp. 1–4 (2019). DOI: 10.1109/ISCAS.2019.8702679
49. Liao, Y., Bin, Y.: To generalise or not to generalise: Towards autoencoders in one-class classification. In: International Joint Conference on Neural Networks. IEEE (2022)
50. Liu, Y., Sun, J., Dong, J.S.: PAT 3: An Extensible Architecture for Building Multi-domain Model Checkers. In: 2011 IEEE 22nd International Symposium on Software Reliability Engineering, pp. 190–199 (2011). DOI: 10.1109/ISSRE.2011.19
51. Maler, O., Nickovic, D.: Monitoring Temporal Properties of Continuous Signals. In: Y. Lakhnech, S. Yovine (eds.) Formal Techniques, Modelling and Analysis of Timed and Fault-Tolerant Systems, Lecture Notes in Computer Science, pp. 152–166. Springer, Berlin, Heidelberg (2004). DOI: 10.1007/978-3-540-30206-3_12

52. Malhotra, P., Big, L., Sheriff, G., Agarwal, P.: Long short term memory networks for anomaly detection in time series. In: European Symposium on Artificial Neural Networks (2015)
53. Manna, Z., Pnueli, A.: The Temporal Logic of Reactive and Concurrent Systems. Springer-Verlag (1992)
54. Mygdalis, V., Iosifidis, A., Tefas, A., Pitas, I.: Graph embedded one-class classifiers for media data classification. Pattern Recognition **60**, 585–595 (2016)
55. Ničković, D., Lebeltel, O., Maler, O., Ferrère, T., Ulus, D.: AMT 2.0: Qualitative and quantitative trace analysis with extended signal temporal logic. International Journal on Software Tools for Technology Transfer **22**(6), 741–758 (2020). DOI: 10.1007/s10009-020-00582-z
56. Ničković, D., Yamaguchi, T.: RTAMT: Online Robustness Monitors from STL. In: D.V. Hung, O. Sokolsky (eds.) Automated Technology for Verification and Analysis, Lecture Notes in Computer Science, pp. 564–571. Springer International Publishing, Cham (2020). DOI: 10.1007/978-3-030-59152-6_34
57. Omer, M., Peled, D.: Runtime verification prediction for traces with data. In: P. Katsaros, L. Nenzi (eds.) Runtime Verification, pp. 148–167. Springer Nature Switzerland, Cham (2023)
58. Paridari, K., O'Mahony, N., El-Din Mady, A., Chabukswar, R., Boubekeur, M., Sandberg, H.: A framework for attack-resilient industrial control systems: Attack detection and controller reconfiguration. Proceedings of the IEEE **106**(1), 113–128 (2018)
59. Pnueli, A.: The temporal logic of programs. In: 18th Annual Symposium on Foundations of Computer Science, Providence, Rhode Island, USA, 31 October - 1 November 1977, pp. 46–57. IEEE Computer Society (1977). DOI: 10.1109/SFCS.1977.32
60. Pnueli, A., Zaks, A.: PSL Model Checking and Run-Time Verification Via Testers. In: FM 2006: Formal Methods, pp. 573–586. Springer, Berlin, Germany (2006). DOI: 10.1007/11813040_38
61. Ruff, L., Vandermeulen, R., Goernitz, N., Deecke, L., Siddiqui, S.A., Binder, A., Müller, E., Kloft, M.: Deep one-class classification. In: International conference on machine learning, pp. 4393–4402. PMLR (2018)
62. Sánchez, C., Schneider, G., Ahrendt, W., Bartocci, E., Bianculli, D., Colombo, C., Falcone, Y., Francalanza, A., Krstić, S., Lourenço, J.M., Nickovic, D., Pace, G.J., Rufino, J., Signoles, J., Traytel, D., Weiss, A.: A survey of challenges for runtime verification from advanced application domains (beyond software). Formal Methods in System Design **54**(3), 279–335 (2019). DOI: 10.1007/s10703-019-00337-w
63. Sohrab, F., Iosifidis, A., Gabbouj, M., Raitoharju, J.: Graph-embedded subspace support vector data description. Pattern Recognition **133**(108999) (2023)
64. Sohrab, F., Raitoharju, J., Gabbouj, M., Iosifidis, A.: Subspace support vector data description. In: International Conference on Pattern Recognition, pp. 722–727 (2018)
65. Tax, D.M., Duin, R.P.: Support vector data description. Machine learning **54**(1), 45–66 (2004)
66. Temperekidis, A., Kekatos, N., Katsaros, P.: Runtime Verification for FMI-Based Co-simulation. In: T. Dang, V. Stolz (eds.) Runtime Verification, pp. 304–313. Springer International Publishing, Cham (2022)
67. Yilin Mo, Kim, T.H.J., Brancik, K., Dickinson, D., Heejo Lee, Perrig, A., Sinopoli, B.: Cyber–Physical Security of a Smart Grid Infrastructure. Proceedings of the IEEE **100**(1), 195–209 (2012)
68. Zapridou, E., Bartocci, E., Katsaros, P.: Runtime Verification of Autonomous Driving Systems in CARLA. In: J. Deshmukh, D. Ničković (eds.) Runtime Verification, pp. 172–183. Springer International Publishing, Cham (2020)
69. Zhou, X., Hu, Y., Jiang, W., Ma, J., Jin, S.: Variational LSTM Enhanced Anomaly Detection for Industrial Big Data. IEEE Transactions on Industrial Informatics **17**(5), 3469—-3477 (2021)

Chapter 10
Advanced Digital Twin Services

Mirgita Frasheri, Till Böttjer, Peter Gorm Larsen, Lukas Esterle and Cláudio Gomes

Abstract This chapter explores Digital Twin (DT) services that help to support decision-making, and the instigation of change in the DT's Physical Twin. We consider "what-if" analysis – the examination in parallel of multiple alternative scenarios – and related topics including design space exploration and fault injection. The potential of the DT in fault diagnosis and predictive maintenance are considered as ways to improve resilience, and predictive maintenance as a means of avoiding failure. We finally consider how to effect change in the PT through reconfiguration.

10.1 Introduction

One of the important capabilities of a Digital Twin (DT) is support for "what-if" simulations. These can use historical data to inform predictions of potential future evolutions of the PT, or applied retrospectively to evaluate the effects that alternative PT configurations (such as different controller parameters) would have had in past situations. Such simulations can take the form of automated sweeps through parameter values to find optimal outcomes using *Design Space Exploration (DSE)* techniques, or they may be created ad-hoc by a human operator. The exploration of scenarios is often focused on possible faults, and how resilient the PT may be. If required, techniques for fault diagnosis, predictive maintenance, reconfiguration, and adaptation can enhance the capabilities of a DT.

These advanced functionalities are explored in this chapter. Sect. 10.2 describes what-if simulations, including enabling techniques such as DSE, as well as techniques like fault injection that enable stakeholder to interrogate the response of the PT to undesirable events. Building on this, Sect. 10.3 delves into fault diagnosis techniques and support for resilience. Thereafter, Sect. 10.4 explores predictive maintenance. Finally, Sect. 10.5 covers the reconfiguration of the PT by the DT in case of errors and unforeseen events.

J. Fitzgerald et al. (eds.), *The Engineering of Digital Twins*,
https://doi.org/10.1007/978-3-031-66719-0_10

10.2 What-if Simulations

A what-if simulation is a data- and computation-intensive simulation used to evaluate the behaviour of a PT in different circumstances and hypotheses [30]. Should unwanted behaviours be detected during the operation of the PT, a what-if simulation enables a human operator or the DT itself to compare interventions, inspecting the potential consequences before applying them. Thereafter, a more informed decision can be made regarding selection of optimal responses. To achieve this functionality, the simulation of different alternatives should be carried out faster than real-time. Additionally, if multiple models are to be simulated together, then co-simulation can be used [19].

In a DT setting, a what-if simulation can access the historical data directly from the DT, potentially synchronise with the PT, and feed the simulation results to other services running in the DT. Ultimately, the insights provided by what-if simulations may provide insights that allow the optimisation of the PT configuration. As with any other simulation, however, if decisions relate to safety, relatively high confidence is required in the fidelity of the simulations.

What-if Simulations for the Incubator

The incubator PT includes a local controller that manages the heater and fan using data from the temperature sensors in the box (Fig. 3.5). Twinpeh were considering a maintenance intervention in the operating incubator, specifically the replacement of the power supply. If they wish to allow the tempeh to continue fermenting, an option is to boost the temperature in the box just prior to the replacement, so that the temperature drifts down when the power is off, but stays in range while the replacement takes place. The control parameter H set the length of time that the heater stays on during the boost, and so a what-if simulations might be used to find its optimal value in this scenario.

The outcome of the simulation can be seen in Fig. 10.1. The red line denotes the largest available time period but with excessive overheating, thereby violating the slight overheat allowance. The green line has the largest available time period compared with the yellow and blue lines, thus represents the value needed for H.

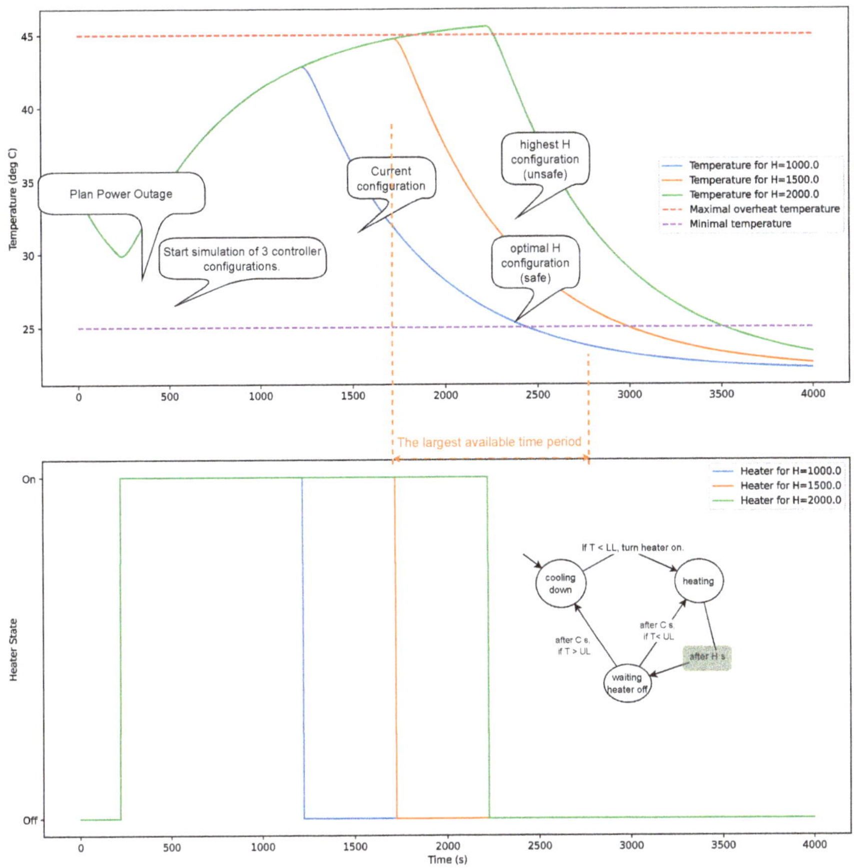

Fig. 10.1: Results of what-if simulations for controller configuration. Reproduction of Fig. 3.11.

10.2.1 Design Space Exploration with Conflicting Objectives

DSE is a technique used to evaluate a set of design alternatives pertaining to a system of interest [12], such that a solution can be found which satisfies requirements such as specific performance characteristics [26]. Each design alternative consists of specific values for the parameters of the design. Through DSE, a (co-)simulation is run for each of the alternatives, with the results being consequently saved for further analysis. Genetic algorithms can be used to evaluate which (co-)simulation should be run next, should the design space be too large for every combination to be simulated [31]. In general, such results would contain information on multiple – possibly conflicting – objectives such as speed, accuracy and energy consumption. A ranking function may be used to select the "best" design if one exists. Otherwise a non-dominated set of designs can be computed and used to calculate the Pareto Optimal front [8].

The principle underpinning DSE – sweeping through a set of alternatives and selecting the best solution – also applies in what-if simulations. However, there is an important conceptual difference between the two techniques. DSE is mostly used in design stages, whereas what-if simulations are used during the operational stages. While DSE aims to find the best design for the system, what-if simulations explore what would happen with the deployed system under different circumstances.

DSE on the incubator

A simple example of a DSE with conflicting objectives has been performed on the incubator. For the purposes of the experiment, two parameters are studied, the heating gap and the controller bound. The former captures how many times the heater should be turned on and off, in other words the actuator effort. The latter is just the integral of the temperature compared to the desired temperature. The results for the DSE experiment are shown in Fig. 10.2. As can be seen these two parameters represent a trade-off, and in the plot one can see what the optimal solutions are represented at the Pareto front which shows the optimal cases allowing a choice of trade-off between actuator effort and controller error.

10.2.2 Fault-injection enabled Digital Twins

Fault Injection (FI) is a validation technique for dependability [1], used for years across different domains, thus becoming a well-established approach for the analysis of the robustness and dependability of a given system. Dependability is an integrating concept, defined as "the ability to deliver service that can justifiably be trusted" [2] There are different attributes to dependability, where the application determines which one(s) are brought to the forefront. They include:

- Availability, which refers to the readiness of the system for usage;
- Reliability, which refers to the continuity of service;
- Safety, which refers to ensuring that no catastrophic events take place; and
- Security, which refers to the composite of confidentiality, integrity and availability.

Early work on FI was conducted by Harlan Mills in 1972 at IBM [36], where the goal was to provide some estimation of a system's reliability. To this end, they counted the number of injected faults that were discovered through testing, as well as the real faults, i.e., those faults that were not injected, but actually found through testing. Later on, in the mid 80s, FI techniques started being used in academia for research on

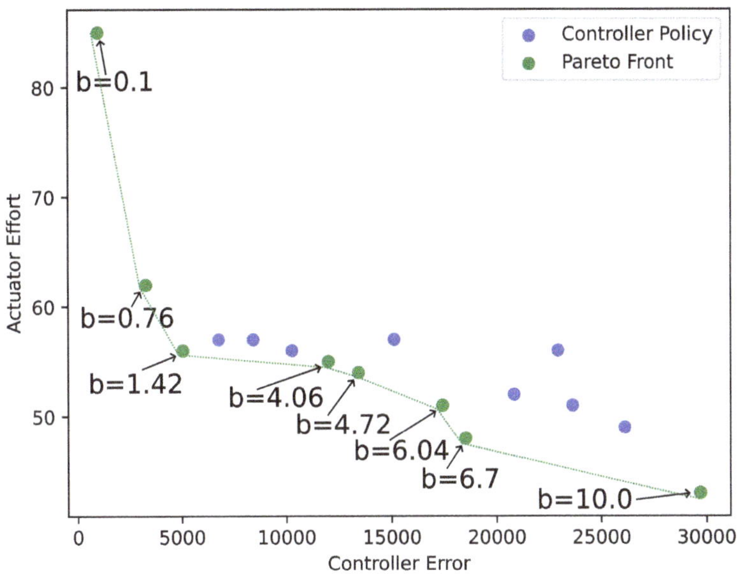

Fig. 10.2: Results of DSE for the incubator case-study with two conflicting objectives.

system testing. Their value is that they provide ways to better understand how faults affect a system, how suitable the adopted fault tolerance mechanisms are in avoiding system failures, and potentially supporting the forecast of faulty behaviour [36].

Faults can happen at the hardware and software levels [7]. Hardware faults can arise in manufacturing or as a consequence of environmental conditions or weaknesses in design. On the other hand, software faults are frequently a result of inadequate design or specification, among others. Faults can propagate and lead to an incorrect state, known as an error. Errors if not detected and dealt with through fault tolerance mechanisms can lead to a failure in the system, which is a deviation from correct behaviour, observable at teh system boundary. The FI concept has been leveraged both in hardware and software [21], as well as in Model-Based Systems Engineering (MBSE), where the response to faults in a system's model is evaluated in simulation [33]. Various simulation tools provide FI capabilities, e.g., Simulink [32], LabVIEW [20], Maestro as an Functional Mockup Interface (FMI)-based co-simulation orchestration engine [17, 25].

Hardware FI consists in physically tampering with the hardware, through such routes as its environment (e.g., with heavy ion radiation), disturbances in power supply, or changing pin values directly, among others. In recent years there has been a shift toward software FI [29], applicable at different levels, from highly intrusive setups that simulate bugs through code changes, to the use of data injections to tamper with the state of different components, and by tampering at the interface points of the components of interest by modifying the inputs and outputs at will. Whichever the level of injection, it is necessary to create adequate fault models, describing what, when and where a fault is to be injected. In this process, aspects such as representation, usability and efficiency are to be addressed properly [29]. Emulation-based FI, where Field Programmable Gate Arrays (FPGAs) are used to save time in comparison to simulation-based approaches and hybrid FI, that combine hardware and software FI, have also been explored [37].

The enhancement of DTs with FI capabilities makes it possible to experiment in a controlled manner with the real system, with minimal intrusion on the hardware side. Thus, it becomes possible to devise more efficient fault tolerance mechanisms, that enable our system to degrade gracefully and not deteriorate in a failure, once an error manifests itself. Different problematic cases can be reproduced, such as the effect of extreme weather conditions (assuming a robot that would operate continuously outside) or the effect of system faults, without the need to actually test the system in such conditions or perform ad-hoc FI on the actual hardware, respectively. Thereafter, the most interesting cases can be tested in the field, with reduced testing costs and risk of damaging the hardware (e.g., [5]). Such FI-enabled DTs provide opportunities for experimentation and testing as the different parts of the hardware become available, and not only on the prototype or finished product. Dealing with hardware brings about many issues that are simply not captured in simulation. The sooner the hardware can be brought in the loop, the earlier we might discover possible issues in the design, setup, etc.

FI can be realised in different ways, but in this chapter the reader is pointed towards FMI-based approaches. Using such an approach allows FI to be applied to any of the models presented in Chap. 5, at the interface points between individual models and components (FMUs). FMI-based FI has been explored previously by Frasheri et al. [17, 16]. In this work, the Maestro co-simulation engine is extended through a plugin to support fault injection of FMUs during run-time[1]. The plugin creates a wrapper around the FMU(s) of interest (Fig. 10.3), which can be configured through an `xml` file. The same principles could also be applied for other (co-)simulation technologies.

Fault Injection on the incubator

In order to simulate the act of removing the lid from the incubator, a fault was injected whereby the air temperature output of the plant FMU was set to provide a lower value than the real one, according to e^{-T_b}. Fig. 10.4 shows the difference between a normal and the fault injected run. In the latter, the predicted value from the Kalman filter is different from the actual real value.

[1] Further details can be found at `https://digital-twin-book.org/`

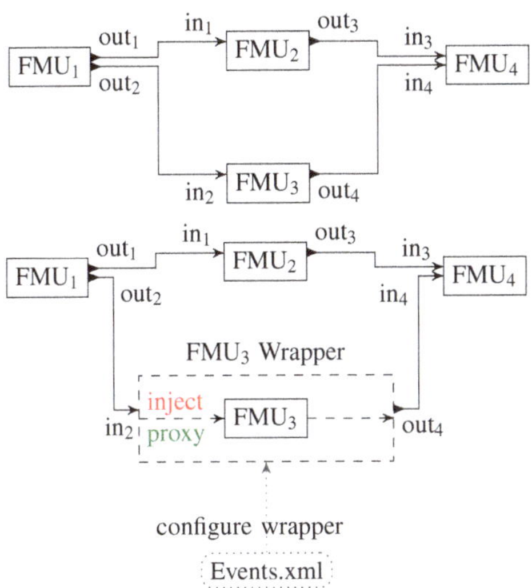

Fig. 10.3: High-level schematic of the connections between FMUs (top) and with the FI mechanism (bottom). From [17].

10.2.3 Runtime Verification of What-if Simulations

Runtime verification, in addition to monitoring the PT, can be also applied in what-if simulations, i.e. at the model level of the DT. In this case, the input trace to the runtime monitor is originated from the (co-)simulated models (see Sect. 5.7.1).

In monitoring what-if (co-)simulations, it is required that the simulation time is frozen every time that the runtime monitor fetches a state from the input trace; an interface-based integration of the runtime monitor into the (co-)simulation is opted, which means that the monitor can be integrated as a separate simulation unit called an FMU in an FMI-based co-simulation. Such an approach has been demonstrated in [34] and seems to be a practical solution. The monitor is updated every time that the `doStep` function of its FMU is invoked by the master algorithm. Monitor updates occur in the right order, for taking into account the updates of all other FMUs in current simulation step, before advancing simulation time to the next time step. Keeping low the overhead of runtime monitoring is of paramount importance. For non-timing properties, if there is no change in the FMUs that affect the verdict, after having executed their `doStep` function, then there is no need for the monitoring FMU to update the monitoring program. Such an optimisation requires a preliminary analysis of the property dependencies on the outputs of the co-simulated FMUs.

The monitoring output(s) may be post-processed to enforce control over the simulation, for example for triggering another scenario for what-if analysis. Alternatively, it may be desirable to detect discrepancies between the trace of a what-if simulation

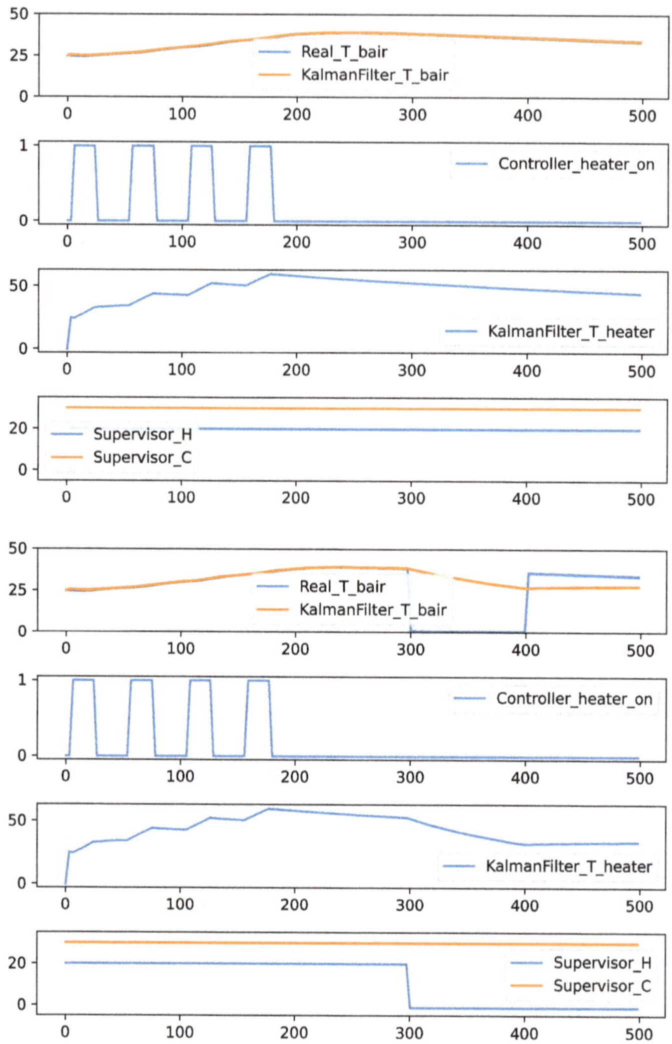

Fig. 10.4: Normal run (top four traces) and Fault Injection (bottom four traces) runs of the incubator.

and the input trace from the PT. This runtime verification task is based on a hyper-property specification [15], but the runtime verification of hyper-properties is still a field with open research challenges.

10.3 Fault Diagnosis and Resilience

Fault diagnosis deals with the detection, isolation and identification of faults. Detection represents the most basic step, that alerts to the fact that there is a malfunction in the system detected by a monitor (see Chap. 9), whereas isolation and identification provide information on the location and aspects like type, shape and size.

Gao et al. classify fault diagnosis techniques as model-based or signal-based [18]. Model-based fault diagnosis monitors the difference (if any) between the observed outputs of a system and those predicted by a model [3]. Gao et al. classify such methods in four groups: deterministic methods, stochastic methods, methods for discrete events and hybrid systems, and methods for networked and distributed systems. Signal-based techniques by contrast extract features of measured signals, as well as using knowledge representing healthy systems, to perform fault diagnosis. These methods are useful for real-time diagnosis and are grouped into time-domain, frequency domain and time-frequency domain techniques.

Symptoms of faults, as well as heuristic fault knowledge, serve as input to a fault diagnosis mechanism [22]. Such symptoms can be analytical, consisting of check-ups of measurable signals or heuristics coming from observations by operators such as noise or visual impressions. Additionally, the history of the system, including information such as load times or occurred faults, can serve as input to the fault diagnosis. The relations between faults and observable symptoms are generally causal. If these relations are not known a priori and therefore cannot be captured analytically, then classification methods that can be experimentally trained can be adopted. Otherwise, the cause-effect relationship can be represented through if-then rules, allowing the adoption of reasoning techniques.

Fault diagnosis is crucial for (industrial) Cyber-Physical Systems (CPSs), especially in the aerospace, automotive and industrial control domains, due to the strict requirements imposed on those systems to operate without faults for long/continuous periods of time [11]. For industrial CPSs, diagnostics techniques fall under the following categories: (i) physics-based and model-driven, e.g., Kalman filters, Markov models, fault trees etc., (ii) data-driven Artificial Intelligence (AI), e.g., neural networks, Machine Learning (ML), fuzzy logic etc. and (iii) Knowledge-based, e.g., Bayesian networks, binary decision trees, etc. In the CPS context, it is also relevant to consider how the CPS can explain the fault diagnostics to a human operator, e.g., by using ML and signal temporal logic as an interpretation language [9].

The curious reader is referred to other examples in the literature such as: anomaly detection support in a DT setting in assembly lines [6] and DT-based fault diagnosis for rotating machinery in smart manufacturing [35].

10.4 Predictive Maintenance

Predictive maintenance is a proactive strategy that uses advanced data analysis techniques and predictive analytics to determine the condition of a PT in order to predict when maintenance should be performed. Its goal is to optimise maintenance activities by addressing potential issues before they lead to equipment failure or downtime [28, 24]. Examples include a DT setting supporting state estimation/prediction in assembly lines [6] and DTs for the estimation of the life expectancy for plastic injection moulds [4], among others. The curious reader is referred to the work of van Dinter *et. al*, for a comprehensive survey covering the literature that has emerged since 2018 [10].

In the case of our incubator system, predictive maintenance can be applied to enhance its reliability and performance by continuously monitoring various parameters and collecting real-time data from the heat bed and fan. Thereafter, predictive maintenance algorithms can be used to analyse the data and identify patterns, trends or anomalies that may indicate potential issues. These algorithms can utilise techniques such as ML, statistical analysis and data mining to identify early indicators of component degradation, wear and tear or abnormalities in the incubator system. For example, the algorithm might detect a gradual increase in temperature variation or a decrease in fan speed over time.

Once potential issues are identified, the predictive maintenance system can generate alerts or notifications to the maintenance team, indicating the specific components or areas of the PT that require attention. This allows maintenance personnel to schedule proactive maintenance activities, such as replacing a worn-out fan or calibrating temperature sensors, before any critical failure occurs. The benefits to the PT are clear. First, the risk of unexpected equipment failures, which can lead to costly downtime or disruptions in the execution of the PT, is reduced. Second, maintenance activities are optimised by eliminating unnecessary preventive maintenance tasks that may not be required, thus reducing maintenance costs. Third, lifespan of the PT is extended by ensuring that components are replaced or repaired at the appropriate time, avoiding premature replacements or unnecessary repairs.

Predictive Maintenance on the Incubator

As a simple example consider the degradation of the fan in the incubator evidenced by higher motor friction, which translates to higher drawn current from the fan. Fig. 10.5 illustrates the gradual increase in current drawn as the fan motor faces more friction, caused by ageing [27, Section 8.9]. Periodic sampling of the current drawn enables the possibility to estimate the time to failure, as higher and higher currents eventually leads to overheating and it is also possible to relate the current with the fan speed using well known motor equations.

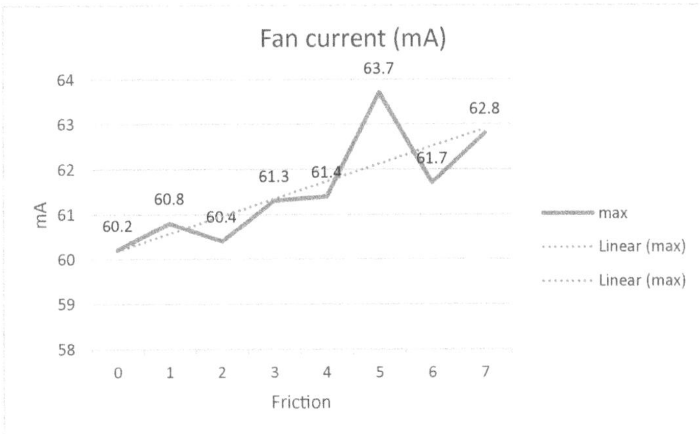

Fig. 10.5: Example of current drawn by fan following increasing levels of friction.

10.5 Re-configuration, Robustness and Optimisation

A key characteristic of a DT is its capability to effect change in the PT when necessary, while relieving the human from the task, e.g., the DT observes an anomaly in the behaviour of the PT, as a result actions are set in motion to (potentially) diagnose the anomaly, bring the PT to a safe state, or reconfigure the PT to adapt to the new circumstances and continue its operation. This process could involve re-optimisation of parameters and update of models, among others.

Consider as an example the case of a mobile robot driving on straight asphalt, that suddenly ends up on irregular terrain. In such case, the tracking capability of the DT would be impacted negatively and the error between predicted and actual positions would exceed the expected bounds. The DT should be able to first identify that there is an issue, and thereafter take action either to change the models, or reconfigure existing parameters, such that the tracking capability can provide the required accuracy. Such a capability enables the whole DT-enabled system to be robust against (inevitable) changes in the environment. More details on the re-configuration of PT using DT are given in Chap. 14.

Re-configuring the Incubator

In tthe incubator (for a detailed description the curious reader is referred to [13]), to support such capability, the Monitor-Analyze-Plan-Execute over a shared Knowledge (MAPE-K) loop [23] was implemented, which is a self-adaptation loop, that refers to Monitor, Analyse, Plan and Execute phases, as well as a Knowledge component. Recall that the purpose of the incubator is the regulation of the temperature inside its insulated box. To show self-adaptation, the work conditions of the PT were changed by opening the lid of the styrofoam box. The plant model, as well as the open-loop

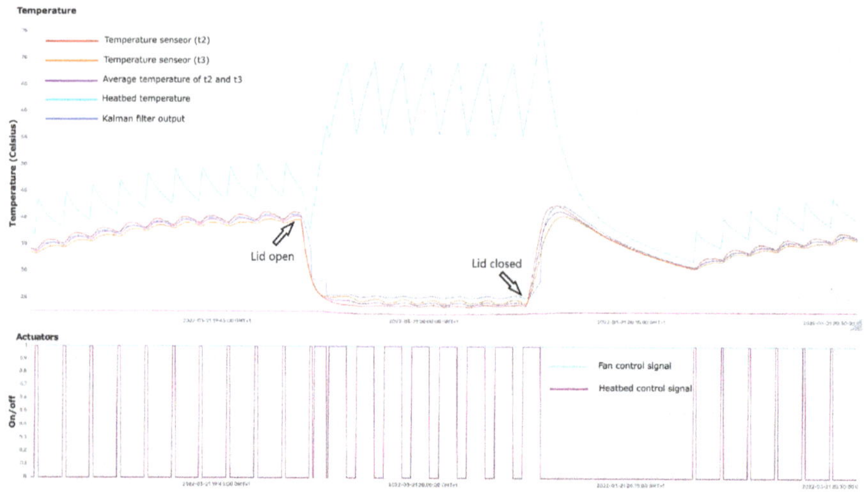

Fig. 10.6: Experimental result of self-adaptation. Adapted from [13].

controller of the incubator are designed to work with the closed lid, thus by opening the lid during operation, both the model and controller are invalidated. Normally, in this case, a human would be needed to recalibrate the model and re-optimise the controller. Since this example consists only of this anomaly, there is no need for an analyse phase, rather only gathering data is required for re-calibration and re-optimisation. For anomaly detection, a Kalman filter is deployed for monitoring and detecting anomalies [14]. Both model and controller are re-calibrated and re-optimised respectively, after getting the data from the InfluxDB database. The new parameters are then calculated and used by the DT to reconfigure the system and send the controller parameters to the PT.

When data is retrieved from the InfluxDB database in the DT, it is necessary to re-calibrate and optimise the model and the controller respectively. After obtaining the necessary parameters, the DT re-configures the model and controller with such parameters and delivers the controller parameters to the PT to be configured (Fig. 10.6). Note that in the initial experiments, the highest air temperature after closing the lid was around 68°C, which exceeded the target temperature of the incubator. This was a result of the optimised control policy that turned on the heater (when the lid was open) at its full power during the total experiment period. Once the lid was closed, the energy was transferred to the air inside the box, causing the overshoot. To address this issue, the maximum amount of energy released into the system was capped by restricting the maximum heating period parameter.

References

1. Arlat, J.: Validation de la sûreté de fonctionnement par injection de fautes: méthode, mise en oeuvre, application. Ph.D. thesis, Toulouse, INPT (1990)
2. Avizienis, A., Laprie, J.C., Randell, B., Landwehr, C.: Basic Concepts and Taxonomy of Dependable and Secure Computing. IEEE Transactions on Dependable and Secure Computing **1**, 11–33 (2004). DOI: http://doi.ieeecomputersociety.org/10.1109/TDSC.2004.2
3. Beard, R.: Failure accommodation in linear systems through self-reorganization (self-reorganization methods for complex linear dynamic systems to compensate for component failures) (1971)
4. Böttjer, T., Ørnskov Rønsch, G., Gomes, C., Ramanujan, D., Iosifidis, A., Gorm Larsen, P.: Data-driven identification of remaining useful life for plastic injection moulds. In: Towards Sustainable Customization: Bridging Smart Products and Manufacturing Systems: Proceedings of the 8th Changeable, Agile, Reconfigurable and Virtual Production Conference (CARV2021) and the 10th World Mass Customization & Personalization Conference (MCPC2021), Aalborg, Denmark, October/November 2021 8, pp. 431–439. Springer (2022)
5. Cheng, K., Wang, Q., Yang, D., Dai, Q., Wang, M.: Digital-twins-driven semi-physical simulation for testing and evaluation of industrial software in a smart manufacturing system. Machines **10**(5), 388 (2022)
6. Chhetri, S.R., Faezi, S., Canedo, A., Faruque, M.A.A.: QUILT: Quality inference from living digital twins in IoT-enabled manufacturing systems. In: Proceedings of the International Conference on Internet of Things Design and Implementation, IoTDI '19, pp. 237–248. Association for Computing Machinery (2019). DOI: 10.1145/3302505.3310085
7. Clark, J.A., Pradhan, D.K.: Fault injection: A method for validating computer-system dependability. Computer **28**(6), 47–56 (1995)
8. Deb, K.: Multi-objective optimisation using evolutionary algorithms: an introduction. In: Multi-objective evolutionary optimisation for product design and manufacturing, pp. 3–34. Springer (2011)
9. Deng, Z., Kong, Z.: Interpretable fault diagnosis for cyberphysical systems: A learning perspective. Computer **54**(9), 30–38 (2021)
10. van Dinter, R., Tekinerdogan, B., Catal, C.: Predictive maintenance using digital twins: A systematic literature review. Information and Software Technology p. 107008 (2022)
11. Dowdeswell, B., Sinha, R., MacDonell, S.G.: Finding faults: A scoping study of fault diagnostics for industrial cyber–physical systems. Journal of systems and software **168**, 110638 (2020)
12. Ernst, R.: Codesign of Embedded Systems: Status and Trends, pp. 45–54. Elsevier (2002). DOI: 10.1016/b978-155860702-6/50006-5
13. Feng, H., Gomes, C., Gil, S., Mikkelsen, P.H., Tola, D., Sandberg, M., Larsen, P.G.: Integration of the MAPE-K Loop in Digital Twins. IEEE (2022). DOI: doi:10.23919/ANNSIM55834.2022.9859489
14. Feng, H., Gomes, C., Thule, C., Lausdahl, K., Sandberg, M., Larsen, P.G.: The Incubator Case Study for Digital Twin Engineering (2021). URL http://arxiv.org/abs/2102.10390
15. Finkbeiner, B., Hahn, C., Stenger, M., Tentrup, L.: Monitoring hyperproperties. Formal Methods Syst. Des. **54**(3), 336–363 (2019)
16. Frasheri, M., Lumer-Klabbers, G., Kvistgaard, J.L., Hausted, J.O., Thule, C., Esterle, L., Larsen, P.G.: Building safer robots: From simulation to hardware deployment. In: 2023 27th International Conference on Methods and Models in Automation and Robotics (MMAR), pp. 63–68 (2023). DOI: 10.1109/MMAR58394.2023.10242467
17. Frasheri, M., Thule, C., Macedo, H.D., Lausdahl, K., Larsen, P.G., Esterle, L.: Fault injecting co-simulations for safety. In: The 5th International Conference on System Reliability and Safety ICSRS'21 (2021)

18. Gao, Z., Cecati, C., Ding, S.X.: A survey of fault diagnosis and fault-tolerant techniques—part i: Fault diagnosis with model-based and signal-based approaches. IEEE transactions on industrial electronics **62**(6), 3757–3767 (2015)
19. Gomes, C., Thule, C., Broman, D., Larsen, P.G., Vangheluwe, H.: Co-simulation: A survey. ACM computing surveys **51**(3), 1–33 (2018)
20. He, D., Hu, N., Wang, M.: Study on real-time fault injection and simulation of mechanic-electronic-hydraulic control system based on amesim and labview pp. 446–450 (2014)
21. Hsueh, M.C., Tsai, T.K., Iyer, R.K.: Fault injection techniques and tools. Computer **30**(4), 75–82 (1997)
22. Isermann, R.: Supervision, fault-detection and fault-diagnosis methods—an introduction. Control engineering practice **5**(5), 639–652 (1997)
23. Kephart, J.O., Chess, D.M.: The vision of autonomic computing. Computer **36**(1), 41–50 (2003)
24. Kim, N.H., An, D., Choi, J.H.: Prognostics and health management of engineering systems. Switzerland: Springer International Publishing (2017)
25. Larsen, P.G., Esterle, L., Fitzgerald, J., Frasheri, M.: Fault Injection in Co-simulation and Digital Twins for Cyber-Physical Robotic Systems, pp. 222–236. Springer Nature Switzerland, Cham (2023). DOI: 10.1007/978-3-031-40132-9_14
26. Larsen, P.G., Fitzgerald, J., Woodcock, J., Fritzson, P., Brauer, J., Kleijn, C., Lecomte, T., Pfeil, M., Green, O., Basagiannis, S., Sadovykh, A.: Integrated Tool Chain for Model-based Design of Cyber-Physical Systems: The INTO-CPS Project. In: CPS Data Workshop (2016)
27. Lynch, K.M., Park, F.C.: Modern robotics. Cambridge University Press (2017)
28. Medjaher, K., Zerhouni, N., Gouriveau, R.: From prognostics and health systems management to predictive maintenance 1: Monitoring and prognostics. John Wiley & Sons (2016)
29. Natella, R., Cotroneo, D., Madeira, H.S.: Assessing dependability with software fault injection: A survey. ACM Computing Surveys (CSUR) **48**(3), 1–55 (2016)
30. Rizzi, S.: What-if analysis. In: Encyclopedia of Database Systems, pp. 3525–3529. Springer US (2009)
31. Rose, M., Fitzgerald, J.: Genetic Algorithms for Design Space Exploration of Cyber-Physical Systems: an Implementation in INTO-CPS. In: H.D. Macedo, C. Thule, K. Pierce (eds.) Proceedings of the 19th International Overture Workshop. Overture (2021)
32. Silveira, A.M., Araújo, R.E., de Castro, R.: Fieev: a co-simulation framework for fault injection in electrical vehicles pp. 357–362 (2012)
33. Svenningsson, R., Vinter, J., Eriksson, H., Törngren, M.: Modifi: A model-implemented fault injection tool pp. 210–222 (2010)
34. Temperekidis, A., Kekatos, N., Katsaros, P.: Runtime Verification for FMI-Based Co-simulation. In: T. Dang, V. Stolz (eds.) Runtime Verification, pp. 304–313. Springer International Publishing, Cham (2022)
35. Wang, J., Ye, L., Gao, R.X., Li, C., Zhang, L.: Digital twin for rotating machinery fault diagnosis in smart manufacturing. International Journal of Production Research **57**(12), 3920–3934 (2019)
36. Yu, Y., Bastien, B., Johnson, B.W.: A state of research review on fault injection techniques and a case study. In: Annual Reliability and Maintainability Symposium, 2005. Proceedings., pp. 386–392. IEEE (2005)
37. Ziade, H., Ayoubi, R.A., Velazco, R., et al.: A survey on fault injection techniques. Int. Arab J. Inf. Technol. **1**(2), 171–186 (2004)

Part IV
Realising Digital Twins

In Part I, Part II and Part III we built towards Digital Twin (DT) services from foundational concepts, models and data. In this part, we move towards the reality of building DTs. Frameworks for DT construction are described in Chap. 11, including the need for bidirectional communication between the DT and its Physical Twin (PT). This also introduces a *Digital Twin as a Service (DTaaS)* platform that allows DTs to be built from reusable components rather than developed from scratch every time. Alongside this, we consider in Chap. 12 a series of research-based case studies, presented in a common systematic framework and offering some insight into how the DT engineering principles introduced so far can play out in practice.

Chapter 11
Realising Digital Twins

Prasad Talasila, Peter Høgh Mikkelsen, Santiago Gil and Peter Gorm Larsen

Abstract Building a Digital Twin (DT) can be a challenge, in part because of the variety of assets, models, data and services that must be marshalled to satisfy the specific needs of stakeholders. In this chapter, we consider the possibility of a generic platform that allows DTs to be built from reusable components and offered as a service to users. We propose such a platform which we call *Digital Twin as a Service (DTaaS)*. The DTaaS is sufficiently generic to support DTs spanning multiple modelling and execution paradigms. The chapter compares DT frameworks and discusses the advantage of treating related DTs as a fleet.

11.1 Introduction

This chapter introduces the reader to technologies that can be used to realise DTs in a more efficient way than building them from scratch every time. In software this is often achieved by offering reusable components such as functions and libraries.

For DTs, numerous frameworks already exist inspired by those developed for the Internet of Things (IoT). In order to provide better support for the DT services explained above in the book (in particular in Chap. 8, Chap. 9 and Chap. 10) we introduce a *Digital Twin as a Service (DTaaS)* platform. This is an open-source development intended to provide all organisations an affordable entry point into engineering their own DTs. The vision of DTaaS and the technologies used to construct it are introduced step by step, and this is compared to several DT applications that are in the literature.

It is worth noting the use of the terms *framework* and *platform*. In this chapter, we regard a framework as a collection of computational resources (such as code packages) from which one might construct a DT. It represents a rather basic technology level accessible mainly to those with the strong software engineering and programming capabilities needed to work with these resources directly. A platform

J. Fitzgerald et al. (eds.), *The Engineering of Digital Twins*,
https://doi.org/10.1007/978-3-031-66719-0_11

offers an interface (an *Application Programmer Interface (API)*) to application developers whose expertise may be more in an application domain and less in the underlying computing environment. In this chapter, we work towards the creation of a platform because our goal is to enable a wide range of potential beneficiaries to build and experiment with DTs. In spite of this, the terms have been used rather interchangeably in the literature to date, and so in our review of frameworks we will make reference to sources that refer to their products as "platforms".

We first take an overview of existing DT frameworks in Sect. 11.2 before considering available technologies for operating a DT either in a cloud setting or using virtualisation (Sect. 11.3). Section 11.4 briefly introduces the idea of a DT as a composition of assets and is followed in Sect. 11.5 by an introduction to how a DT (and its assets) can be configured. The Sect. 11.6 explains the creation of DT classes and instances which help with the management of Physical Twins (PTs) and its many instances manufactured and used. We are then in a position to give an overview of a DTaaS platform that we are in the process of constructing (Sect. 11.7). We pay particular attention to the management of DT execution depending on the available infrastructure (Sect. 11.8). The current status of our DTaaS platform is described in Sect. 11.9[1]. Then Sect. 11.10 presents an overview of DT platforms reported in the literature in comparison to our own. Finally, taking the idea of DT composition a step further, we consider the realisation of DTs for fleets of deployed PTs (Sect. 11.11).

11.2 Digital Twin Frameworks

Many DT frameworks already exist in open source and more traditional commercial forms, but gaining a coherent overview of them requires a great deal of time and insight. In this section we build on an existing survey of open-source DT frameworks [9], providing pointers to existing frameworks and summarising strengths, weaknesses and possibilities.

The demand of ready-to-use DT frameworks and platforms has grown with the interest in DT technology. However, DT frameworks vary from domain to domain and feature to feature, although their core has been mostly built upon IoT infrastructure and connectivity layers with add-ons such as integrated monitoring, fault diagnosis, modelling, simulation, optimisation and visualisation. Many of them are adapted for specific domains and groups of feasible applications in mind. They vary from commercial to open-source solutions with the possibility of being integrated with external technologies for DTs [31, 22].

It is worth highlighting the efforts of industry to standardise DT frameworks, such ISO standard 23247:2021 *Automation systems and integration – Digital twin framework for manufacturing* [13] stating basic functionalities and services that a DT and

[1] Updates can be followed at `https://github.com/INTO-CPS-Association/DTaaS`.

the DT framework should include as minimum requirements. Asset Administration Shell (AAS), the German initiative to standardise Industry 4.0 and DTs [43], represents another important contribution from industry. The DT community and their open-source contributions have also driven the adoption of DT technology and moves towards the establishment of common foundations [3]. Overall, commercial solutions offer a straightforward approach to getting started with DT frameworks, while the open-source solutions offer the Small and Medium-sized Enterprises (SMEs) a chance to undertake DT projects with limited resources.

A recent survey [19] performed a comparative analysis of three popular DT platforms: Amazon Web Services (AWS), IoT Greengrass, Microsoft Azure DTs and Eclipse Ditto-Hono-Vorto. The platforms were compared in seven dimensions, covering 13 requirements and seven quality characteristics. Microsoft Azure DTs contains the Digital Twin Definition Language (DTDL)[2] a definition language that offers a common schema for data models and DT structures. Twined[3] is also a definition language, that similarly to DTDL.

AAS provides a generic meta-model for aligned with industrial standards and has several open-source implementations developed by different entities including Eclipse BaSyx[4], PYI40AAS[5], SAP I4.0 AAS[6], NOVAAS[7] and AASX Package Explorer[8]

Eclipse Ditto[9] is a similar and alternative approach to AAS, although not specifically in the industrial domain. It provides a meta-structure for IoT and Web of Things applications and supported by a flexible set of communication tools.

In relation to 3D-enabled DTs, iTwin[10] provides an open-source toolkit that can be seamlessly integrated with external software. The Unity[11] gaming and 3D physics rendering engine enables users to create interactive 3D visualisations of their DTs. It offers a commercial toolkit with physics engines to perform simulation and deploy high-fidelity 3D models. Unity also offers free license for students.

In the geospatial data domain, TerriaJS[12] and Digital Twin Cities Centre (DTCC) Platform[13] offer a toolkit for geospatial-enabled DTs, where the PT can be displayed on maps and from there pattern analysis in the geospatial context can be performed.

---

[2] https://github.com/Azure/opendigitaltwins-dtdl

[3] https://github.com/octue/twined

[4] https://www.eclipse.org/basyx/

[5] https://git.rwth-aachen.de/acplt/pyi40aas

[6] https://github.com/SAP/i40-aas

[7] https://gitlab.com/novaas/catalog/nova-school-of-science-and-technology/novaas

[8] https://github.com/admin-shell-io/aasx-package-explorer

[9] https://www.eclipse.org/ditto/

[10] https://www.itwinjs.org/

[11] https://unity.com/

[12] https://terria.io/

[13] https://gitlab.com/dtcc-platform

Name	Creator / Maintainer	Domain	License	Main focus	Strengths	Additional features
Microsoft Azure DT & DTDL	Microsoft	General / IoT	Commercial / DTDL open-source	Large-scale applications	Robustness, flexibility, facility	Support, extended modelling and computing features, DTDL compatibility
AWS IoT Greengrass	AWS	General / IoT	Commercial	Large-scale applications	Robustness, flexibility, facility	Support, extended modelling and computing features
Eclipse Ditto	Eclipse	General / IoT	Open-source	Integrating devices with solutions	Communication infrastructure	Several supported protocols and available SDKs
AAS Implementation	Various	Industrial	Open-source	Industry 4.0 applications	Common semantics and standards, industrial support	Available SDKs, compatibility with industrial protocols
iTwin	Bentley Systems	Infrastructure	Open-source	3D-enabled infrastructure DTs	Built-in visualisations	Support, client-side web applications, examples with third-party software
Unity	Unity Technologies	Video games	Commercial	3D, 2D, VR and AR applications	Built-in physics engine	Multiple file sources / Cross-platform
TerriaJS	Australian Government	Geospatial Data	Open-source	Web-based geospatial applications	Robust, flexible, modular	High support for geospatial data sources
DTCC	Chalmers University of Technology	Geospatial Data	Open-source	Smart City applications	Multiple visualisation engines	
CPS-Twinning	SBA Research	CPSs	Open-source	State replication for PEs	supports standard specifications and physical models, provides simulated environment	Compatible with AutomationML documents
Twined	Octue Ltd.	General / IoT	Open-source for the definition language	Formatting, validation and structure of DT schemas and data	Useful to validate DT specifications	
INTO-CPS Co-simulation Framework	INTO-CPS Association	Co-simulation	Open-source	Simulation and model-based applications	FMI standard compliant, high-fidelity models	Provides useful FMU blocks for DT deployment

Table 11.1: Features of DT frameworks. Updated from [9].

In relation to model-driven approaches, CPS-Twinning[14] includes AutomationML documents in the core of the solution to deploy the DTs in simulated networks. It is intended to cover physical simulations and the INTO-CPS Association[15] provides a toolkit for co-simulation using high-fidelity modelling and FMU blocks that can be used in the DT context with certain modifications.

Lehner et al. [18] propose an architectural extension to integrate behavioural models into DT platforms and thus enable simulation-based services in DTs that are developed using those platforms, with for example, FMU models. Such an approach brings the potential to bridge the gap between DT platforms and model-driven frameworks for DTs.

The DT frameworks mentioned above (as at the time of writing) are summarised in Table 11.1.

11.3 Cloud and Virtualisation Technologies

A DT platform needs to support the execution of many DTs, but the computational resource requirements of DTs are diverse. Some DTs may be capable of being executed on a low-capability devices such as Raspberry Pi [28]; on the other hand, large scale city DTs [36] require execution on multi-cloud environments. However, innovations in computing environments, such as virtualisation, containerisation, cloud and fog computing technologies hold out some promise here.

Virtualisation refers to the creation of one or more virtual hardware and software resources on top of a physical computer [39]. A classic example is the process isolation feature of an Operating System (OS). All the processes managed by an OS are isolated by default. In addition, an OS provides virtualisation of hardware resources such as disks, processors and network connections. Thus virtualisation is inherently built in modern computer hardware from the processor to Cloud OS. There have been innovations in creating virtualised OS Application Programming Interfaces (APIs) so that applications built for one OS can execute on another. In the same manner one may create a virtualised computer providing a custom processor, memory, network and disk resources on a base computer. The virtualised processor can have a different instruction set architecture from that of the physical processor [11]. This ability to create a virtual environment enables execution of one or more application programs isolated at the level of the OS or the hardware.

Virtualisation technologies have created a fertile ground for *containerisation*, which refers to the bundling of an application with the libraries and files it needs into a package that is able to run on a range of infrastructures [24]. The process control groups feature of Linux OS is an example of containerisation feature build natively

[14] `https://github.com/sbaresearch/cps-twinning`

[15] `https://into-cps-association.readthedocs.io/en/latest/`

into OS [34]. Popular innovations like Docker containers[16] and the Open Container Initiative[17] have led to container execution environments like Kubernetes[18]. Kubernetes and the ecosystem of related technologies facilitates creation and execution of applications built using multiple docker containers. Kubernetes-related technologies also provide scalability to a containerised application. Thus a DT created for a Kubernetes environment can execute and scale in a seamless manner. Large scale DTs benefit significantly from using application architectures supported by Kubernetes [5].

A parallel innovation to containerisation is cloud computing [6]. Cloud computing offers virtualisation and containerisation technologies at scale. In addition, cloud computing services provide different levels of services ranging from Hardware as a Service to API as a Service [33]. In the context of DTs, cloud computing technologies can play a vital role in managing data received from PTs. Cloud computing services can also integrate with on-premise computers thus providing seamless scalability and on-demand availability of services. Such availability is key to enabling on-demand execution of DTs.

The rise of IoT technology has greatly enabled instrumentation of PTs so that data of interest can be extracted. Such IoT sensors and the low-capability devices that manage them can be labeled as edge devices. The data of interest is often beyond the transmission capabilities of networks. In addition, there is a distinct need for in-situ analysis and decision making. Hence the need for fog computing arrives. Fog (computing) is both a virtualised and non-virtualised computing paradigm that provides networking, storage and computation services amid cloud servers and IoT devices [27]. Placement of DTs either at the edge or in the fog increases the speed of response at the price of less available computing resource. It is possible to use a mix of edge, fog and cloud computing resources for execution of DTs, as we will describe later in this chapter.

11.4 Digital Twin Composition

In general it is easier to create DTs from reusable assets than build from scratch [1, 38]. As indicated in Sect. 3.6, the DT's data, models and software tools are reusable assets.

Relatively basic functions such as those acting on and managing data (in the Data & Data Management layer introduced in Sect. 1.3.2) can be reusable assets. An example might be a function that converts units on incoming temperature data (e.g., from Fahrenheit to Celsius). For such elementary functions, standard libraries could be written or reused from programming frameworks. In case of more complex data

[16] `https://www.docker.com/`

[17] `https://opencontainers.org/`

[18] `https://kubernetes.io/`

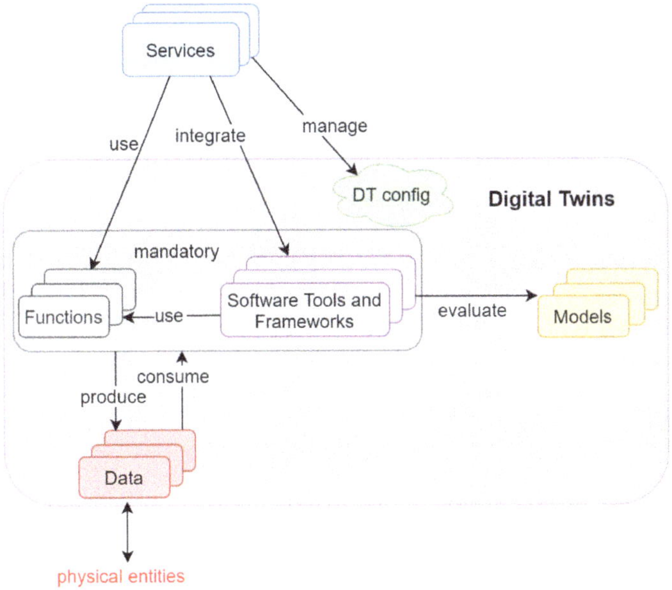

Fig. 11.1: Reusable assets composed for creation of DTs. The relationship between the assets is labelled on the arrows between the assets. Adapted from [40].

processing needs, dedicated tool chains such as Apache Spark[19] can be used. Such data processing tool chains are also part of the Data & Data Management layer.

DTs implemented as distributed software need to rely on internal services for efficient communication between different executing components. In addition, there are services that are reusable across DTs. It is meaningful to make the DT platforms host these services. The DTs needing these platform services can access them on demand. The DT services described in Chap. 3 are best hosted as platform services.

The software tools and frameworks evaluate models with data while utilizing functions and services. To be precise, tools are software implementations of domain-specific algorithms that are capable of evaluating related models, for example finite element models. The tools themselves need an execution environment (OS, processor architecture etc.). The responsibility of making the execution environment available to tools and thereby to a DT falls on the DT platforms.

A potential relationship between the elements that are eligible for composition is shown in Fig. 11.1. The functions and tools are mandatory assets. Among the two, functions can be reused by both services and tools. The additional packages and libraries used by software tools is an example of function reuse by software tools

[19] https://spark.apache.org/

and frameworks. The models are evaluated / executed by tools and frameworks with the use of data.

11.5 Digital Twin and Physical Twin Configuration

In this section there will be focus on different ways in which a DT can be configured. This includes considering different computer representations for structuring the information. If you come from a background in physical disciplines, the details of this may not be in your main area of interest. It is enough to think of this as a computer-readable way to express the configuration.

11.5.1 Requirements

The section above has explored the conceptual dependency between the components from which a DT may be built. The user requirements for a DT lead to the selection of components. Conceptually, we may think of a DT in terms of asset selection as outlines in Equation 11.1.

$$D_t :\{D^*, M^+, (FT)^+, S^*, SM^*, \}C_{dt} \tag{11.1}$$

where D denotes data, M denotes models, F denotes functions, T denotes tools, S denotes services, SM denotes components from service management layer, C_{dt} denotes DT configuration and D_t denotes the DT itself. The $*$ is a notation for zero or one more instances of a component while + is a notation for one or more instances of a component. This book focuses on model-centric DTs. Thus at least one model is used for the construction of a DT. DTs that use co-simulation will normally have multiple Functional Mockup Units (FMUs) thereby using multiple models in one DT, as described in Sect. 5.7.1. The DTs can also be used during prototyping in the early stages of product development [41], during which data from PT is not yet available.

Fig. 11.2 shows a conceptual mapping of a DT configuration and its relationship to the DT itself [40]. The configuration contains four sub-parts, namely configurations for the assets, PT, platform services and execution infrastructure. The asset configuration refers to selection and customisation of the assets selected for a DT. It is not mandatory for each of the assets to have a configuration of its own, but often they do, for example the parameters of reduced order and co-simulation models.

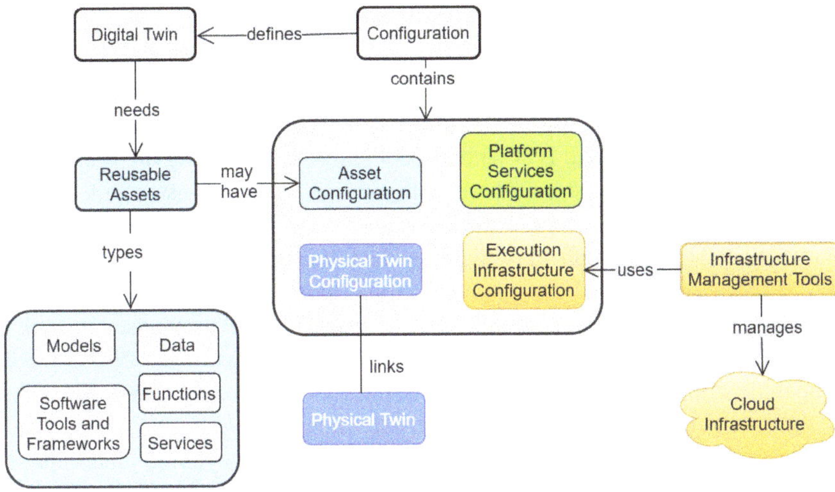

Fig. 11.2: A conceptual model mapping of DT configuration and DT. The configuration provides enough information for execution of DT created from reusable assets and linking DT with its PT.

The PT configuration specifies the required information for two way communication between a DT and its PT. Sometimes this information originates from the PT (e.g., current on-board controller being used on PT) and sometimes from the DT (e.g., selection of one on-board controller from many available).

The platform services configuration refers to linking of components from the service layer with the DT. Users often interact with DTs only via platform services; thus the option of linking DTs with platform services is desirable. Finally, the execution infrastructure configuration refers to the computing infrastructure on which a DT will be executed. Here infrastructure implies the physical location of the computer(s) on which the DT can be executed. There are general purpose infrastructure configuration languages such as Topology and Orchestration Specification for Cloud Applications (TOSCA) [21] which specifies a standard for configuration of software over cloud computing infrastructure. This configuration is used by infrastructure management tools such as MiCADO [17]. If a DT gets executed at one location, it may be called as a centralised DT. It is also possible to see DTs as potentially executable at multiple locations, one of them being near the PT, such as at the edge. The DTs which are executed at multiple locations may be called as distributed DTs [20]. This section deals only with configuration of centralised DTs, but the same configuration principles apply to distributed DT twins as well. Section 11.8 provides further detail on the execution of DTs.

11.5.2 Existing Digital Twin Configuration Formats

The existing DT frameworks define the skeleton for creating, running and maintaining DTs and typically the users must provide the specific configuration, structure, or schema for their PTs and DTs with a specific format, following the framework's architecture and design pattern. These configuration files allow users with minimal knowledge of programming or the internal framework's functionalities to deploy DTs relatively quickly. Users only need to follow the framework's guidelines to construct the configuration files and there are even some tools that help users to construct DT configuration files in a more user-friendly way. However, users need to stick to the configuration file template of the framework they use, which is not necessarily compatible with other frameworks. It means, the DT assets can be reused only within the same tool environment or its supported extensions.

Some common configuration formats are listed below, some of which may be used by the frameworks listed in Table 11.1:

XML: The eXtensible Markup Language is widely used in software systems. It can be used for defining the structure or schema of the DTs when these are defined by a metamodel. Some of the DT frameworks (and their core dependencies) that use XML include AAS and AutomationML.

JSON: This object notation format is also widely used in software systems and is easier for humans to read and write than XML. It may be used for defining the structure or schema of DTs when these are defined by a metamodel and also for defining parameters required by the DT and the frameworks. Some of the DT frameworks (and their core dependencies) that use JSON include Azure DTDL, Eclipse Ditto, Twined, AAS and the INTO-CPS Co-simulation framework.

GeoJSON/CityJSON: These specialisations of JSON are designed for representing geographical and 3D city features respectively. They may be used for containing static data and structures of geospatial DTs. Some of the DT frameworks (and their core dependencies) that use GeoJSON or CityJSON include TerriaJS and DTCC.

RDF: The Resource Description Framework is used for representing linked data on the semantic web. It may be used for defining astructure for DTs as a metamodel or following the structure of other metamodels. It is usually used along with ontologies, and it enables the representation of static data, parameters and semantic relationships required by the DT. The Semantic Micro Object Language (SMOL) language [15][20] is an example of a DT framework that uses RDF as the configuration format.

Other configuration formats, such as Yet Another Markup Language (YAML), are used for lifting serialised data and may be used for DT configuration files, although they are not mentioned here with a particular example.

[20] `https://smolang.org/index.html`

A minimal example for the configuration file of the incubator with JSON in Eclipse Ditto is as in Listing 11.1.

Listing 11.1: "Configuration file for the incubator using Eclipse Ditto"

```
{
  "definition": "AU-DT:DigitaltwinIncubator:1.0.0",
  "attributes": {
    "manufacturer": "AarhusUniversity"
  },
  "features": {
    "fan": {
      "properties": {
        "automatic": true,
        "state": "off"
      }
    },
    "heater": {
      "properties": {
        "automatic": true,
        "state": "on"
      }
    },
    "temperature": {
      "properties": {
        "value": 20.8,
        "unit":"Celsius"
      },
      "desiredProperties":{
        "value" :35
      }
    }
  }
}
```

Listing 11.2: "XML schema of the incubator using the AAS framework"

```
<?xml version='1.0' encoding='UTF-8'?>
<aas:aasenv>
  <aas:assetAdministrationShells>
    <aas:assetAdministrationShell>
      <aas:idShort>Incubator_AAS</aas:idShort>
      <aas:identification idType="IRI">urn:dtexamples.into-cps.
          Incubator_AAS</aas:identification>
      ...
    </aas:assetAdministrationShell>
  </aas:assetAdministrationShells>
  ...
  <aas:submodels>
    <aas:submodel>
      <aas:idShort>OperationalData</aas:idShort>
```

```
    <aas:identification idType="IRI">urn:dtexamples.into-cps.Incubator.OperationalData</aas:identification>
    <aas:submodelElements>
      <aas:submodelElement>
        <aas:property>
          <aas:idShort>Temperature</aas:idShort>
          <aas:semanticId>
            <aas:keys>
              <aas:key idType="IRI" local="false"
                type="GlobalReference">
                urn:dtexamples.into-cps.Incubator.OperationalData.Properties.Variables.Temperature</aas:key>
            </aas:keys>
          </aas:semanticId>
          <aas:valueType>float</aas:valueType>
        </aas:property>
      </aas:submodelElement>
      <aas:submodelElement>
        <aas:submodelElement>
          <aas:operation>
            <aas:idShort>heating_operation</aas:idShort>
            <aas:semanticId>
              <aas:keys>
                <aas:key idType="IRI" local="false"
                  type="GlobalReference">
                  urn:dtexamples.into-cps.Incubator.OperationalData.Operations.heating_operation</aas:key>
              </aas:keys>
            </aas:semanticId>
          </aas:operation>
        </aas:submodelElement>
      ...
</aas:aasenv>
```

Even though the configuration files are different, they are normally used either as a structure for a specific metamodel the DT requires to follow or as a parameterisation for the correct initialisation of the DT. The state of the DT can also be stored in a file, but it requires that the file contains serialised objects, which can then be transferred from and to the running object of the DT.

Listing 11.1 shows the very simplistic way of representing the DT in Eclipse Ditto. From that file, an object representing the DT *Incubator* is made available on the Eclipse Ditto platform. Similarly, Listing 11.2 lists the XML schema of the incubator following the AAS metamodel. This schema can then be uploaded to one of the multiple AAS implementations to create on object representing the DT *Incubator* and be connected to other services or channels within the AAS environment.

11.5.3 Digital Twin Configuration Template

The importance of DT configuration C_{dt} has been described so far with the help of two examples. Based on the DT configuration definitions provided in Equation 11.1 and the accompanying explanation provided for Fig. 11.2, we can revise the two DT configuration templates. Listing 11.3 contains the revised specification expressed in YAML format.

Listing 11.3: "A YAML template specification for a DT configuration."

```yaml
version: 0.1

assets:
 config:
  location: string [file, url]

compute:
 host: enum [docker, isolated host, shared host]
 software-id: dockerid / osid

dt_services:
 internal:
  rabbitmq:
   host:
   port:
   credentials:
     username:
     password:
   topic:
  ....

 external: #integration with external services
  service_name:
   ports:
   hostnames:
  ....

physical_twin: #info comes from PT to DT
 config:
  location: string [file, url]
  ....
```

This configuration template is necessarily generic and thus any DT platform needs to provide automation for configuration generation, verification, modification and deployment. There are auto-generated keys such as service access credentials. The DT platform must auto-generate these while satisfying the relevant security standards.

The minimal DT can be specified only using *assets[config]* file and compute sections. This minimal configuration is sufficient to run a digital twin. The *phys-*

ical_twin section is only needed for a real-time feedback loop with the PT. The *services* part need to be filled only if internal or external services are required. An example configuration for the incubator use case is given in Listing 11.4

Listing 11.4: "The YAML specification for the DT configuration of the Incubator."

```
version: 0.1

assets:
  config:
    location: incubator.conf

compute:
host: docker
software-id: ubuntu22:latest

dt_services:
  internal:
    rabbitmq:
    host: rabbitmq.foo.com
    port: 5473
    credentials:
        username: incubator
        password: xxxxx
    topic: incubator

external: #integration with external services
  influx:
    ports: [8086]
    hostnames: [influxdb.cloud-provider.com]

physical_twin: #info comes from PT to DT
config:
    location: pt.conf
```

The configuration indicates that the incubator DT should be run in a docker container running an Ubuntu 22 image.

The incubator needs both the RabbitMQ and the InfluxDB services. The configuration indicates that RabbitMQ is an internal service provided by the DT platform. InfluxDB is an external service to which the DT has access. In the case of external services, the DT platform only needs to provide network access; there is no need to know the credentials of external services. The *pt.conf* indicates the Secure SHell (SSH) access details for connecting to the PT. The incubator DT can connect to the PT using this kind of information.

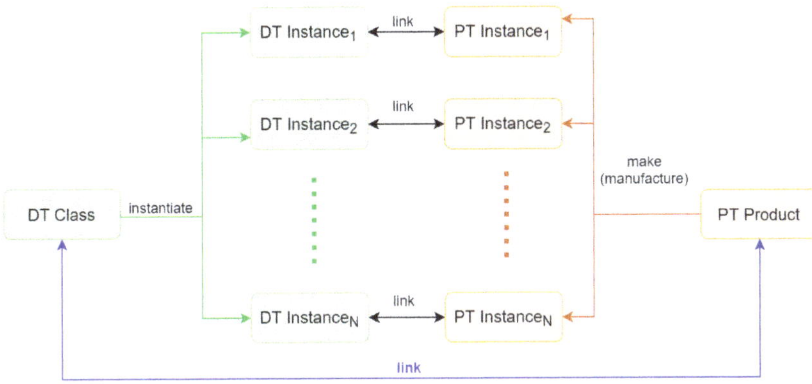

Fig. 11.3: Mapping of DT class and instances to the corresponding PT model and PT instances.

11.6 Digital Twin Class and Instances

The preceding sections have highlighted the value of building DTs from reusable assets (Sect. 11.4) and a configuration format to create them (Sect. 11.5). It is possible to enhance the value of DTs by following three guidelines during the management of DTs:

Adaptability: A DT is likely created, customised, executed and used by more than one user. Thus there is a need to modify/update a DT during the period of its use.

Reuse: In many domains, multiple PTs will be manufactured as instances of the same or similar designs, and often using components at the PT and design levels. An example is the physical design and embedded software of incubator using which many incubator PTs can be made. Similar reuse is desirable even in DTs. If a DT can be used to create many similar DTs, then it is advantageous. The DT that is used to create similar DTs can be referred to as a DT template. We thus envisage the reuse of both DTs and their component parts.

Concurrency: In domains where there are multiple PTs built from similar instances, each operational PTs is likely to benefit from having a dedicated DT, and so multiple similar DTs may be running concurrently. We envisage this with Twinpeh's incubator: many incubator PTs will be operational, each with a DT and Twinpeh (at company level) may well wish to use such a multi-DT to mimic the situation of many PTs running concurrently.

These principles argue for the ability to create many concurrent instances of a DT to accompany their partner PTs. To realise this, we turn to the concepts of object-orientation from software engineering, which draws a distinction between objects and classes. In an object-oriented view, we see a system as being composed of interacting objects, each of which is an individual but many of which share common features. Each incubator is different, for example, but they all have a size and shape, thermal performance parameters determined by for example their materials, controller, etc. In the object-oriented framework, these common features can be gathered into a class which serves as a template form which more useful objects can be created. We will use the term "class" and "template" interchangeably; similarly the terms "object" and "instance" shall be used interchangeably. These can safely be used as synonyms in the context of DTs.

The DT configuration template specified in Sect. 11.5.3 can be viewed as a definition of a DT class. This template binds DT assets, platform services, potential PTs and a suitable execution environment. The binding is at the level of selection but not the use. This template is still not completely reusable. Even a class cannot be directly used in a programming language. There might be further configuration required for assets, for example parameters for co-simulation models, error thresholds for numerical algorithms. There is also a need to specify authentication and authorisation details for services and execution environments. These details are specific to each execution instance of a DT class. These specific details are used for construction of one of many instances of a DT from one DT class. Thus the DT class (template) is used to create any DT objects (instances).

A similar situation exists on the PT side as well. A PT will also have one PT product model (a PT class) and many PT instances. An incubator can be a product model (a PT product) from which many incubator PTs are manufactured. Each of the manufactured PTs become PT instances (typically with a unique identification). Fig. 11.3 shows the relationship between DT classes and instances to their corresponding PT product and PT instances. A DT class is often developed in tandem with a PT product. Thus, a DT class is intricately tied to PT product. Each manufactured PT instance shall have its linked DT instance. These two instances evolve in tandem and become on instance of DT enabled system.

11.7 DTaaS: Reference Architecture for Digital Twin Platforms

In this section we outline the vision of the DTaaS platform. The stakeholders of a DT and their requirements have been outlined in Chap. 4. Some requirements, for example those related to manufacturing, are not relevant to the DT platform. However, some roles and capabilities that are more specific to the engineering of DT-enabled systems (outlined in Sect. 4.8) would be expected to give rise to particular requirements on the platform, examples of which are summarised in Table 11.2.

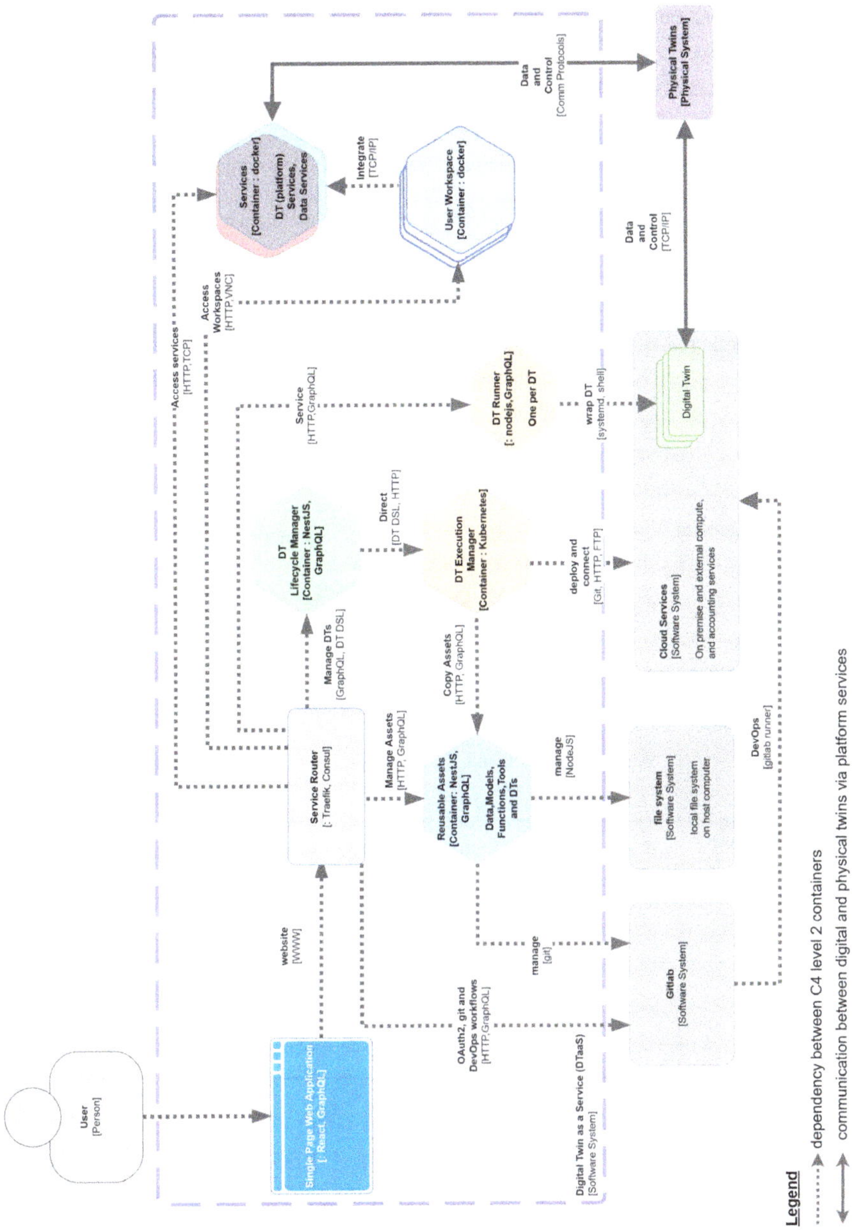

Fig. 11.4: System architecture of DTaaS platform. Updated from [40].

Table 11.2: DT-enabled systems engineering roles and requirements for DT platforms.

Role	Requirements
PT design	1. Access historic data from on field PTs to improve the design of future PTs. 2. Prototype PTs from available PT components to make design decisions. 3. Create reusable PT and DT assets.
PT manufacturing	1. Save test data from PT prototypes for future use. 2. Link PT with prototype DT running on DT platform and interact with both. 3. Tweak the prototypes of either PT or DT to perform what-if analysis.
Model management	1. Create and maintain complementary models for DTs. 2. Compose models for system of systems. 3. Execute DTs with different models to assess the suitability of a model for a PT-DT combination.
PT maintenance	1. Use DT services to derive insights on the current state of PT. 2. Create monitoring, predictive maintenance services on top of DT. 3. Configure PT and DT based on the recommendations provided by DT services.
DT architecture	1. Compose, configure DT and required services. 2. Perform analysis and decision making on one or a fleet of DTs. 3. Integrate DTs with available internal and external services to satisfy requirements of other roles.
Calibration	1. Configure/reconfigure DT based on the state of PT and its environment. 2. Create useful, potentially reusable calibration services for DTs.

Fig. 11.4 shows a simplified overview of the DTaaS architecture using the C4 notation[21]. The system has been divided into complementary containers to promote scalability. Each container fulfils one or more of the requirements listed in Table 11.2. The containers have also been separated based on the DT elements introduced in Sect. 1.3.2 and elaborated in Sect. 3.6, specifically physical entities, data & data management, models & model management, DT services, each of which has a dedicated container. The data processing and storage container is a collection of

[21] `https://c4model.com/`

data-related services / technologies integrated into the DT platform. The data storage technologies such as databases or data lakes come under data-related services. The PTs inherently generate time-series data; thus data processing technologies such as Apache Spark and InfluxDB are good candidates for including inside the data processing and storage container. The PTs have direct but secure access to this container, thus enabling direct storage of data.

The reusable assets container is responsible for performing all the tasks related to asset management in the DT platform. This container must have components to provide version control, traceability and potentially authoring of DT assets. There is a wide array of tools for creating and using DT assets, especially models.

The platform services can be structured in containers that provide the high level services that face DT users, such as monitoring, visualisation, calibration and re-configuration. Services like Jupyter notebooks[22] have a rich ecosystem of data visualisation and data science services. In addition to standard services available from free and commercial software providers, the DT platforms must enable users to start their own services. The services provided by users can help one or more DTs fulfil their requirements. A good example is that of predictive maintenance service for a fleet of DTs that can integrate with the Information Technology (IT) systems of the manufacturer. The service itself might be unique to DTs corresponding to one PT model. Yet this service can be used by all the DTs running within the DTaaS platform.

One of the crucial containers is the DT Lifecycle Manager component which manages the DT through its creation, deployment, maintenance and termination [40]. The support for DT life cycle can be either inherently built into a DT at the time of its construction. Otherwise, DT platform must provide a standard interface to DTs to support life cycle operations.

Among the other elements of the reference architecture, the DT Execution Manager container will have components responsible for receiving the configuration of a DT and executing on a cloud and virtualisation infrastructure, and is discussed in greater depth in Sect. 11.8. Finally, the generic components like a service router, a website and authentication provide unified user interface and necessary security to the users of the DT platform.

[22] https://jupyter.org/

11.8 DTaaS: the DT Execution Manager

The primary responsibility of an execution manager is to support on-demand execution of DTs on the available compute infrastructure in a manner that is transparent to the DT's users. The execution manager has to fulfil this responsibility for both centralised and distributed DT structures. This section explores design decisions than can be made for implementation of an execution manager.

11.8.1 On-demand Execution

The foremost requirement of an execution manager is the on-demand execution of DTs. Without such a capability, all the available DTs need to be run all the time. This results in inefficient use of compute resources.

The execution manager can only satisfy the on-demand execution requirement if it has access to the necessary compute resources. The cloud and virtualisation technologies explained in Sect. 11.3 play a key role here. The user / complementary DT services supply the DT configuration provided in Listing 11.3 to the execution manager, and this needs to be converted into the equivalent API invocation of public and private cloud providers to provide the infrastructure needed.

There have been two broad approaches taken to generate the configuration for the cloud compute infrastructure, both of them illustrated in Fig. 11.5. Note that, although the connection of a PT to its matching DT is shown directly, these are only logical connections. The actual connections have to pass through the DT platforms.

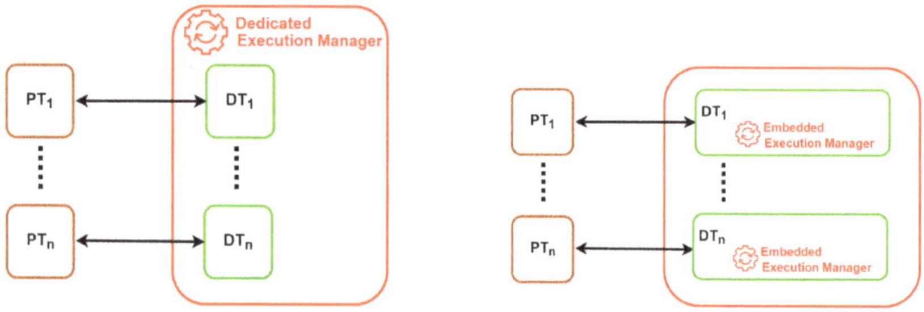

(a) A common execution manager (b) Embedded execution manager

Fig. 11.5: Approaches to integrating execution manager into DT platforms.

One approach is to allocate the responsibility of provisioning IT infrastructure solely to the execution manager, as shown in Fig. 11.5(a). In this case, DTs are only concerned with the execution of DT assets as per the asset configuration provided.

The DTs are not aware of the infrastructure on which they are run. Thus DTs can be deployed on a variety of compute infrastructures, e.g., docker containers, Kubernetes, Terraform, public and private cloud providers. The downside is the limitation on the packaging of DTs. The DTs cannot be packaged as anything but reusable assets (objects) linked through the DT configuration and cannot have notions of packaging like Dockerfile, Helm manifests for Kubernetes, Terraform templates, Ansible playbooks, TOSCA application description templates, etc.

If the DT needs to have complete flexibility in terms of execution environment, an embedded execution manager is needed. This is similar in design to the dedicated execution manager except that it only serves one DT. For example, a DT could have an Ansible playbook to deploy specific parts of DT assets to external compute infrastructure. Such innate awareness of compute infrastructure enables DTs to optimise their use of compute infrastructure. The what-if analysis of DTs is a good use case for the embedded execution manager. If the user demands what-if analysis, an embedded execution manager can receive the request and then execute multiple scenarios on a distributed compute infrastructure acquired temporarily, collate results and release the acquired infrastructure. The embedded execution manager indirectly extends the capabilities of a DT platform. The requirements of a DT platform can then be scaled down to that of executing "lightweight DTs". These DTs are lightweight only platform's perspective in that the main work demanded by the DTs is performed outside of the platform itself. Thus the platform indirectly achieves cost-effective scalability.

The embedded execution manager can only be permitted to configure the compute infrastructure resources that it has authorisation to. It is important for the DT platforms to prohibit the embedded execution manager from changing the system configuration of the DT platform.

The embedded and the dedicated execution manager approaches are two ends of a spectrum. It is possible to have both these approaches supported within one DT platform. The dedicated execution manager can deploy DTs under different configurations. For example, one dedicated execution manager can deploy a DT in the form of a docker container or a Kubernetes application or a Virtual Machine (VM). A few of DTs supported by DT platform, for example a VM, can have its own embedded execution manager in the form of docker-compose files executable within the VM. The platform-wide dedicated execution manager will be unaware of such DT-specific capabilities.

11.8.2 Execution Isolation

The second requirement is the isolation of DTs while they are being executed. This is both for safety and security reasons. In abstract terms, a DT can be thought of as an OS process. Process isolation is a key requirement of every OS. A DT can be executed from a single binary as a single/multiple process(es), or a collection of programs executed as multiple processes. The Linux control groups feature also allows process grouping, which in turn allows containerisation of DTs. A DT can also be executed as a Kubernetes application which is executed as a set of linked docker containers. The isolation at execution time is a requirement irrespective of the scale and variety in DT execution scenarios.

One way to achieve execution isolation is to provide only pre-defined execution environments. For example, a DT software platform can demand that all DTs must only be executable within a Kubernetes cluster. Such a demand requires (i) the DT to be specified as a collection of docker containers and (ii) the DT to be configured as a Kubernetes application. Similarly a DT software platform can demand that a DT must always run in validated VMs that are disconnected from the Internet. If a DT were to be executed in such an isolated environment, it could only be packaged as a fully self-contained executable program which does not require access to the Internet in order to provide value to users.

The need for isolation indirectly demands deliberate specification of resources available to a DT in its execution. An a priori specification is the only way to limit one DT from resource starving another that is executing in the same environment. For example, a DT can greedily consume process, memory and GPU resources to perform its what-if analysis, degrading the performance of other DTs running on the same computer.

Good security practices demand that a program (and by extension a DT) adheres to the principle of least privileges. A valid DT configuration must implicitly or explicitly specify the Role-Based Access Control (RBAC) for a DT on a DT software platform. An execution manager must clearly enforce the security privileges specified in the DT configuration. The security restrictions are another form of isolation on DTs in its execution.

11.8.3 Distributed Execution Manager

The execution manager must scale with the requirements of DT software platforms. An execution manager which is capable of executing all the available DTs on one computer suffer from a scalability problem. It has previously been solved using a "big iron" server computing or vertical scalability approach. However, this is an expensive solution that is not scalable. An economic alternative is horizontal scalability, in which multiple low-capacity computers are used to distribute the work using a tiered approach [33].

A variation of the tiered approach is the manager-worker design pattern in which a manager distributes tasks to workers who complete the work and return the results to the manager [8]. In the context of DT software platforms, a DT to be executed is one task. The execution manager sends the details of one DT to a worker and suggests execution. The worker takes upon the responsibility of executing the DT. There can be multiple hierarchical levels of workers who receive and execute DTs. At the lowest level, one worker can be given responsibility for execution of one DT. This worker may be referred to as *runner*. A runner is only responsible for managing one DT. A next higher-level worker can manage multiple runners. Fig. 11.6 shows a two-level distributed execution manager in whcih an execution manager running in the cloud is managing multiple workers deployed in the fog (computing networks that are of intermediate capacity). Their capability lies between the low-capacity edge computers and high-capacity cloud data centers. The workers deployed in the fog are managing the runners deployed on computers running on the premises, i.e., on-site of PTs. A worker in collaboration with the execution manager takes responsibility for managing multiple runners.

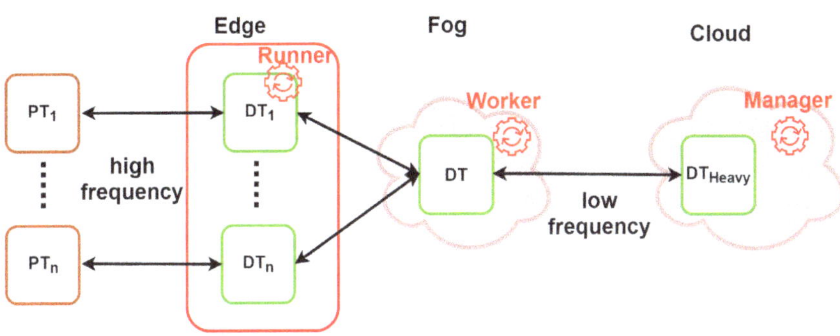

Fig. 11.6: Execution manager supporting a distributed DT.

A tiered approach is a natural fit for distributed DTs in which components of DTs are executed at more than one location. Fig. 11.6 shows the scenario of a distributed DT being deployed at three locations: cloud, fog and edge. For the sake of simplicity, the term "component" will not be used in this section, but it must be understood that a part or whole of a DT is executed at a location on the network.

An edge DT runs on a computer that is physically located next to the PT such as an edge DT running on a smaller, resource-constrained computer [29]. Although it is tempting to make edge DTs work as a data collection point, the real value of edge DTs lies in the ability to interact with the PT at high frequency. A DT running in a fog has more compute resources at its disposal. Thus, a fog DT can potentially run more accurate models and enable higher quality decisions. In addition, it is possible for a fog DT to work as a hierarchical DT over multiple edge DTs. Hierarchical DTs play a crucial role in the construction of DTs for Systems of Systems (SoS) [32]. A cloud DT can be the most accurate representative model of a PT (or a set of PTs in the case of a hierarchical DT). A cloud DT can run highly accurate models on cloud computing resources to provide more accurate results to both users and PTs. The only downside is the significant distance and network delay involved to complete one feedback loop between the DT and the PT. Imperfections in communication networks such as variable delay, data loss etc. must be considered. The execution manager is best situated in the cloud. In the context of a distributed execution manager, each edge DT is managed by one runner while fog DTs are managed by workers while the execution manager takes the overall responsibility for the execution of centralised/distributed DT.

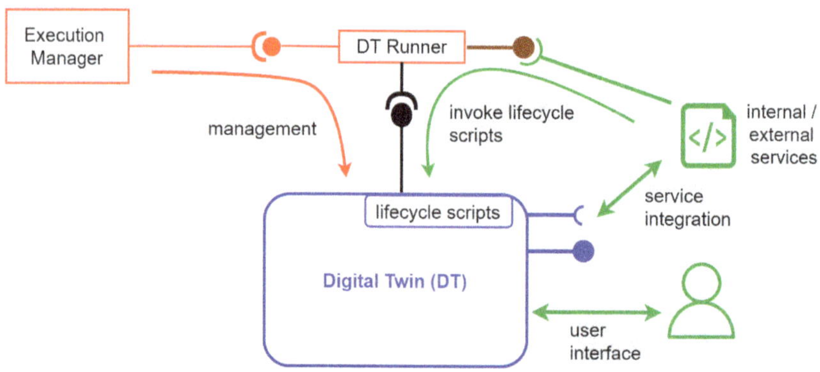

Fig. 11.7: The role of a DT Runner in a DT software platform.

A DT runner component can extend the capabilities of the execution manager. There will be many instances of runners in a live DT software platform. The lifetime of a runner closely matches that of the DT it manages. One possible design for a DT runner is illustrated in Fig. 11.7. The runner relieves the top-level manager from the low-level responsibility of managing each DT. At the same time, the runner can also enhance the accessibility of a DT to its users. The execution manager launches one runner instance on the available compute infrastructure and provides it with details of the DT(s) for which the runner should take responsibility. The runner then fetches the required DT assets, configures compute resources (memory amount, initialises specified service access credentials, etc.) and waits for service requests from either users or other processes. The runner also provides a management interface to the execution manager and a service interface to others. The execution manager can ask a runner to terminate a DT. The terminate command leads to termination of a DT and release of the resources used. The DT under consideration can be in any life cycle phase and may prefer to be active. Nevertheless the runner terminates a DT and releases all the resources allocated to it.

Most DTs do not have inherent support for life cycle phases. The life cycle scripts are executable programs which act as a guided interface between a DT and its users. The scripts can be written in any programming language. Users can invoke them to change the life cycle phase that a DT is in. There are many DTs that do not offer an API interface to users. Even for the DTs that offer API interfaces, there may not be support for life cycle phases. The user-facing side of DT runners provide an API that invokes life cycle scripts. This API can be invoked by either internal or external services which want to integrate a DT with their own software systems. The users expect to interact with APIs of the DTs; the API can either be the DTs or the one provided by the runner. Thus the execution manager and infrastructure must support network-level access to this API from outside of DT platforms.

11.9 Prototype Implementation

A prototype DTaaS software platform has been built to demonstrate the utility of our proposed architecture. Fig. 11.8 shows this implementation prototype at the time of writing.

The single page web application has been developed using the React[23] and GraphQL[24] technologies. The prototype has two-level user authentication and authorisation mechanisms in place. The mTLS standard certificates have been used to verify user identity while Gitlab has been used as OAuth2 authorisation provider.

The reverse proxy technologies such as an Apache HTTP server[25] and Traefik[26] are suitable for providing service router-cum-gateway functionality. The service router can perform dynamic updates of the background containers or microservices that it connects to, using Consul[27] for service discovery. In this prototype both an Apache HTTP server and a Traefik server have been used as the service router. The dynamic configuration updates are based on updates to configuration files used by these services.

A dedicated microservice has been developed to manage reusable DT assets. This microservice provides a unified interface to two storage backends, namely local files and Gitlab files. The NestJS[28], HTTP and GraphQL technologies have been used for developing this microservice.

---

[23] `https://react.dev/`

[24] `https://graphql.org/`

[25] `https://httpd.apache.org/`

[26] `https://traefik.io/traefik/`

[27] `https://www.consul.io/`

[28] `https://nestjs.com/`

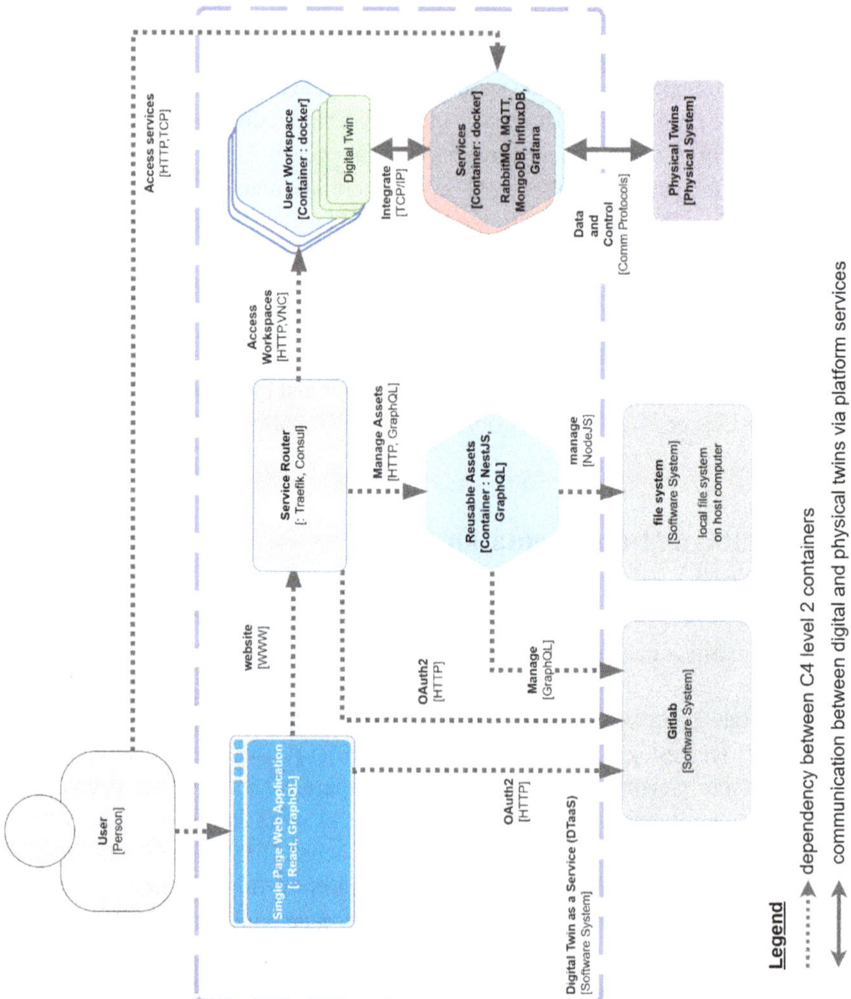

Fig. 11.8: Demonstration of the Incubator and other DTs on a distributed deployment of DTaaS.

Two data processing and storage technologies, namely InfluxDB[29] and MongoDB[30] have been integrated into the DTaaS software. The Incubator DT and its PT use InfluxDB for data storage while MongoDB has been used by DTs utilising JSON

[29] https://www.influxdata.com/

[30] https://www.mongodb.com/

documents for data storage. The RabbitMQ[31] and Eclipse Mosquitto[32] communication brokers have been integrated and provided as platform services for all the DTs hosted on the software platform. These two support the most widely used IoT and DT communication protocols. Both the Incubator DT and other DT prototypes use these two communication brokers for the PT-DT interaction. Grafana[33] is a visualisation and monitoring software; it is hosted as a DT service in the DTaaS.

All users have dedicated workspaces. Each of these workspaces has powerful VS Code and Jupyter lab tools installed on them. These can be used for the DT configuration, management and graphical interaction. The users run their DTs in workspaces. These DTs communicate with PTs either directly or indirectly via the communication services hosted by the DTaaS. The user workspaces also have support for graphical DTs. In addition, the workspaces come with pre-installed machine learning and data science frameworks.

This prototype implementation showcases the utility of separating the DT platform components into the ones suggested in the DTaaS reference architecture (see Fig. 11.4).

11.10 Support for DT Services

Examples of DT platforms found in the literature are summarised in Table 11.3. The proposed DTaaS software architecture and the prototype implementation support visualisation, state estimation, monitoring and anomaly detection. As can be seen from the table, such a combination of forms of support is not available at the time of writing in the other platforms reported. This support in DTaaS is feasible because of the flexible architecture that can integrate complementary services (including off-the-shelf services) available both inside and outside the DTaaS software.

The incubator example can be used to illustrate a typical approach taken to support the four stated features. The incubator DT needs InfluxDB and RabbitMQ services. The Incubator DT communicates with the PT via the RabbitMQ server. The InfluxDB is used to record data from PT. The RabbitMQ service has been integrated into the platform while an external influxDB database has been used. The visualisation of measurements has been done using InfluxDB and Grafana dashboards. The Kalman filter state estimator, monitor and anomaly detector are bundled into the incubator DT. In addition, the monitoring and anomaly detection services can be built on top of the InfluxDB and Grafana services.

---

[31] `https://www.rabbitmq.com/`

[32] `https://mosquitto.org/`

[33] `https://grafana.com/`

Table 11.3: Comparison with other DT platforms.

Paper	Specific Area	Vis'n	State Est.	Monitoring	Anomaly Det.
Bilberg, Malik [4]	Smart Factory	x			
Chhetri et al. [7]	Assembly Line		x	x	x
Karadeniz et al. [16]	Ice Cream Machines	x		x	
Mandolla et al. [23]	Aircraft	x		x	
Min et al. [26]	Petrochemical Factory				
Shangguan et al. [37]	Industrial Robot	x			
Ruohomaki et al. [35]	3D Energy Mapping	x			
DTaaS [40]	Multiple domains	x	x	x	x

11.11 Fleet Analysis

Increasingly, smart products like Twinpeh's Incubator tend either to be shipped with local DTs that could run, for example, on a user's smart phone, or are connected to corporate DTs on the supplier's digital premises. As a result, it is increasingly important to think about whole fleets of DT-enabled systems. The goal of a fleet-level DT is the orchestration of many DT-enabled systems and/or other components.

Fleets can be thought of as SoSs, or more specifically as Systems of DT-enabled Systems. There is a large body of research and practice in SoS engineering which addresses challenges arising from the relative independence of ownership and management among the individual constituent systems, and the degree to which they are set up to cooperate, acknowledge participation in a wider system, or receive direction. This SoS character is present in fleets of DT-enabled systems, bringing challenges that range from ensuring privacy and protection of Intellectual Property to the need to offer vertical and horizontal integration of DTs [25].

Fleets of DTs can be managed either in a decentralised way or in a centralised way [42]. The decentralised option requires methods for managing distributed DTs, which were covered in the previous section. On the other hand, the centralised alternative can be covered, among others, with methods like composition of DTs [14]. Others have proposed the concept as aggregation of DTs [12].

The composition of DTs is particularly relevant for large-scale systems where multiple entities co-operate or are related to one another to some extent, which means that the systems are 1) complex and 2) highly dynamic. Common approaches for including composition of DTs include ontologies [2] and micro-service architectures [30]. A modelling approach for composable DTs has been proposed, where multiple independent DTs can be modelled and then implemented as a composed system [10]. The approach takes advantage of ontologies to define semantic relationships, including the *isComposedOf* relationship, which defines the relation from a larger DT to a smaller DT that composes the larger one. Fig. 11.9 illustrates a modelling approach for composing DTs. The technical implementation and example are available[34] where a case study with two robotic arms is used and the robotic arms and grippers are composed into the larger system, i.e., a manufacturing cell. This way, the larger system has access to the smaller DTs' components, such as the models, operations and attributes, making the larger system a centralised manager of the system of DTs.

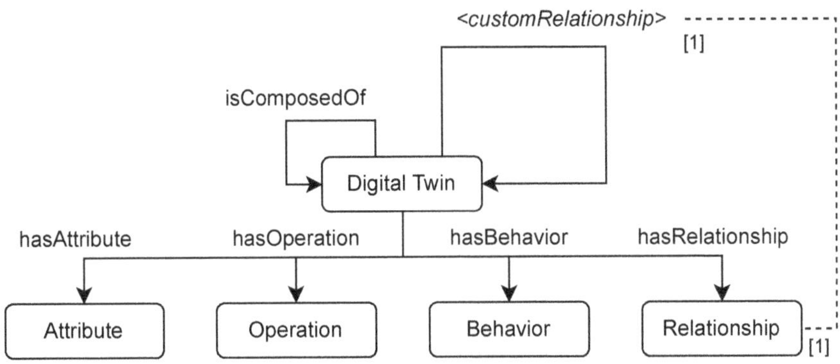

Fig. 11.9: Diagram of composition of DTs using an ontological model [10]. The dashed line refers to custom DT-2-DT semantic relationships and their dependency to individuals of the Relationship class.

Fleet analysis requires access to the state of DTs and information exchanged between multiple PTs and DTs; access to state of DTs is an additional benefit. DT services (see Sect. 3.6.4) include trend analysis, monitors and calibration. These three services would be expected to provide particular value for a fleet. The PT designer and model producer can use the results of services that use fleet data to provide insights for next generation of PTs. The PT maintenance expert, calibration expert and manufacturing engineers can utilise trend analysis and calibration to roll out mass updates to fleet of PTs. The fleet analysis can supplement the traceability analysis on DT assets. Finally, a DT architect can use analysis and decision making services to iterate upon the existing and new DTs.

[34] `https://github.com/sagilar/DigitalTwin_Composition`

A fleet of Incubators

For the Twinpeh company it makes sense to consider how to learn from the fleet of DT-enabled incubators that they will have access to. Since the incubator PTs do not interact with each other in the physical world there is no need to consider connections between the PTs. In this case there is potential to use machine learning for the fleet as a whole, for example. This could support predictive maintenance (see Sect. 10.4) or learn more about how the physical Incubator products are actually being used, and thereby improve the controllers implemented in the current PTs, as well as inspiring future products for the market.

References

1. et al, V.Z.: Industrial digitalization in the Industry 4.0 era: Classification, reuse and authoring of digital models on digital twin platforms. Array p. 100176 (2022). DOI: https://doi.org/10.1016/j.array.2022.100176
2. Andryushkevich, S.K., Kovalyov, S.P., Nefedov, E.: Composition and application of power system digital twins based on ontological modeling. In: IEEE International Conference on Industrial Informatics (INDIN), pp. 1536–1542 (2019). DOI: 10.1109/INDIN41052.2019.8972267
3. Barnstedt, E., Lin, S.W., Boss, B., Malakuti, S., Scientist, S., Clauer, E., Van Schalkwyk, P., Isaacs, D., Martins, T.W.: Open Source Drives Digital Twin Adoption. IIC Journal of Innovation pp. 1–16 (2021). URL https://arc.aiaa.org/doi/pdf/10.2514/6.2012-1818
4. Bilberg, A., Malik, A.A.: Digital twin driven human-robot collaborative assembly **68**(1), 499–502 (2019). DOI: 10.1016/j.cirp.2019.04.011
5. Burns, B., Beda, J., Hightower, K., Evenson, L.: Kubernetes: up and running. O'Reilly Media, Inc. (2022)
6. Buyya, R., Yeo, C.S., Venugopal, S., Broberg, J., Brandic, I.: Cloud computing and emerging it platforms: Vision, hype, and reality for delivering computing as the 5th utility. Future Generation Computer Systems **25**(6), 599–616 (2009)
7. Chhetri, S.R., Faezi, S., Canedo, A., Faruque, M.A.A.: QUILT: Quality inference from living digital twins in IoT-enabled manufacturing systems. In: Proceedings of the International Conference on Internet of Things Design and Implementation, IoTDI '19, pp. 237–248. Association for Computing Machinery (2019). DOI: 10.1145/3302505.3310085
8. Dean, J., Ghemawat, S.: Mapreduce: simplified data processing on large clusters. Communications of the ACM **51**(1), 107–113 (2008)
9. Gil, S., Mikkelsen, P.H., Gomes, C., Larsen, P.G.: Survey on open-source digital twin frameworks–A case study approach. Software: Practice and Experience (December 2023), 1–32 (2024). DOI: 10.1002/spe.3305
10. Gil, S., Mikkelsen, P.H., Tola, D., Schou, C., Larsen, P.G.: A Modeling Approach for Composed Digital Twins in Cooperative Systems. In: 2023 IEEE 28th International Conference on Emerging Technologies and Factory Automation (ETFA), pp. 1–8. IEEE (2023). DOI: 10.1109/ETFA54631.2023.10275601
11. Goldberg, R.P.: Architecture of virtual machines. In: Proceedings of the Workshop on Virtual Computer Systems, pp. 74–112. Association for Computing Machinery, New York, NY, USA (1973). DOI: 10.1145/800122.803950

12. Human, C., Basson, A.H., Kruger, K.: A design framework for a system of digital twins and services. Computers in Industry **144**(June 2022), 103796 (2023). DOI: `10.1016/j.compind.2022.103796`
13. International Organization for Standardization: Automation systems and integration — Digital twin framework for manufacturing, ISO 23247:2021(E) edn. International Organization for Standardization, Geneva, Switzerland (2021). URL `https://www.iso.org/standard/78743.html`
14. Jia, W., Wang, W., Zhang, Z.: From simple digital twin to complex digital twin Part I: A novel modeling method for multi-scale and multi-scenario digital twin. Advanced Engineering Informatics **53**(July), 101706 (2022). DOI: `10.1016/j.aei.2022.101706`
15. Kamburjan, E., Klungre, V.N., Schlatte, R., Johnsen, E.B., Giese, M.: Programming and Debugging with Semantically Lifted States. In: Lecture Notes in Computer Science (including subseries Lecture Notes in Artificial Intelligence and Lecture Notes in Bioinformatics), vol. 12731 LNCS, pp. 126–142. Springer (2021). DOI: `10.1007/978-3-030-77385-4_8`
16. Karadeniz, A.M., Arif, I., Kanak, A., Erguen, S.: Digital Twin of eGastronomic Things: A Case Study for Ice Cream Machines. In: 2019 IEEE International Symposium on Circuits and Systems (ISCAS), pp. 1–4 (2019). DOI: `10.1109/ISCAS.2019.8702679`
17. Kiss, T., Kacsuk, P., Kovacs, J., Rakoczi, B., Hajnal, A., Farkas, A., Gesmier, G., Terstyanszky, G.: Micado - microservice-based cloud application-level dynamic orchestrator. Future Generation Computer Systems **94**, 937–946 (2019)
18. Lehner, D., Gil, S., Mikkelsen, P.H., Larsen, P.G., Wimmer, M.: An architectural extension for digital twin platforms to leverage behavioral models. In: 2023 IEEE 19th International Conference on Automation Science and Engineering (CASE), pp. 1–8 (2023). DOI: `10.1109/CASE56687.2023.10260417`
19. Lehner, D., Pfeiffer, J., Tinsel, E.F., Strljic, M.M., Sint, S., Vierhauser, M., Wortmann, A., Wimmer, M.: Digital Twin Platforms: Requirements, Capabilities, and Future Prospects. IEEE Software **39**(2), 53–61 (2022). DOI: `10.1109/MS.2021.3133795`
20. Leng, J., Wang, D., Shen, W., Li, X., Liu, Q., Chen, X.: Digital twins-based smart manufacturing system design in Industry 4.0: A review. Journal of manufacturing systems **60**, 119–137 (2021)
21. Lipton, P., Palma, D., Rutkowski, M., Tamburri, D.A.: TOSCA Solves Big Problems in the Cloud and Beyond! IEEE cloud computing (2018)
22. Liu, M., Fang, S., Dong, H., Xu, C.: Review of digital twin about concepts, technologies, and industrial applications. Journal of Manufacturing Systems **58**, 346–361 (2021). DOI: `https://doi.org/10.1016/j.jmsy.2020.06.017`. Digital Twin towards Smart Manufacturing and Industry 4.0
23. Mandolla, C., Petruzzelli, A.M., Percoco, G., Urbinati, A.: Building a digital twin for additive manufacturing through the exploitation of blockchain: A case analysis of the aircraft industry **109**, 134–152 (2019). DOI: `10.1016/j.compind.2019.04.011`
24. Menage, P.B.: Adding generic process containers to the linux kernel. In: Proceedings of the Linux symposium, vol. 2, pp. 45–57. Citeseer (2007)
25. Michael, J., Pfeiffer, J., Rumpe, B., Wortmann, A.: Integration Challenges for Digital Twin Systems-of-Systems. In: Proceedings - 10th IEEE/ACM International Workshop on Software Engineering for Systems-of-Systems and Software Ecosystems, SESoS 2022, vol. 1, pp. 9–12 (2022). DOI: `10.1145/3528229.3529384`
26. Min, Q., Lu, Y., Liu, Z., Su, C., Wang, B.: Machine Learning based Digital Twin Framework for Production Optimization in Petrochemical Industry **49**, 502–519. DOI: `10.1016/j.ijinfomgt.2019.05.020`
27. Naha, R.K., Garg, S., Georgakopoulos, D., Jayaraman, P.P., Gao, L., Xiang, Y., Ranjan, R.: Fog computing: Survey of trends, architectures, requirements, and research directions. IEEE access **6**, 47980–48009 (2018)
28. Picone, M., Mamei, M., Zambonelli, F.: WLDT: A general purpose library to build IoT digital twins. SoftwareX **13**, 100661 (2021)

29. Picone, M., Mamei, M., Zambonelli, F.: A flexible and modular architecture for edge digital twin: Implementation and evaluation. ACM Transactions on Internet of Things **4**(1), 1–32 (2023)
30. Preuveneers, D., Joosen, W., Ilie-Zudor, E.: Robust Digital Twin Compositions for Industry 4.0 Smart Manufacturing Systems. In: Proceedings - IEEE International Enterprise Distributed Object Computing Workshop, EDOCW, pp. 69–78. IEEE (2018). DOI: 10.1109/EDOCW.2018.00021
31. Qi, Q., Tao, F., Hu, T., Anwer, N., Liu, A., Wei, Y., Wang, L., Nee, A.Y.: Enabling technologies and tools for digital twin. Journal of Manufacturing Systems **58**(PB), 3–21 (2021). DOI: 10.1016/j.jmsy.2019.10.001
32. Ricci, A., Croatti, A., Mariani, S., Montagna, S., Picone, M.: Web of digital twins. ACM Transactions on Internet Technology **22**(4), 1–30 (2022)
33. Rimal, B.P., Choi, E., Lumb, I.: A taxonomy and survey of cloud computing systems. In: 2009 fifth international joint conference on INC, IMS and IDC, pp. 44–51. Ieee (2009)
34. Rosen, R.: Resource management: Linux kernel namespaces and cgroups. Haifux, May **186**, 70 (2013)
35. Ruohomäki, T., Airaksinen, E., Huuska, P., Kesäniemi, O., Martikka, M., Suomisto, J.: Smart City Platform Enabling Digital Twin. In: 2018 International Conference on Intelligent Systems (IS), pp. 155–161. DOI: 10.1109/IS.2018.8710517
36. Schrotter, G., Hürzeler, C.: The digital twin of the city of Zurich for urban planning. PFG–Journal of Photogrammetry, Remote Sensing and Geoinformation Science **88**(1), 99–112 (2020)
37. Shangguan, D., Chen, L., Ding, J.: A Hierarchical Digital Twin Model Framework for Dynamic Cyber-Physical System Design. In: Proceedings of the 5th International Conference on Mechatronics and Robotics Engineering, ICMRE'19, pp. 123–129. Association for Computing Machinery. DOI: 10.1145/3314493.3314504
38. Shao, G., Helu, M.: Framework for a digital twin in manufacturing: Scope and requirements. Manufacturing Letters **24**, 105–107 (2020)
39. Smith, J.E., Nair, R.: The architecture of virtual machines. Computer **38**(5), 32–38 (2005)
40. Talasila, P., Gomes, C., Mikkelsen, P.H., Arboleda, S.G., Kamburjan, E., Larsen, P.G.: Digital Twin as a Service (DTaaS): A Platform for Digital Twin Developers and Users. In: 2023 IEEE Smart World Congress (SWC), pp. 1–8 (2023). DOI: 10.1109/SWC57546.2023.10448890
41. Tao, F., Cheng, J., Qi, Q., Zhang, M., Zhang, H., Sui, F.: Digital twin-driven product design, manufacturing and service with big data. The International Journal of Advanced Manufacturing Technology **94**, 3563–3576 (2018)
42. Wang, Z., Gupta, R., Han, K., Wang, H., Ganlath, A., Ammar, N., Tiwari, P.: Mobility digital twin: Concept, architecture, case study, and future challenges. IEEE Internet of Things Journal **9**, 17452–17467 (2022). DOI: 10.1109/JIOT.2022.3156028
43. Ye, X., Hong, S.H.: Toward Industry 4.0 Components: Insights into and Implementation of Asset Administration Shells. IEEE Industrial Electronics Magazine **13**(1), 13–25 (2019). DOI: 10.1109/MIE.2019.2893397

Chapter 12
Case Studies in Digital Twins

Bentley James Oakes, Houxiang Zhang, Lars Ivar Hatledal, Hao Feng, Mirgita Frasheri, Michael Sandberg, Santiago Gil and Cláudio Gomes

Abstract We present four case studies of Digital Twins (DTs) through a DT description framework, covering 14 fundamental characteristics for each DT and its Physical Twin (PT). These characteristics uncover the diversity of the case studies, providing a deeper understanding of how these DTs interact with the real world, a sense of how they were built, and the challenges that were encountered. Following these examples, we advocate for the DT community to report on their own case studies using such a description framework. Having a common structure to experience reports, combined with explicit details for each characteristic, can both open further research avenues such as empirical studies on digital twins, and reveal best practices within the community.

12.1 Introduction

Even though the DT approach is continuing to evolve in academia and industry, it is often difficult to understand exactly what capabilities and structure a DT may have. Case studies in experience reports often do not fully report on all relevant details. This not only obscures the DT engineering process, but also does not allow the complexity of the DT-enabled system to be revealed.

Past work has partially addressed this reporting issue by presenting an initial set of 14 characteristics for describing DTs in experience reports [28, 30]. These characteristics, summarised in Sect. 12.2, attempt to capture the fundamentals of a DT in a conceptual framework and diagram, such that readers can quickly assess and compare the DT and its capabilities to other DTs. The hope is that this more detailed reporting framework allows for the better understanding of case studies and consequently faster DT research and development.

In this chapter, we examine four case studies through their description using these characteristics. This chapter thus provides two contributions: a) clear examples of

J. Fitzgerald et al. (eds.), *The Engineering of Digital Twins*,
https://doi.org/10.1007/978-3-031-66719-0_12

how to describe a DT using the characteristics from this reporting framework, and b) a structured format to allow readers to better understand the capabilities and conceptual architecture of the DTs presented.

The case studies described are as follows. The first three are implemented at Aarhus University (AU), while the fourth is at the Norwegian University of Science and Technology (NTNU):

- The Tempeh Incubator (Sect. 12.3);
- Robotti and Desktop Robotti (Sect. 12.4);
- Cooperative Robotic Manufacturing (Flex-cell) (Sect. 12.5);
- The Research Vessel Gunnerus (Sect. 12.6).

12.2 Summary of Characteristics

This section summarises the 14 characteristics of the DT description framework from [28, 30]. We refer the reader to these articles for further explanation of each characteristic.

1. **System-Under-Study (SUS):** The system (essentially the PT), its environment, and any agents (computational or human) present who may interact with the system.
2. **Acting Components:** Any additions and modifications to the SUS which enables communication from the DT to the SUS.
3. **Sensing Components:** Any additions and modifications to the SUS which enables communication from the SUS to the DT.
4. **Multiplicities:** How many systems and DTs are involved in the DT ecosystem and their relationships.
5. **Data Transmitted:** The data transmitted from the SUS to the DT.
6. **Insights/Actions:** The information from the DT to agents in the SUS, or the automatic controlling actions from the DT to the SUS.
7. **Services:** The activities that the DT is used for. This could also be termed the *capabilities* of the DT. The *insights* and *actions* of the DT originate from these services.
8. **Enablers:** Computational components which take models and data, and support the services of the DT.
9. **Models and Data:** The input and output for the enabler components, with some data coming from the SUS.
10. **Constellation:** The conceptual relationships within the DT of the models/data, enablers and services.
11. **Time-scale:** The time-scale of the communication between the DT and SUS, and the computation within the DT. This includes data, insights, actions and any present simulations.

12. **Fidelity Considerations:** For each component/service of the DT, how they correspond to the SUS. For example, whether a reduced-order model was built to balance fidelity versus performance [29].
13. **Life-cycle Stages:** The stages of the SUS (*ideation*, *realisation*, *utilisation*, etc.) for which the DT is utilised. If the scope of the SUS changes, this should also be reported.
14. **Evolution:** The evolution of the DT throughout its development.

Note that this DT description framework has been recently extended to 21 characteristics in [12] by merging three DT frameworks, including the one above, to present a unified characterization of DTs. [12] adapts the material found here to provide a detailed description for all 21 characteristics for the Flex-cell use case. As well, the Incubator and Robotti DTs found in this chapter are also summarized in tabular form.

12.3 The Tempeh Incubator

The tempeh incubator example introduced in Sect. 3.2 has been used through this book as a representative example of a system to be controlled by a DT. The business context involving the Twinpeh company has been slightly fictionalised, but a real incubator and DT have been built at AU. Here, we return to the real incubator DT to provide an introductory example showing the usage of the DT description framework. We summarise the incubator case here, and also provide links to those sections of the book which contain additional context for each characteristic of the incubator.

12.3.1 Incubator DT Overview

The purpose of the PT is to maintain a specific temperature within the incubator's chamber to support the fermentation of tempeh. The primary purpose of the incubator's DT is to enhance the functionality of the PT through the DT services. As mentioned in Sect. 3.5, these services support business goals including, but not limited to:

1. Visualising the incubation process.
2. Conserving energy, for example, by turning off the controller when the lid is opened.
3. Signalling when a batch of tempeh is ready.
4. Indicating when equipment requires maintenance.
5. Detecting subpar batches of tempeh.

6. Measuring and suggesting control actions to improve the quality of tempeh.

For the co-simulation of the flex-cell we use Maestro and the RabbitMQFMU [7]. A graphical representation of the incubator PT and its DT in Fig. 12.1. In this figure, the PT is on the left, while its DT is on the right. The following sections will refer to this figure to explain the conceptual components and characteristics of the incubator and its DT.

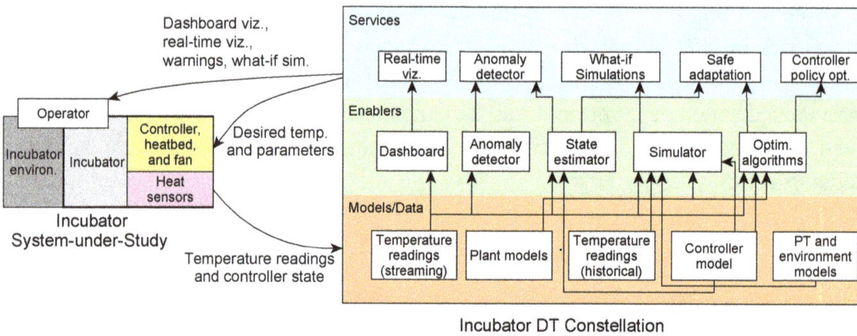

Fig. 12.1: The incubator system and its Digital Twin.

12.3.2 Fundamental Characteristics

This section presents the incubator case study through the 14 fundamental DT characteristics as introduced in Sect. 12.2.

12.3.2.1 C1: System-under-study

The SUS includes the incubator PT along with its environment. The elements of the SUS are described in Table 12.1, organised hierarchically.

The main interactions between the incubator and its environment are:

- the room is assumed to be colder than the incubator. It therefore acts as a heat sink, contributing to cooling down the incubator's box and the air inside the incubator.
- the user can open the lid to check on the tempeh incubator. When the lid is open, the cooling-down process is accelerated.

Table 12.1: Components of the incubator SUS.

	Part	Description
Plant	Power supply	Powers electrical components
	Fan	Circulates air throughout box
	Heater	Raises the temperature of surrounding air
	Electrical switching board	Allow controller to turn on/off heater/fan
	Temperature sensors	Two inside box, one outside
	Styrofoam™ box	With lid, to contain components and air
	Air volume	
	Tempeh	Object to be temperature-controlled
Controller	Running on Raspberry PI	Controls the fan and heater
Environment	Surface	Where the incubator is placed
	Surrounding air	Heat sink
Agents	User	Manual operation of the lid

12.3.2.2 C2: Acting Components

Two components were added and/or updated to allow the DT to interact with the SUS. First, the controller of the incubator was updated to connect to a secure RabbitMQ server (see Sect. 7.4), for checking whether there are requests to update the controller parameters and desired temperature. Second, a safety layer was coded into the controller that discards non-essential values for parameters and/or desired temperatures. This mitigates errors in the DT from propagating to the incubator and creating an unsafe situation, such as raising the temperature too high.

12.3.2.3 C3: Sensing Components

To allow the PT to communicate directly with the DT, the controller was modified to add two abilities. The first ability is to connect to a secure RabbitMQ server and upload the controller and incubator state. This includes the heater state, the temperatures measured and all other necessary information as mentioned below. The second ability is to recover when the network connection drops, to improve the robustness of the DT-PT communication.

12.3.2.4 C4: Multiplicities

Currently, there is only one DT connected to one incubator. In the future, it is anticipated that multiple incubators could be controlled by one DT. Note that in this scenario, the heat produced by one incubator may affect another one. This would require a more sophisticated composition approach for analysis and control over a fleet of incubators (see Sect. 11.11).

12.3.2.5 C5: Data Transmitted

The data sent periodically to the DT by the controller is comprised of: a) the temperature of each sensor, b) the average temperature used by the controller to make a decision, c) the time it takes for each control cycle, d) the current state of the controller (e.g., waiting or cooling down, as in Fig. 12.2), and e) the values of each of the controller parameters.

12.3.2.6 C6: Insights/Actions

As one *insight*, the incubator DT provides a dashboard for performing monitoring of the incubator's state, such that the operator understands the incubator's behaviour at a glance. This dashboard serves as a sink for all things that can be visualised related from the different DT services. For instance, the Lid Opening Service will produce data that is visualised in the dashboard (see Sect. 8.4). Other insights include *real-time visualisation* of the incubator, *warnings* about anomalous behaviour, and *what-if simulation* results for the user.

The automatic *actions* that the DT performs on the PT depend on the controller version being used. There are different versions of the controller, but generally the actions send the parameters to be set on the controller, as well as the desired temperature. The simplest controller version, with a state machine model as introduced in Sect. 5.6.1, can have its lower and upper bounds of temperature set by the DT. It can also be commanded to shutdown. A more advanced version of the controller contains waiting states that allow for finer tuning of its behaviour in the presence of more powerful heating elements. This version is illustrated in Fig. 12.2. In this advanced version the DT can configure, in addition to the lower and upper bounds, the parameters C and H in seconds. The above actions can be leveraged to, for example, turn off the controller when the lid is detected to be open.

12.3.2.7 C7: Services

The DT currently provides the following services for the incubator:

- *Real-time visualisation* for the user to understand the incubator's current state at a glance (see Sect. 8.4).
- An *anomaly detector* which alerts the user to any anomalous behaviour that may interfere with proper incubation, such as the lid being open for too long (see Sect. 9.4).
- Providing *what-if simulations* such that the user can enter in proposed conditions and examine the resulting behaviour (see Sect. 10.2).

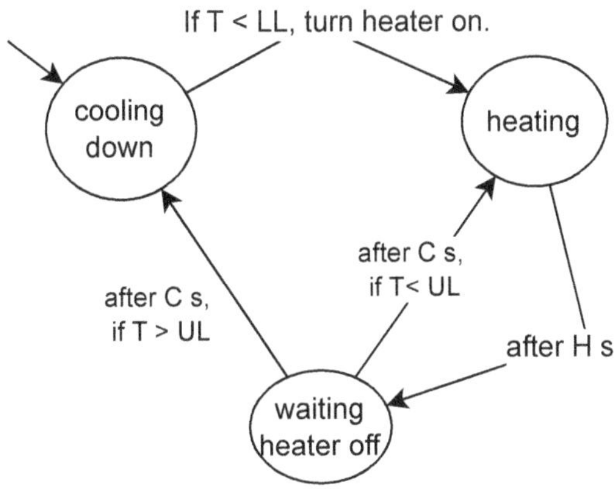

Fig. 12.2: State machine of the controller containing waiting states.

- *Safe adaptation when the lid is opened* to adjust the controller parameters when the lid has been opened for too long (Sect. 10.5). This saves energy, and can prevent unsafe situations where the heater runs at maximum power.
- *Controller policy optimisation when PT changes*. When a component of the system is changed, such as the heater or power supply, the DT will detect that the incubator physical models used are no longer a valid representation of the PT configuration. A re-calibration of such models will be triggered, as well as re-optimisation of the control parameters (see Sect. 10.5).
  - One adaptation sub-process will re-calibrate the models of the PT if the dynamics have been detected to change.
  - A second adaptation sub-process will then re-compute the optimal control policy if the dynamics have changed.

12.3.2.8 C8: Enablers

There are several enablers that support the incubator DT services:

- A *simulator*, which is able to simulate the behaviour of the incubator forward in time and perform what-if experiments.
  - There are two options: a) the *incubator* itself can be simulated for self-adaption purposes, or b) the combination of the *controller and the incubator* can be simulated for control policy optimisation.
- An *anomaly detector* that can determine what is the normal behaviour of the incubator, what is anomalous behaviour, and can alert the user to anomalies.

- A *state estimator* for the incubator which predicts the current state (open/closed lid).
- *Optimisation algorithms* that are used for calibration purposes and the optimisation of the controller behaviour.
- The real-time *dashboard* for displaying information/visualisations.

12.3.2.9 C9: Models and Data

Before detailing the models, we summarise the main quantities of importance about the system, as well as their symbolic representation and units:

- T_b – Average temperature of the air volume inside the box (°C).
- T_h – Temperature of the heating element (°C).
- T_r – Room temperature (°C).
- H_{Heater} – State of the heater (on, presented as 1, and off represented as 0)
- UL – Controller temperature upper limit (°C)
- LL – Controller temperature lower limit (°C)
- C – Cooldown timer (s)
- H – Heating timer (s)

The incubator has the following models for the incubator, which we organise according to their focus.

Plant Models

2 Parameter Model: A simple scalar Ordinary Differential Equation (ODE) representing the plant dynamics. It takes as inputs T_r, H_{Heater}, and returns as an output T_b. The parameters include the heat transfer coefficient between the air volume inside the box and the room, and the heat capacity of the air volume. The initial condition of this model is the initial temperature of the air volume $T_b(0)$.

4 Parameter Model: A plant dynamic model (also an ODE), introduced in Sect. 5.4.1, that takes as inputs T_r, H_{Heater}, and returns as an output T_b and T_h. The parameters include the same parameters as the 2 parameter model plus the heat transfer coefficient between the heater and the air volume as well as the heater capacitance. The initial conditions are: The initial temperature of the heater as well as the air volume.

ANN Model: An Artificial Neural Network (ANN) based model, that takes as inputs the room temperature, the state of the heater, the previous average temperature of the air volume inside the box, and returns as outputs the average temperature of the air volume inside the box.

ANN State Estimation Model: An ANN based model, that takes as inputs the room temperature, the state of the heater, the previous average temperature of the air volume inside the box, and returns as outputs the average temperature of the air volume inside the box as well as the temperature of the heating element.

Controller Model

A controller model (State machine), introduced in Sect. 5.6, that takes as inputs the average temperature of the air volume T_b and outputs the state of the heater T_h. The parameters include the temperature upper and lower limits. The initial condition represents the initial state of the controller (typically cooling down).

Physical Twin Model

The hybrid automata model introduced in Sect. 5.7.2. The input is the room temperature, and the set of parameters is the union of the 4 Parameter model and the controller state machine model.

Environment Model

A simple algebraic model $T_r = A\sin(Bt) + C$ representing the temperature of the room as a function of the time (in seconds). The parameters are A, B, C.

PT + Environment Model

The model introduced in Sect. 5.7. Regarding the data, we first look at the historical data (versus the streaming data from the PT [31]), exemplifying one of the datasets of the incubator, and subsequently we look at the data used to configure the DT services.

An example dataset can be found online[1], which includes a description of the goal of the experiment conducted, the author, the configuration of the system, a log of the experiment, the CSV files containing the raw data, and plots of the results. The CSV files contain all the data sent from the PT to the DT.

The data used to configure the DT services is stored in configuration files, that are loaded when the services are launched. Such configuration file contains the following sections:

RabbitMQ: Contains the configuration parameters of the access to the routing queue server, for instance, the IP address, port, username, password, etc.

InfluxDB: Contains the configuration for the access to the time series database including the url, the access token, and so on.

[1] `https://digital-twin-book.org/`

Physical Twin: Contains the initial parameters of the controller.

Digital Twin: This is further subdivided into different sections containing service specific configurations, and the different sets of parameters for each model.

Interfaces to Models

In this case study two interfaces are been used to interact with the models: the Functional Mockup Interface (FMI) has been used for co-simulation studies involving the above models (recall Sect. 5.7.1); as well as an FMI-like interface implemented in the python programming language. Examples of models implementing each interface can be found online.[2 3]

12.3.2.10 C10: Constellation

Fig. 12.1 shows how the models/data, enablers, and services provide information to each other. We can consider the top-level services as being supported by a particular "slice" through the enablers and models/data which can be inferred by the information flow arrows in Fig. 12.1.

12.3.2.11 C11: Time-Scale

The DT for the incubator operates at different rates:

- **Slower-than-real-time:** The anomaly alerts and dashboard information provided to the user may not be acted upon in real-time.
- **Real-time:** Dashboard updates, fan/heat-bed control/adaptation, temperature readings.
- **Faster-than-real-time:** The simulator is able to simulate into the future.
  - This must be performed for both re-calibration, and for controller optimisation.
  - The simulator must be able to run hundreds of simulations within a few seconds to support these services.

[2] https://sites.google.com/view/fm2021tutorialdt/home

[3] https://digital-twin-book.org/

12.3.2.12 C12: Fidelity Considerations

The dashboard does not need to be extremely precise for the user. The temperature propagation model on the other hand, must be very accurate (*high-fidelity*) as the incubator temperature must be simulated into the future. It is also important that the *validity frame* [36] of the incubator models must encompass the control policies that are going to be simulated.

In this section we discuss the different procedures we have followed to quantify the accuracy of each of the models produced above.

Plant Models

2 and 4 Parameter Models These two models were calibrated following the procedures introduced in Sect. 6.4.3: experiments were conducted with an empty incubator, with a fixed control signal that just turns ON the heater for specific periods of time. A plot of the example dataset is available online[4] can be seen in Fig. 12.3, and a plot of the simulation results for each of the models can be seen in Fig. 12.4. As can be seen from the plots, for that particular control profile and room temperature conditions the 4 parameter model is very close to the data, whereas the two parameter model is not so accurate for the prediction of the temperature evolution. Nevertheless this model could be used as a very fast predictor of how much energy is required over a long period of time.

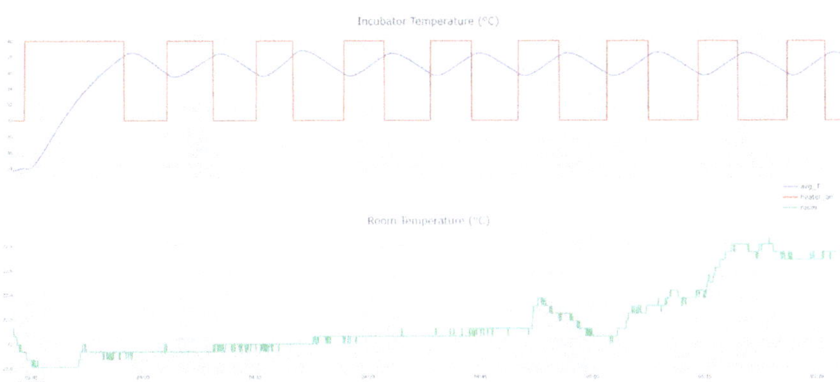

Fig. 12.3: Example dataset used for calibration of the ODE models for the Incubator.

[4] `https://digital-twin-book.org/`

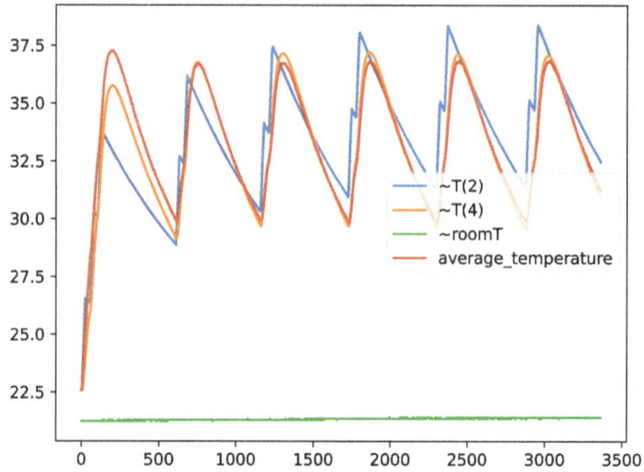

Fig. 12.4: Simulation results for the 2 and 4 parameter Incubator plant models.

ANN Model The calibration procedure of this model was very similar to the calibration procedure of the 2 and 4 parameter models. The model is a simple neural network of a single layer that can take the previous N temperature measurements of the incubator air volume and heating states, and outputs the next temperature estimate (the estimated value of the temperature after 3 seconds, which corresponds to the sample interval in the data set.). It is worth noting that what this model is calibrated to do is to perform one step prediction. This is very different than the use case of the ODE models which are given to a solver and can predict a temperature evolution for many steps. The size of the first layer depends on the parameter N and is given by $2N$. The results of the one step prediction are shown in Fig. 12.5 for $N = 2$. For this model to be used for multistep prediction each temperature predictor needs to be fed back into the model with an iterative evaluation. The errors caused by each prediction quickly accumulate and cause large divergences in multi step predictions, as is illustrated in Fig. 12.6.

ANN State Estimation Model A neural network based model, that takes as inputs the room temperature, the state of the heater, the previous average temperature of the air volume inside the box, and returns as outputs the average temperature of the air volume inside the box as well as the temperature of the heating element.

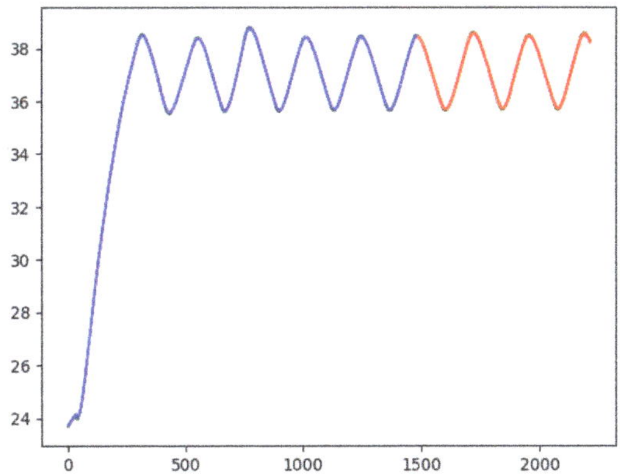

Fig. 12.5: One step prediction results for the Incubator ANN model. The real data set is underneath the blue and red lines. The red line represents data that has not been used to train the neural network demonstrating its ability to generalise.

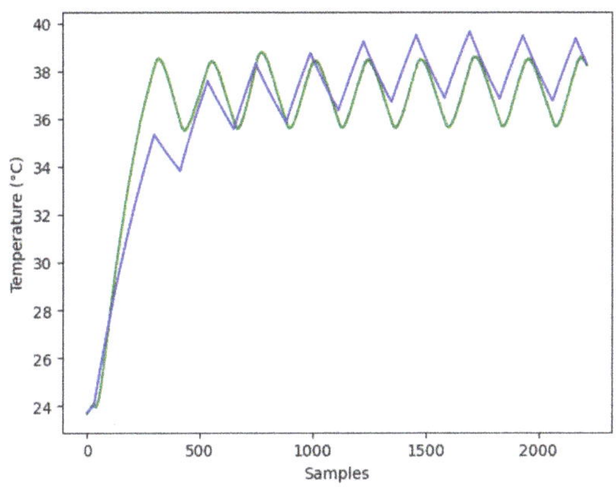

Fig. 12.6: Multi step prediction results for the ANN model (blue line), compared to the data set (green line). This is an example of a bad usage of this Incubator model, because even if each step prediction error is small, those small errors are being fed back into the network and getting amplified.

Controller Model

The parameters of this model can be determined exactly from inspection of the controller code. Its calibration is therefore trivial.

Remaining Models

The remaining models represent compositions of the above models and therefore do not need to be calibrated. However, it is worth discussing their accuracy. As can be seen in the Fig. 12.7, the predicted temperature follows a similar shape as a temperature in the dataset but these two signals become misaligned over time. This divergence occurs because small inaccuracies in the temperature measurements will cause small inaccuracies in the time at which the controller transitions are taken. The difference in the controller transition time in turn leads to the temperature dropping earlier or later than the real temperature in the data set, therefore exacerbating the differences between the two signals. These inaccuracies accumulate overtime leading to total divergence between the two signals. This is one of the main reasons one needs to use the data fusion techniques introduced in Sect. 7.6.

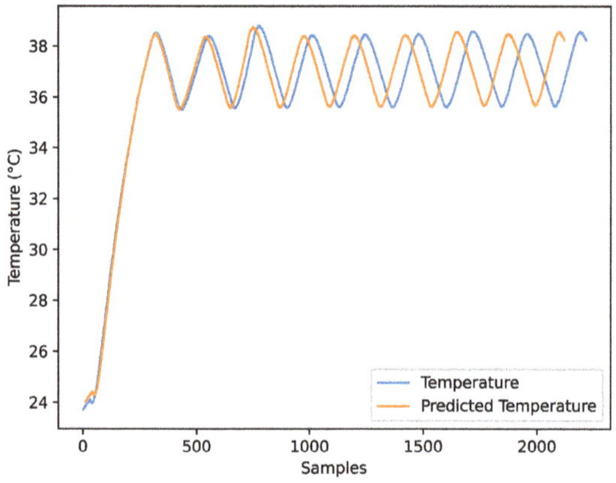

Fig. 12.7: Example co-simulation results and comparison with dataset for the Incubator. The figure shows an example state divergence which is a common phenomenon in models.

12.3.2.13 C13: Life-cycle Stages

All the services provided by this DT are for the *operation* life-cycle stage [35] of the incubator.

12.3.2.14 C14: Evolution

The evolution of the incubator DT is reported in [5, 4]. It can be summarised as follows, and is illustrated in Fig. 12.8.

1. The relevant physics were identified before building the first version of the PT.
2. From this preliminary plant models were built with rough guesses for the parameters.
3. This enabled us to answer questions such as how powerful and heater was necessary for our incubator.
4. The first prototype of the incubator was built.
5. Lab experiments were carried out to refine the estimates for the parameters of the models and compare the different plant models.
6. A controller model was built and the parameters of the controller model were tuned in co-simulation together with a plant model.
7. The controller code was deployed into the physical controller of the incubator and tested.
8. The visualisation service was the first DT service to be deployed.
9. Subsequently the state estimation and normal detection services were implemented.
10. The self adaptation loop was then implemented using the models to optimise the controller whenever the lid was opened.

It is worth highlighting that the preliminary DT services have enabled us to reach new insights which affected the implementation of subsequent DT services, thus highlighting one of the big benefits of DTs and their incremental development.

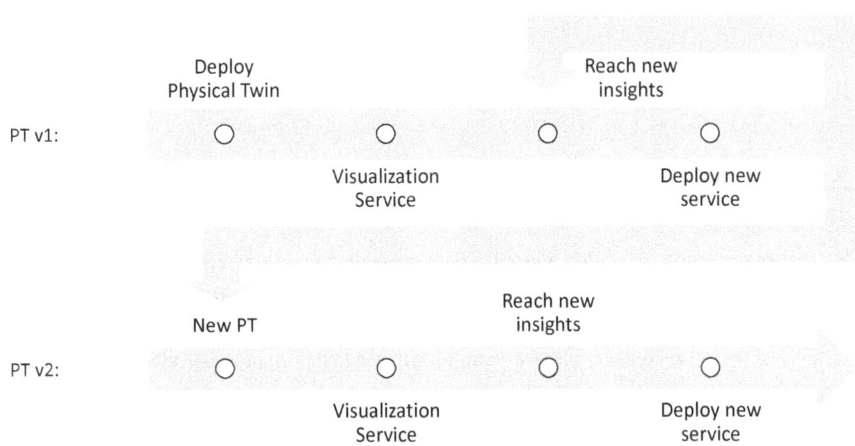

Fig. 12.8: Illustration of the main development steps in the incubator case study, as well as one of the main causes for evolution of the physical twin.

The challenge that arises in the incremental development of DTs is that there are limits to the sensors and actuators of the PT. These in turn set a limit on how advanced the future DT services can be. In the incubator case study, we can highlight this phenomenon by looking at the CO_2 emissions of a healthy tempeh fermentation, displayed in Fig. 12.9. As can be seen, there is significant evidence that such a sensor can bring enormous value to the incubator PT, because it provides an earlier indicator of the temperature of the tempeh rising ahead of healthy fermentation. Note that this sensor was not included in the initial version of the incubator, but it is being included in the second version that is under construction.

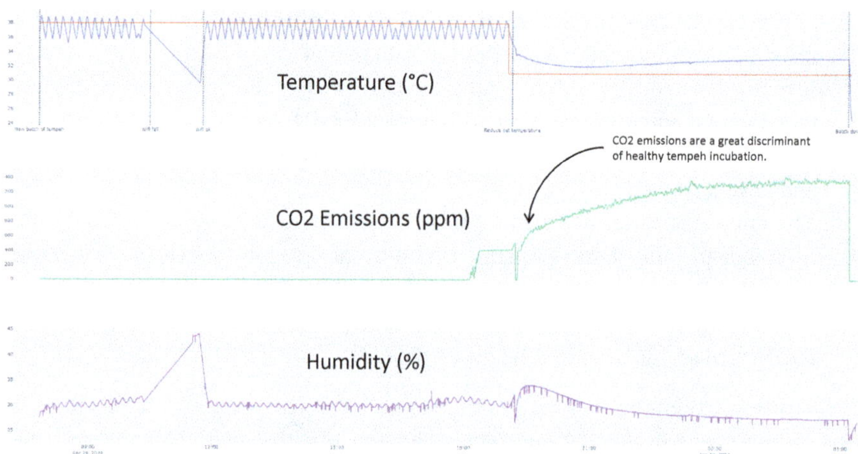

Fig. 12.9: Plot of the data set providing evidence that CO_2 sensor is a great discriminant of healthy tempeh fermentation.

12.3.3 Summary and Future Work

The incubator study provides a small yet representative system for engineering DTs. As the controller must account for the temperature propagation throughout the box, and react to the open lid, the incubator represents some of the challenges seen in twinning CPSs. The source code for the incubator DT is open-source[5], and we invite collaboration on developing further services.

In the short-term, we are enhancing the incubator DT. First, we are further developing visualisations for the incubator based on the Godot game engine. These visualisations will allow the user to control the incubator directly from the visu-

[5] `https://digital-twin-book.org/`

alisation. Second, we are adding new sensors to the incubator. This includes the carbon-dioxide (CO_2) sensor mentioned above to record the release of CO_2 as the tempeh ferments. This will open up new possibilities for tracking fermentation progress, although new models will have to be developed.

12.4 The (Desktop) Robotti

The Desktop Robotti (DR) is a small physical model of the agricultural field robot Robotti developed by AgroIntelli (Fig. 12.10), for which a DT is being built using the technologies described in previous chapters. The role of the DR, and in this case the PT, is to allow a variety of tests to be carried out easily in the lab before these can be scaled up and performed in the field. The purpose of the DT is to provide services from monitoring to reconfiguration, and to enable the early testing of features intended to be part of the PT itself in the future. More specifically, the DT would be able to follow the behaviour of the robot, e.g., by predicting its location, as well as receiving its actual location from a live feed. Such functionality enables monitoring, which in turn supports the reconfiguration and adaptation of the robot in the case of anomalies. The DT would also be able to predict potential collision zones

Fig. 12.10: Real Robotti [6] (top) and the DR (bottom).

of its robot with other robots or objects in the environment and respond adequately to prevent such collisions. The setup makes it possible to perform different experiments in a variety of scenarios, including safety-related ones, with the DR in the loop, with or without fault injection as well as model swapping. The latter consists in updating models (e.g. FMUs) at run-time without stopping and restarting the DT and PT system. The DR DT/PT system is comprised of a physical robot, a co-simulation that runs on a PC, as well as ROS2 nodes and a RabbitMQ server for communication between DT and PT (Fig. 12.11).

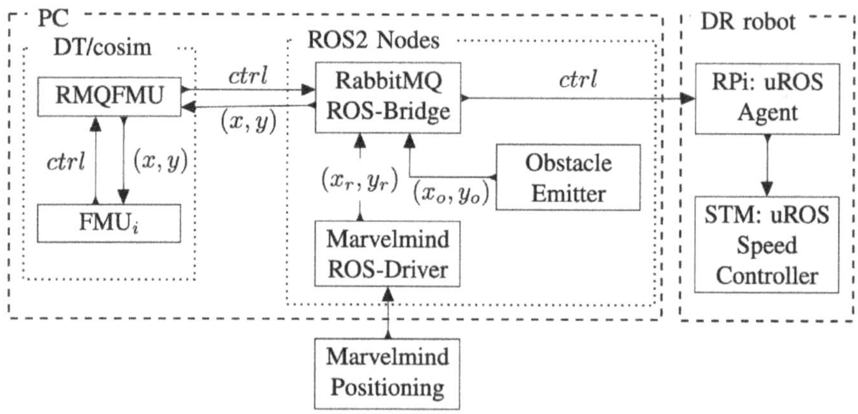

Fig. 12.11: The DR DT/PT system is distributed across the robot platform and the PC containing the co-simulation components, as well as the involved ROS2 nodes. `(x,y)` refer to position coordinates, whereas `ctrl` corresponds to control commands sent to the robot. Figure taken from [9].

12.4.1 Desktop Robotti DT Overview

As mentioned above, the DT for the DR is intended to cover functionalities from monitoring to reconfiguration, and it specifically covers the following capabilities: *parallel operation, fault injection with hardware in the loop*, and *runtime dynamic model swapping*. A graphical representation of the DR and its DT is given in Fig. 12.12 with the DR on the left side and the DT on the right. The following sections will refer to this figure to explain the conceptual components and characteristics of the DR and its DT. For the co-simulation of the flex-cell we use Maestro and the RabbitMQFMU [7].

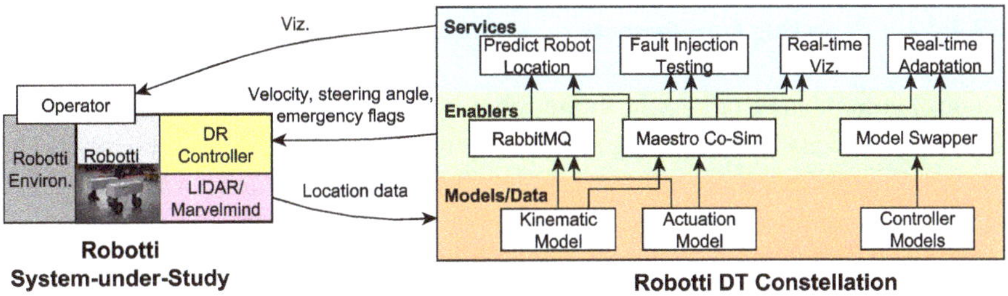

Fig. 12.12: The Desktop Robotti System-under-Study and its DT.

12.4.2 Fundamental Characteristics

This section presents the DR case study through the 14 fundamental DT characteristics as introduced in Sect. 12.2.

12.4.2.1 C1: System-under-study

The PT in this case is the DR mobile robot. The environment could be any space in which the robot is placed to drive, but is typically a structured and known laboratory. The DR has not yet been tested in an outdoor environment, where it can be subject to different weather conditions and terrain types. A human operator is also present to tele-operate and otherwise control the robot if needed. The PT is a car-like steered mobile robot, equipped with two on-board processors – a Raspberry Pi (RPi) 4B and an STM32 arm cortex-M4 based Micro-Controller Unit (MCU) (refer to the right block in Fig. 12.11). The RPi controls the steering motors and communicates with the STM32 using micro-ROS and a serial connection. Meanwhile, the MCU acts as a speed controller and can be in a driving or halted state, with the robot's movement being determined by transmitting a target speed and steering angle. The robot's position is determined by an indoor positioning system using four stationary beacons and a mobile beacon placed on the robot, which operates via a serial connection to an external modem. Lidar readings can also be used for additional information. On the PC, dockerized ROS2 nodes and a RabbitMQ server facilitate data exchange between the co-simulation and the ROS2 network. Messages relevant to the co-simulation are converted and published over RabbitMQ by a dedicated ROS2 node, with incoming messages being received and published as ROS2 messages via a subscriber node on the RPi.

12.4.2.2 C2: Acting Components

Control commands can be sent to the robot by an operator or the DT itself. These consist of velocity and steering angle used by the actuation module in the robot. The DT sends control commands if it detects an anomaly in the form of an emergency stop. The messages are sent through the RMQFMU, a RabbitMQ client encapsulated in an FMU, thereafter forwarded to the RabbitMQ server. A node on the DR side reads data from the RabbitMQ server and the forwards them to the actuation module. Note that the RMQFMU is not a custom solution for the DT of the presented robot, but rather a generic solution to bridge an FMI-based co-simulation with the external world.

12.4.2.3 C3: Sensing Components

The DR can be set up with the MarvelMind indoor positioning system, by using four beacons that serve as references for the one beacon mounted on the robot. In addition, it is possible to read data from a 2D Lidar (Rplidar A1) in the form of planar scans of the surroundings, that can be used to generate a map using the navigation stack package in ROS. As such, the robot could be tracked with respect to an initial position. In either case, the data is published to the RabbitMQ server, and transmitted to the RMQFMU residing inside the co-simulation. The RMQFMU exposes the data through its outputs, which is thereafter used to set the inputs of the interested FMUs.

12.4.2.4 C4: Multiplicities

Currently there is one DT per PT, but initial experiments have been made with two independent PTs. In this case, the DT would monitor the collision of the two (or more) robots (see Sect. 11.11) and apply a mitigation strategy by potentially triggering a mission re-plan. As an example, consider two robots that are tasked to traverse a series of rows in a field. Consider the DT predicting that these robots will collide in the i_{th} row. Then a re-plan could be triggered such that one robot continues in the i_{th} row, whereas the other robot traverses another row in the series.

12.4.2.5 C5: Data Transmitted

The robot can be operated outside of the DT by a user, through ROS topics, by providing velocity and steering commands. The DT communicates with the PT through the RabbitMQ server, and the data transmitted, which is packaged in a JSON format, could consist of velocity, steering angles and (potentially) emergency flags. In this case the RMQFMU is added to the co-simulation-based DT serving as a bridge to the external/PT world. An example of a control command (speed and steering angles) sent to the robot by publishing to a ROS topic through the command line is given below:

```
$ ros2 topic pub -1 /robotti1/move_cmd ackermann_msgs.msg/
    AckermannDrive "{msg.speed = '20', msg.steering_angle =
    '0.8'}"
```

where a message of type *ackermann_msgs.msg/AckermannDrive* is published to the */robotti1/move_cmd* topic. On the other hand, an example of a JSON message published from the DT, i.e., from the RMQFMU, consisting in an emergency stop for the robot is given below:

```
{
 "time": "2023-06-19T13:50:07.390+00:00",
 "em-stop-flag": "True"
}
```

where the *time* field is the timestamp associated with the message, and the *em-stop-flag* field is a Boolean flag, set to *True* to make the robot stop in case of emergency.

Should there be network disturbances (as discussed in Chap. 7) the DT/PT system will become out-of-sync. In order to deal with such circumstances, the Orchestration Synchronisation Administration (OSA) FMU [8] could be added to the co-simulation constellation. This would be able to degrade the DT to a DS by disabling the sending function of the RMQFMU, as well notifying a user of the occurrence. If new messages arrive in a bunch, it would be possible for the DT to skip to the latest messages, if the loss of messages in between can be tolerated by other models in the DT.

12.4.2.6 C6: Insights/Actions

Insights: The system can be coupled with RVIZ tool providing a visual for the location of the robot as shown in Fig. 12.13. In this case the visualisation is in 3D as described in Sect. 8.4, and physical models of the robot could also be added if they are available. The RVIZ interface offers insights into different variables, e.g., robot pose and location etc.

Actions: The DT can send control commands, or affect the behaviour of the robot, e.g., limiting its maximum speed (the third scenario described in Sect. 12.4.2.10), depending on the placed monitors.

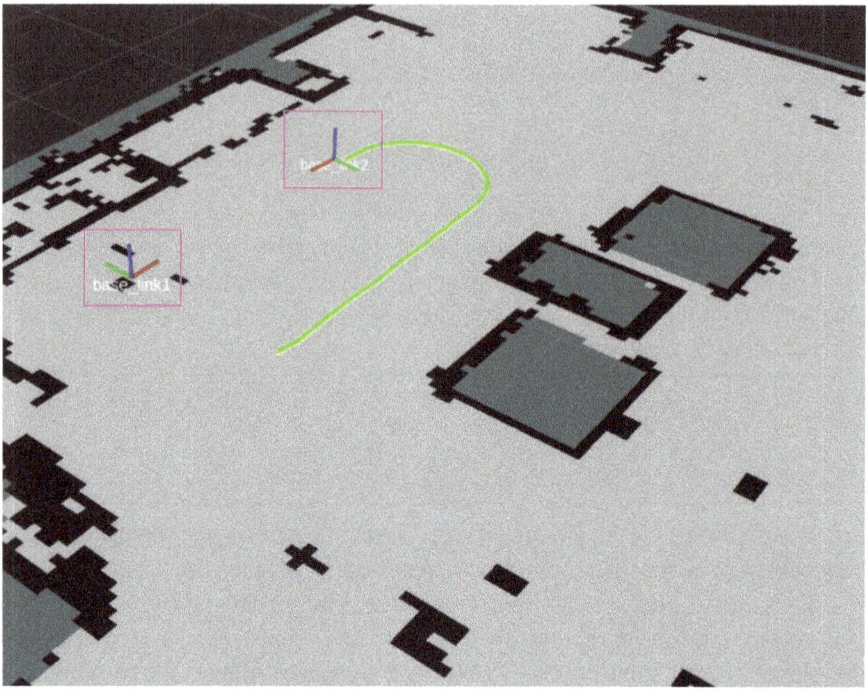

Fig. 12.13: Rviz visualisation of two DRs, one stationary, and one being tele-operated by the user (marked with the magenta rectangles). The green line depicts the path that has already been traversed by the robot.

12.4.2.7 C7: Services

The DT can be configured to support the following scenarios (independently):

1. Parallel operation – Runs kinematics/actuation models on the DT to predict robot location, thereafter used to compare with the live data. This service falls into monitoring and prediction already discussed in Chap. 9 and Chap. 10.
2. Fault injection with hardware in the loop – inject monitors or the RMQFMU data broker to perform different fault scenarios with hardware in the loop. This service falls under what-if simulations enabled with Fault Injection (FI) explored in Chap. 10.
3. Runtime model swapping – update at runtime different FMUs, like the RMQFMU, or a controller FMU. This service falls under reconfiguration capabilities during run-time covered in Chap. 10.

These scenarios are discussed further in the *Constellation* characteristic below.

12.4.2.8 C8: Enablers

The following elements are needed to support the DR DT:

1. Monitors of robot movement and distance to obstacles, that enable the DT to take action if the PT is too close to an obstacle.
2. The RMQFMU, the AMQP based rabbitMQ client encapsulated in an FMU, that enables the two way communication between DT and PT.
3. The Maestro extension for specifying FI scenarios, that allows experimentation with hardware in the loop with fault injected input/outputs.
4. The Maestro extension for model swapping, that enables swapping FMUs in the DT during run-time when new ones are made available from a user.
5. The OSA mechanism able to identify and take appropriate measures when the DT/PT are not in sync, e.g., due to network degradation, allowing operation as a DS as discussed in Sect. 1.2.2 until the network is restored.

12.4.2.9 C9: Models and Data

A kinematic and actuation model have been used to track the position of the robot. This is captured in Fig. 12.14, the bottom block, depicting how the control commands, desired speed and angle, are fed to the actuation model, which outputs the actuation acceleration, as well as angle and speed to the kinematic model. The latter computes the new position for the robot, in the form of (x,y) coordinates. The models are in the form of equations, encapsulated in python scripts, each encapsulated within a separate FMU.

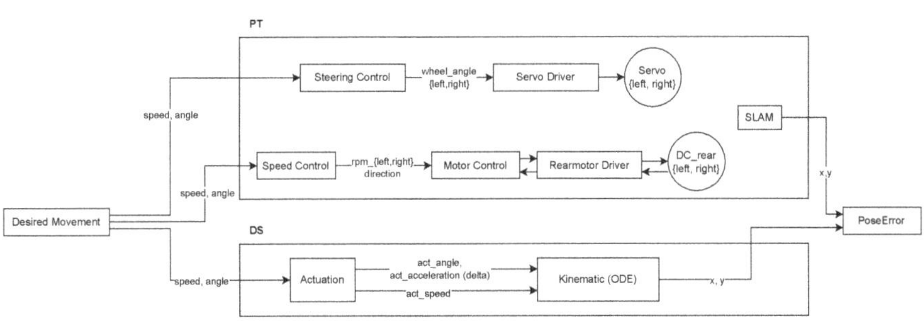

Fig. 12.14: Parallel operation scenario with the DT DR system.

The kinematics of the DR are captured by a bicycle model assuming virtual wheels in the centres of the front and rear axles, thereafter translated for the real wheels of the robot at the front and back (the reader is referred to [18] for the details of the translation). A kinematic model is physics-based (see Sect. 5.4), and uses difference equations, computing current/future state from a past state (new position is computed from a past position). The bicycle model takes as input the velocity υ and steering angle θ, and produces the change in pose as follows:

$$\dot{\psi} = \upsilon / R, \tag{12.1}$$

$$\dot{x} = \upsilon \cdot cos\psi, \tag{12.2}$$

$$\dot{y} = \upsilon \cdot sin\psi, \tag{12.3}$$

where $\dot{\psi}$, $\dot{x}$, $\dot{y}$ represent the change in heading, x, y positions respectively.

The actuation model captures the behaviour of the DC motors, modelled as a first order system. Since DC motors are controlled by some input voltage, resulting in a mechanical output, there is no direct correlation to speed. As such a calibration procedure is carried out to compute the calibration coefficients for the model (see Chap. 6). To this end, a Proportional-Integral regulator was implemented on a microcontroller to manage the speed of DC motors within the Desktop Robotti. The selected regulator imparts a type 1 system characteristic, ensuring a steady-state error in response to step inputs. Subsequently, the model, predicated upon a sampling rate of 60 Hz, undergoes discrete-time conversion and is seamlessly incorporated into the co-simulation platform, with attention given to considerations of stability and phase margins.

12.4.2.10 C10: Constellation

In the following, the DT/PT setup is shown for each scenario.

Scenario 1: Parallel Operation

The components of the system in this experiment are shown in overview in Fig. 12.15. The most significant elements here are the models used in the co-simulation environment to estimate where the robot is, given initial conditions, velocity and steering angle. The results of short experiments performed with the DR are shown in Fig. 12.16, where it is possible to visually compare the actual and predicted positions of the robot.

Scenario 2: Fault Injection

Two fault injection experiments are performed to simulate issues in the components (e.g., monitors) as well as corruption in the data, which would lead to inaccurate

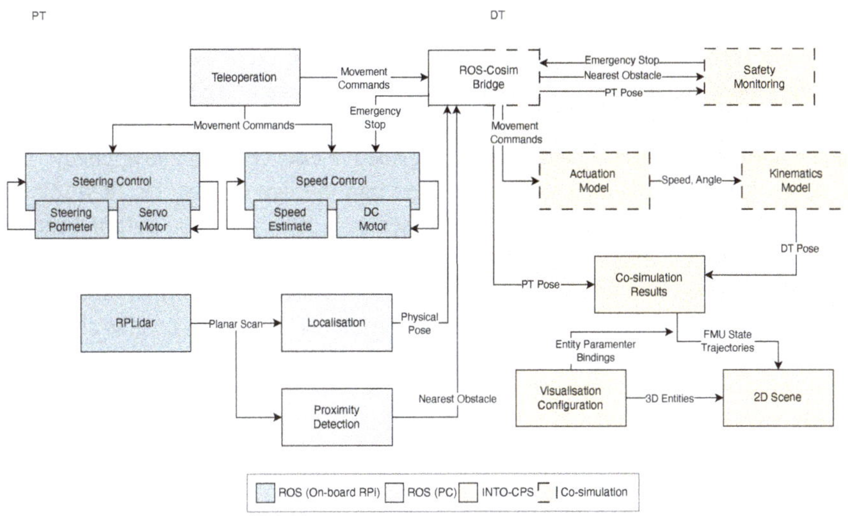

Fig. 12.15: DT/PT system for the parallel operation scenario. The shaded area represents the DT. Figure taken from [26].

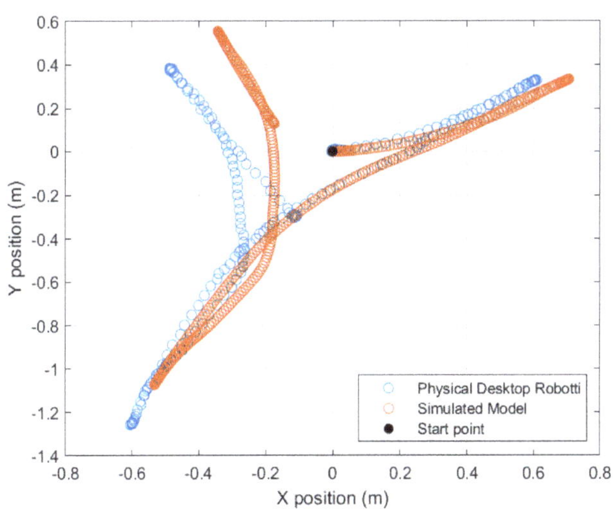

Fig. 12.16: Actual and predicted positions of the DR during experiment. Figure taken from [26].

evaluations in the monitors. The monitors used here check the distance to a virtual obstacle placed in the path of the DR, and set an emergency stop flag if the distance drops below a configurable threshold. Note that there is a redundancy of monitors to support fault-tolerance, one at the PT and the other at the DT level, referred to as PTM and DTM respectively. Should the PTM fail, the DTM might still be able to send an emergency stop if the PT gets too close the obstacle. To perform tests under faulty conditions, the PTM is encapsulated as an FMU and placed within the DT, such that the fault injection techniques described in Sect. 10.2 can be applied.

In the first experiment, a fault is injected into the data broker, i.e., RMQFMU, signifying bad data coming from the PT (see Fig. 12.17). As the data is corrupt, both monitors (DT and PT ones) will not output an emergency stop when needed. However, it is possible to imagine a situation in which correct information comes to the DT from other sources, e.g., an operator or other DTs, and can be thus used from the DTM to produce the desired output (the emergency stop). This source of correct information is captured by the Correct Information (CI) component.

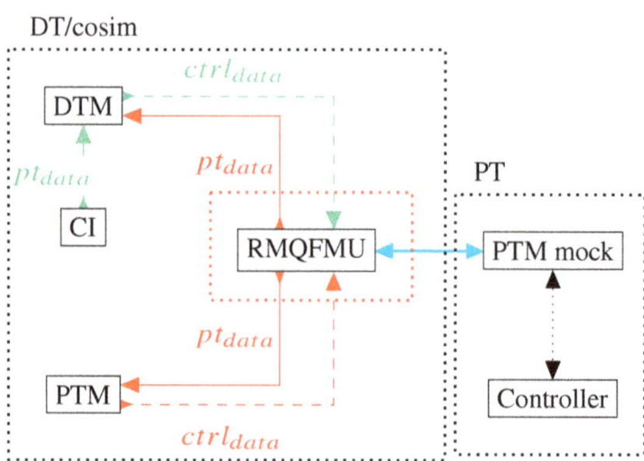

Fig. 12.17: Fault injecting the data broker (RMQFMU) to test the fault tolerance of the DT through the DT Monitor (DTM) when a source of correct information is available (CI). Figure taken from [9].

In the second experiment, it is desirable to test the fault tolerance capabilities of the DTM, should the PTM fail due to some internal crash. Similarly to the previous experiment, to perform fault injection on the PTM within the DT ecosystem, the monitor first needs to be encapsulated as an FMU. Thereafter, the PTM can be included in the DT, and its output can be tampered with by the fault injection mechanism. The faulty behaviour consists of an emergency stop not being issued when the distance to the obstacle is less than a predefined threshold. The system is depicted in Fig. 12.18, where the DTM is the DT monitor, and the PTM is the PT monitor temporarily placed in the DT for testing.

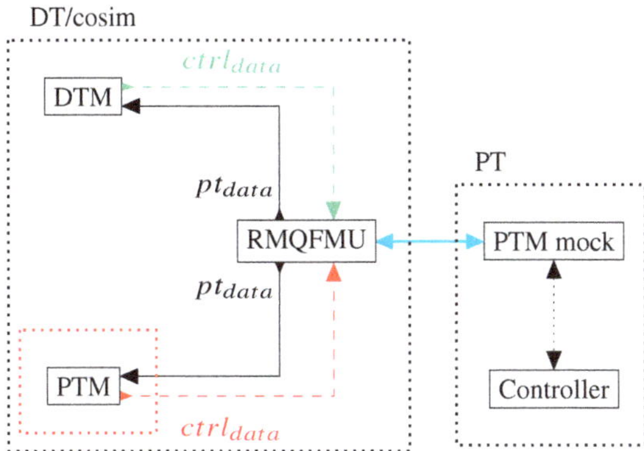

Fig. 12.18: Fault injecting the PT Monitor (PTM) to test the fault tolerance of the DT through the DTM. Figure taken from [9].

Scenario 3: Model Swapping

A *model swapping* technique enables the DT to reconfigure itself when new models become available, without the need to shut down the system and restart it as a whole. The basic DT setup consists of the data broker (RMQFMU), an actuation, and a kinematic model. Two examples of run-time swapping are considered below.

In our first example, a new FMU, specifically the OSA FMU that deals with the detection of time discrepancies between DT and PT, should be inserted. Note that this action would change the existing structure of FMUs. Furthermore, in order to support

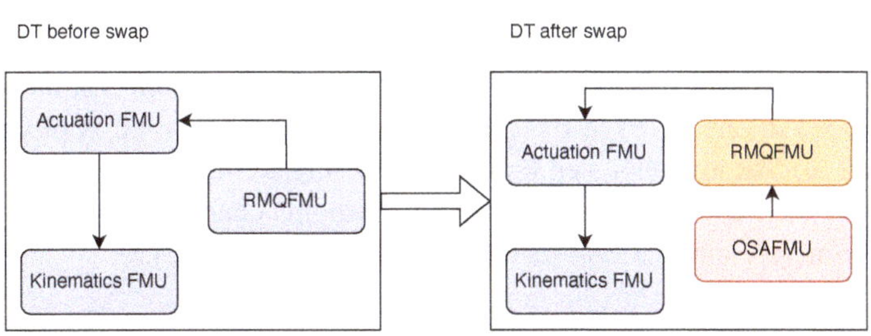

Fig. 12.19: Model swapping by replacing the RMQFMU with a newer version and introducing a new FMU (OSA) with time discrepancy detection capabilities.

the new functionality provided by the OSA FMU, the RMQFMU should be updated as well to include the feature of disabling sending of data if a time discrepancy is discovered. In other words, allowing the transition from DT to DS as discussed in Sect. 1.2.2 for as long that the discrepancies are present. The constellation of the system is depicted in Fig. 12.19. The results of this swap under different conditions are visualised in Fig. 12.20, specifically the values of the steering angle that is fed to the RMQFMUs. The different conditions refer to different stepping and swapping conditions placed for the RMQFMUs. Specifically, when the step condition is satisfied, the inputs of the FMU to be swapped become active. This is useful when these inputs should be synchronised to the outputs of other FMUs. When the swap condition is satisfied, the outputs of the FMU to be swapped are activated, thus the swapping is complete, with the old FMU being discarded. From the tests it can seen that in order to swap in a safe way, i.e., without discrepancies in the steering angles, the new RMQFMU should be stepped only when its initial output matches that of the old RMQFMU. Note also that the new RMQFMU will not have access to the messages already acknowledged by the RabbitMQ server, which are being internally processed by the old one clearly visible in Fig. 12.20a.

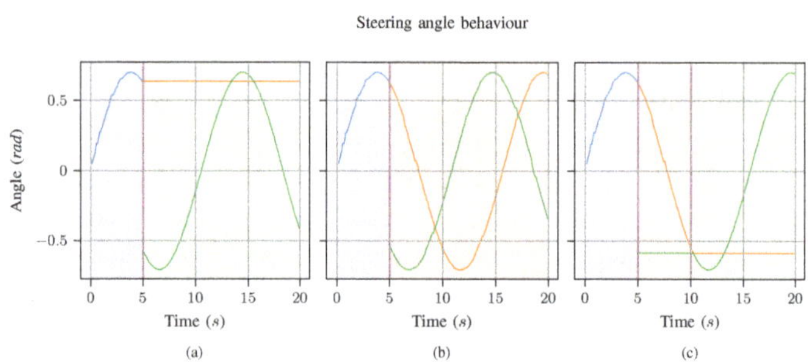

Fig. 12.20: (a) In the first run the FMUs are swapped instantly at time 5s. At that time the old RMQFMU (blue and orange lines) is not stepped any longer (orange line), whereas the new one (green line) starts being stepped. (b) In the second run, a condition is placed for swapping the FMUs, that corresponds to when the steering angles produced by the FMUs match. Note that this never happens, the sinusoids are displaced in phase. (c) In the third run, a stepping condition is placed on the new FMU, so that it cannot step until the outputs (steering angles) between FMUs match. The swap condition is the same as the stepping condition, therefore the swap happens when the outputs have matched. Figure taken from [2]

In our second example, an existing controller is updated with a new one that applies a speed cap to the speed provided by a user (Fig. 12.21). As can be seen in Fig. 12.22, the new controller is already available at time 6s, however the swap only takes place after the robot has reduced its speed under a predefined safety threshold. After that time, even if the user inputs a higher speed than the cap, the new controller will not allow said speed to be propagated to the robot. It will send the cap itself.

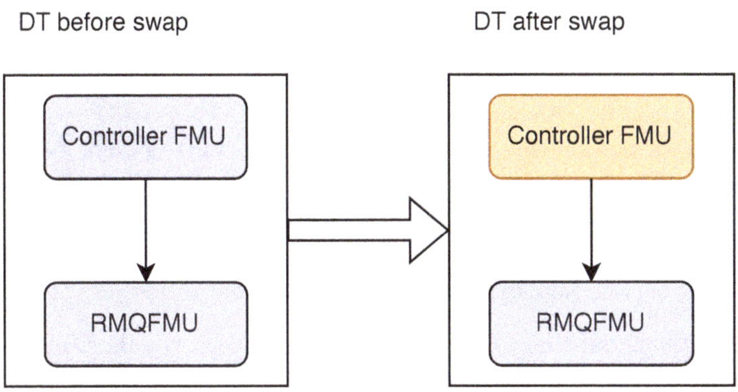

Fig. 12.21: Swapping controllers for the DR.

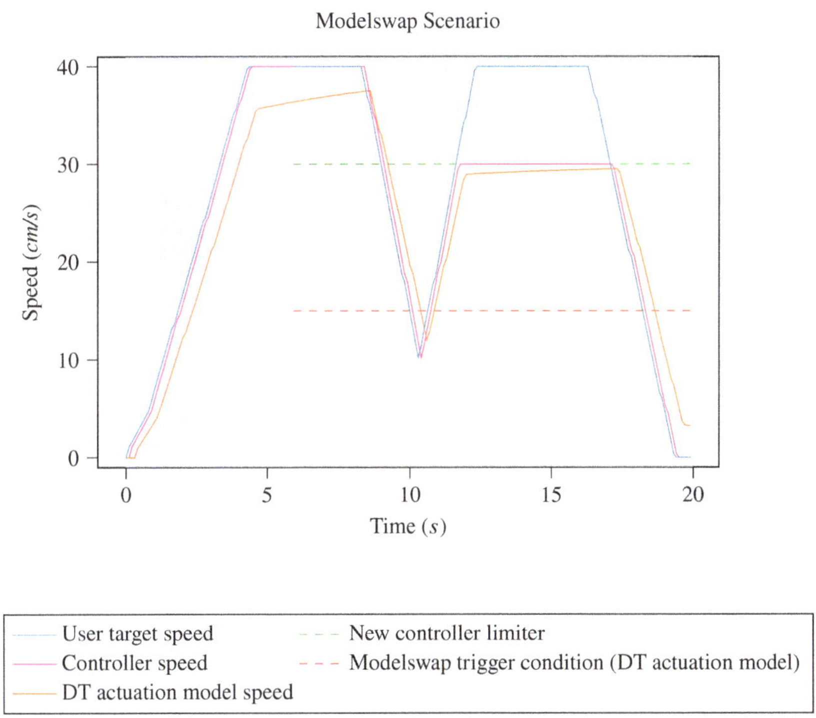

Fig. 12.22: Model swapping by replacing an existing controller with a new one that allows placing a cap on the speed of the DR. Figure taken from [2].

12.4.2.11 C11: Time-Scale

The model inside the DT can run faster than real-time, and while it is expected that it does so, there are currently no guarantees that it will. The speed of execution can be halted or slowed down by the rate of data coming in the DT through the RMQFMU. This is due to the RMQFMU requiring a constant feed of data in order to step. Any delay in this feed will leave the RMQFMU in a waiting mode until a certain timeout is reached, thus resulting in the whole DT being stuck.

In more detail, the Robotti DT operates at these time-scales:

- **Slower-than-real-time:** The services that need to go over the DT Manager and generate an output based on PT and DT data.
- **Real-time (best effort):** control commands, data recording, and visualisation.
- **Faster-than-real-time (best effort):** co-simulation based DT.

12.4.2.12 C12: Fidelity Considerations

This case requires medium to high-fidelity models, as they would be used to predict anomalies and divergences between predicted and actual robot positions, and possibly collision zones. If the models are not of sufficient fidelity, the DT cannot be trusted to send commands to the robots, as it might try to apply mitigation strategies where they are not needed, e.g., by stopping the robots as a precaution. On the other hand, it might react too late or not at all to dangerous situations, e.g., collisions, that would cause damage to the robots themselves, and perhaps others situated nearby.

With respect to this, a known issue is present in the kinematic model described earlier. There, the change in angle is assumed to happen instantaneously, which has contributed to the divergences observed in Fig. 12.16.

12.4.2.13 C13: Life-cycle Stages

The DT can be used throughout the *development* (consider different levels of FI simulations, depending on what hardware is available), and through *operation* (consider that FMUs used only for testing can be removed in the deployed DT, leaving only what is needed for the operation).

12.4.2.14 C14: Evolution

The evolution of the DR DT has been reported in several publications. The initial attempts, focused on the setup of the parallel operation [26]. Thereafter, the enhancement of the DT with fault injection capabilities was explored with the DR in the loop [10, 9]. The DT/PT system is affected by the quality of the communication, or lack thereof, which has an impact on the execution of the DT. To tackle this issue, the DT was extended with time discrepancy detection capabilities, allowing it to degrade to DS when there was a degradation in communication as discussed in Sect. 1.2.2 [8]. Finally, the DT was extended further with a model swapping capability, enabling it to swap models at run-time under configurable conditions [2].

12.4.3 Summary and Future Work

The DR case-study has shown proof-of-concept DTs for a mobile robot, covering functionalities such as: (i) parallel operation, where the DT continuously predicts where the robot should be as well as having access to the actual position data, thus laying the groundwork for monitoring and fault detection; (ii) fault injection, where different components in the DT can be injected with faulty values allowing different kinds of experiments with hardware in the loop, as close to the operational stage as possible; (iii) model swapping, where the DT is able during run-time to swap components, given that certain conditions – specified by the user – are satisfied.

The main challenge is related to the simulation-reality gap, and the development of high-fidelity models. Future work will focus on this aspect, as well as enhancing the DT to deal with a multi-robot system, providing functionalities like the evaluation and mitigation of collision zones between robots during deployment. Additionally, fault detection and diagnosis shall be explored, with relevant techniques to be implemented in the DT such that scenarios where, e.g., a robot gets a flat tire, can be identified and handled properly. Finally, the fault injection framework could be extended to allow for fault injection at the DT and the PT levels simultaneously, allowing for richer experimental setups for testing the adopted fault tolerance mechanisms. Fault injection could also be used in the context of what-if simulations, by providing possibly useful insights to the fault diagnostics.

12.5 The Flex-cell

The flex-cell is a manufacturing cell composed of two robotic arms used for cooperative assembly. A key feature is that the two arms are different and from diverse manufacturers. As an academic DT engineering case study, it provides opportunities for researchers to explore:

- composable DTs;
- cooperative systems with synchronous motions and coupled models;
- multiple abstraction levels of attributes and operations of both robots and grippers;

- complex system with kinematic and dynamic models for the behavioural aspects in the cooperative setup; and
- multiple robotics applications that include different operations, such as pick-and-place, peg-in-hole, optimisation in assembly, payload estimation, online recalibration and collision detection/avoidance.

As shown in Fig. 12.23, the PT is composed of four main assets: a Kuka LBR iiwa 7 robotic arm; a UR5e robotic arm; an OnRobot RG6 gripper; an OnRobot 2FG7 gripper. The flex-cell DT's purpose is to assist the PT with the features listed above. The applications that are run on the flex-cell are based on a 2D cooperative assembly where the scope is related to robot positioning. For this. the real workspace (x, y, z) of the flex-cell plate is discretised to (X, Y, Z), where the discrete positions use the plate holes, creating a grid of 16x24 holes separated by 5mm each; the Z axis is also discretised using a 5mm distance between points.

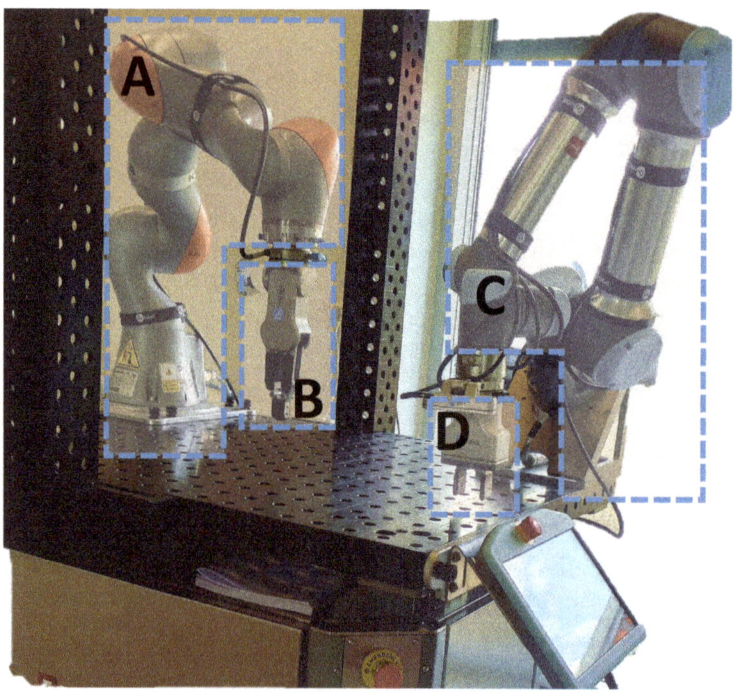

Fig. 12.23: Flex-cell and its components: robotic arms Kuka LBR iiwa 7 (A) and UR5e (C), and grippers OnRobot RG6 (B) and 2FG7 (D).

12.5.1 Flex-cell DT Overview

The flex-cell DT has been approached from a composable perspective using the modelling approach proposed in [11], therefore, the four main assets, i.e., robotic arms and grippers, have a sub-DT representation in the final composed DT. More precisely, using such a modelling approach, each asset is represented as a DT defined by *Attributes*, *Operations*, *Behaviours*, and *Relationships*. In this approach, composition is achieved by the particular relationship *isComposedOf*, which requires the subject (the parent, i.e., the composed DT) and the objects (children, i.e., smaller DTs). For the flex-cell case study, the flex-cell DT instance *isComposedOf* the DT instances of the Kuka LBR iiwa 7, the UR5e, the gripper OnRobot RG6, and the gripper OnRobot 2FG7.

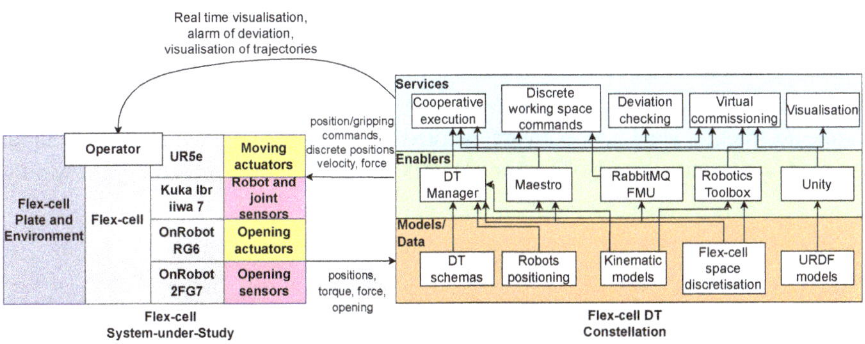

Fig. 12.24: Constellation of the flex-cell DT.

Fig. 12.24 illustrates the constellation of the flex-cell DT. Only the two robotic arms are provided with simulation models, while the grippers only contain the static and structural information of their PTs. Moreover, the simulation models for the robotic arms only contain the kinematic models; these are created using the Robotics Toolbox [1] and embedded into FMUs via the UniFMU tool [23].

For the co-simulation of the flex-cell we use Maestro and the RabbitMQFMU [7]. The purpose of the co-simulation and the RabbitMQFMU is to enable synchronised motions of both robots in the real and virtual settings. In this sense, RabbitMQFMU replicates the commands of the actual controller to the robots during runtime. Fig. 12.25 shows a representation of the co-simulation setup of the flex-cell to execute synchronous and cooperative commands in the flex-cell system.

We use the *DT Manager* approach proposed in [24] to handle the connectivity between PT and DT and the game engine Unity[6] along with Unified Robotics Description Format (URDF) models for the robots and grippers to generate the

[6] `https://unity.com/`

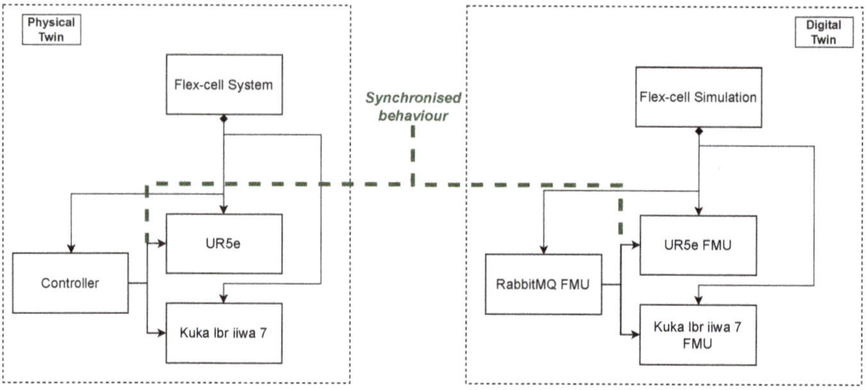

Fig. 12.25: Co-simulation setup of the flex-cell DT.

visual representation of the DT of the flex-cell shown in Fig. 12.26. The DT Manager provides an interface to administer DTs through (i) abstracting PTs and DTs at the same abstraction level, i.e., both are handled as *Twins* and (i) providing the methods to read, write and execute operations on PTs and simulation units straightforwardly.

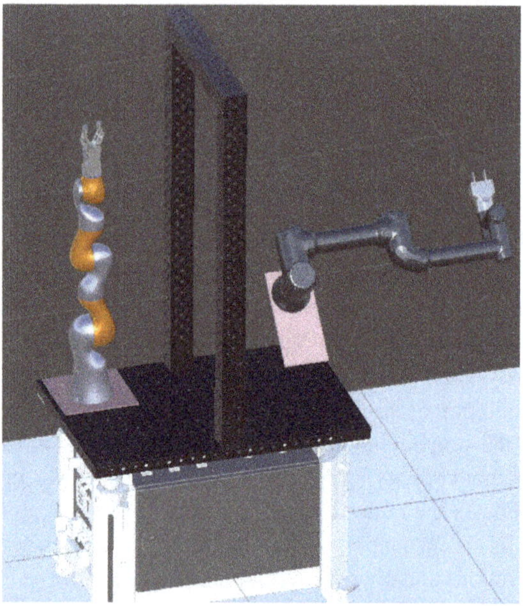

Fig. 12.26: Flex-cell graphical representation in Unity.

12.5.2 Fundamental Characteristics

This section presents the flex-cell case study through the 14 fundamental DT characteristics as introduced in Sect. 12.2.

12.5.2.1 C1: System-under-study

The *flex-cell system* itself is the manufacturing cell PT described above with the two robotic arms with attached grippers. The *environment* is a plate with 16 × 24 holes where there are bricks and pins/pegs to place on the holes. The environment also has a safety system, which is activated in case of a collision or an abnormal condition. The safety system also has a proximity sensor which is enabled to stop the robots if anyone gets close to the working space. Each robotic arm has a "teach pendant" allowing an *agent*, a human operator, can manipulate the robots or set them to work with remote control for automatic execution. The sequences to assemble a shape are executed remotely from a PC by the human operator.

12.5.2.2 C2: Acting Components

The acting components in the system are the two robotic arms, the two grippers, and the safety system. These are provided with network interfaces, which allow the execution of actions with their particular APIs, either through Transmission Control Protocol (TCP) for the robotic arms or ModbusTCP for the grippers. The commands sent to the robotic arms are processed by their internal controllers to perform different motion types, such as cartesian, joint, linear, and circular movements, and update their parameters, such as speed, acceleration, etc. The teach pendants can also be used for manually moving the robots and updating their configurations. Similarly, the commands to the grippers are processed by their internal controllers, where the human operator can update the default values of width/opening of the gripper and the gripping force. The safety system simply stops the robots if there is a collision or a person tries to get close to the working space.

12.5.2.3 C3: Sensing Components

Regarding the sensors, the robotic arms contain several internal sensors from which it is possible to collect data, such as current position, target position, torque, current, voltage, speed, acceleration and so on. Similar to the acting components, the assets are provided with network interfaces to allow the collection of data with their particular APIs, either through TCP (arms) or ModbusTCP (grippers). Some of the variables for the robot contain their angular and cartesian value, and there are some variables that are specific for each joint and some that are for the whole robotic arm. We can collect 117 and 31 variables for the UR5e robot and the Kuka lbr iiwa 7 robot respectively, although not all of them are used. For the grippers, it is possible to collect force and width/opening values. The safety system has a proximity sensor and internal flags in case any of the robotic arms collide.

12.5.2.4 C4: Multiplicities

The flex-cell DT is addressed with the composable modelling approach presented in [11]. Therefore, the DT-enabled system contains seven DT instances, namely, one DT per asset in the flex-cell system, two intermediate composition of robotic arm + gripper (one for the Kuka LBR iiwa 7 + OnRobot RG6 and one for the UR5e + OnRobot 2FG7), and a final composition of the flex-cell system which composes the two intermediate compositions. Through this modelling approach, the multiple DTs and their corresponding PTs can be handled at the higher hierarchy level, in this case, the flex-cell DT, by disaggregation. Although there are multiple instances of DTs, there are no multiplicities of the same DT class since the flex-cell system is built with different assets.

12.5.2.5 C5: Data Transmitted

The transmission of data between PT and DT and vice-versa is carried out by the DT Manager approach [24], which provides the methods to read, write, and execute operations on the endpoints of PTs and DTs. More precisely, reading from an endpoint is possible through the method `getAttributeValue(s)`, writing to an endpoint is possible through the method `setAttributeValue(s)`, executing operations on an endpoint is possible through the method `executeOperation`. These three methods enable the bidirectional communication between PTs and DTs as follows: The method `getAttributeValue(s)` provides the interface to read data from the twins and the methods `setAttributeValue(s)` and `executeOperation` provide the interface to write data to and execute operations on the twins. In terms of sending data to PTs, the method `setAttributeValue(s)` is preferred to send *insights*, such as parameter updates, whereas the method `executeOperation` is preferred to execute direct actions or commands. These methods work on both PTs and DTs since both are categorised at the same abstraction level, so they are covered by the same class. It is therefore possible to use these methods on both equally, depending on the variables and operations they actually have. For the PT, any of the available variables can be transmitted, and the same with the different available move commands. For the DT, this is limited to what the models support and which variables they can provide. With the kinematic models that we use, it is possible to get position variables and execute angular and cartesian motions. However, speed, acceleration and torque data are not available for those specific models.

12.5.2.6 C6: Insights/Actions

As *actions*, the DT Manager can perform moving commands on both the PT and the DT. The DT provides the joint angles to the PT when the move commands are based on the flex-cell discrete working space. Similarly, if DT services are added on top of the DT Manager, *insights* are generated, such as for example, deviation checking or discrete collision detection in the simulation. The visualisation with Unity (see Fig. 12.26) also provides insight into how the trajectory will be performed by the real robots. Refer to Sect. 8.4 for further information on visualisation.

12.5.2.7 C7: Services

The services that the DT currently provides for the flex-cell are:

- *Cooperative execution* that can perform moving commands on both robots on both physical and virtual versions synchronously.
- *Trajectory visualisation* to help the user understand how the robots are going to move for a certain move command in the flex-cell working space (see Sect. 8.4 for further information on visualisation).
- *Discrete flex-cell working space-related commands* that translate the target (X, Y, Z) position in the flex-cell to joint angles of the robots.
- *Deviation checking* is used to compare that the physical and the virtual robots are moving similarly as theoretically expected (see Sect. 9.4 for further information on *anomaly detection*).
- *Virtual commissioning* where the DT is decoupled from the PT to perform *what-if* experiments and validate on the virtual flex-cell before deployment to the physical one (see Sect. 10.2 for further information on *what-if simulations*).

12.5.2.8 C8: Enablers

The following *enablers* support the flex-cell DT services:

- The DT Manager to provide the access and interfaces to PT and DT, and to the DT services.
- The Robotics Toolbox and UniFMU to encapsulate the kinematic models of the robotic arms.
- The urinterface, kukalbrinterface, ModbusTCP, and MQTT to access the physical robots and grippers and bridge the communication to external networks.

- Maestro and the RabbitMQFMU to perform the synchronous and coupled co-simulations.
- Unity, ZeroMQ and URDF are joint enablers for displaying visualisations.

12.5.2.9 C9: Models and Data

The flex-cell DT contains the following:

Data models. There is a data model for each robotic arm and gripper, which handles the *attributes* and *operations* of the DT using schemas. The data models are also extended with the definitions for *behaviours* and *relationships* as presented in [11].

Structural models. There is a structural model for each asset in the flex-cell system, which are used for visualisation. These models use the URDF format, which contains the descriptions of the rigid body, links, and joints.

Behavioural models. Additionally, for the robotic arms there are behavioural models as follows: (i) kinematic models using the Denavit-Hartenberg parameters; (ii) dynamic models with the Newton-Euler formulation (not being used). There is also a calibrated dynamics model for the UR5e robotic arm generated with the AURT tool [27]. The data of interest is contained in the DT schemas and the available outputs from the behavioural models in use, which are related to the positioning of the robotic arms.

12.5.2.10 C10: Constellation

We can think of the services offered in terms of vertical slices through the layered structure shown in Fig. 12.24. In a scenario for *cooperative execution*, *deviation checking*, and implicitly *discrete working space commands*, Fig. 12.27 shows one of the experiments of running coupled cooperative motions along with the co-simulation. The figure shows the joint positions over time of the two robots. The experiment consists of moving the robots between two different discrete waypoints. The simulation is initialised from zero values, but the real system starts from its previous position. Therefore the models and the real robots converge after the models have reached the first target position.

In the scenario for *virtual commissioning and visualisation*, Fig. 12.26 shows the graphical representation of the flex-cell system in Unity. The Unity app embeds the URDF models and the incoming messages from either the flex-cell simulation or real flex-cell are translated into motions, enabling the visualisation of the system as a DT. When using the simulated flex-cell, this setup is also used for virtual commissioning in which it is possible to see the trajectories of the robotic arms before deployment.

In the scenario for *deviation checking*, a monitoring function is used that checks for deviations periodically. This function takes as inputs the DT and PT, the variable

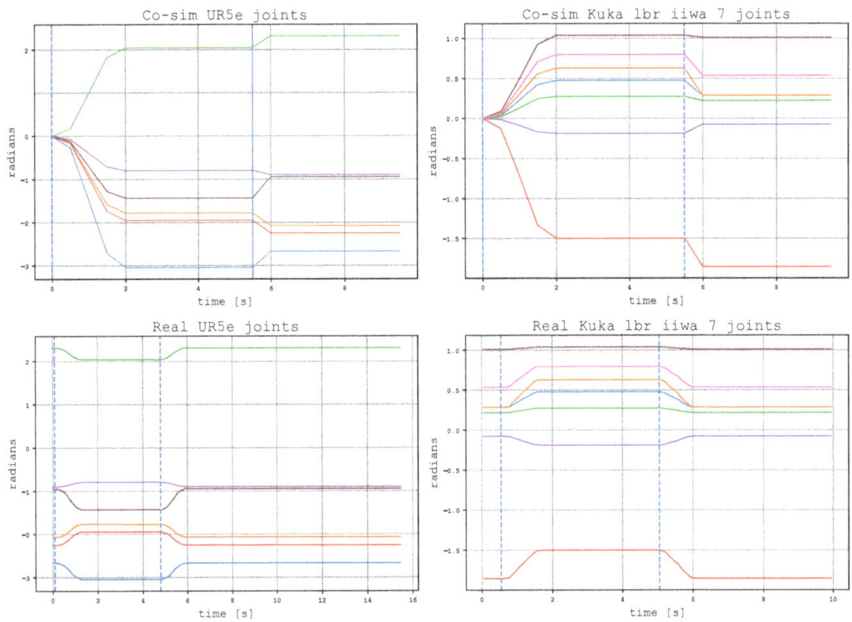

Fig. 12.27: Plot of the joint positions of the robotic arms in the cooperative co-simulation setting. The vertical dashed lines indicate the invocation of moving commands. Each coloured line represents the angle of one of the robot's joints.

of interest, and the tolerance, and returns *true* when there is a deviation in the variable between the two objects given the tolerance, otherwise *false*.

12.5.2.11 C11: Time-Scale

The DT for the flex-cell operates at different rates:

- **Slower-than-real-time:** The services that need to go over the DT Manager and generate an output based on PT and DT data.
- **Real-time:** Control commands, data recording, and visualisation.
- **Faster-than-real-time:** For virtual commissioning, and so the DT is decoupled from the PT to run independent simulations and scenarios.

12.5.2.12 C12: Fidelity Considerations

Medium- to high-fidelity models are required to deliver good performance for the provided services, especially those related to the positioning of the robotic arms.

Note that insufficient fidelity may cause the robots to collide with and damage static objects like the flex-cell structure, with other dynamic objects like the other robotic arm or grippers, and with humans.

As downsides, there are the following:

- The kinematic models do not include the grippers as attached on the end-effector of the robotic arms.
- The trajectory generation with the kinematic models provides certain time behaviour based on an assumed motion speed. However, it does not consider the actual dynamics of the robots, and so, it can have medium to high time offsets. Higher fidelity models for the actual dynamics may overcome this problem.

12.5.2.13 C13: Life-cycle Stages

The services provided on top of the flex-cell DT include the *design*, *manufacturing*, and *service* life-cycle phases for DT applications proposed in Liu et al. [25]. The manufacturing and service phases are administered by the DT Manager and the DTaaS platform (see Sect. 11.7). For the design phase, the virtual commissioning service of the DT is used, i.e., decoupled from the PT, to validate changes to the applications to be run in the physical system.

12.5.2.14 C14: Evolution

The flex-cell case study has evolved considerably during development. Some of the evolution steps are reported in [11, 13]. The most representative milestones are as follows:

1. The DT schema-based object models for robotic arms and grippers were designed.
2. The object models were coupled to the DT Manager architecture.
3. The visualisation with Unity and the URDF models were created.
4. The composable representation of the flex-cell was established.
5. The kinematic and dynamic models for the robotic arms were designed.
6. The space and mapping model for the flex-cell was created.
7. The execution of online PT-to-DT movements with individual non-cooperative motions was achieved.
8. The first experiments of applications for the flex-cell discrete workspace using the DT as the virtual commissioning mechanism were successful.
9. The skill-based engineering approach was integrated to disaggregate the broader term *operations* into *device primitives*, *skills*, and *tasks*.
10. The kinematic models were embedded into FMUs.

11. The co-simulation with Maestro and RabbitMQFMU was added to provide simulation capabilities of the coupled cooperative motions.
12. Synchronous cooperative experiments are successful and used for deviation checking.
13. The flex-cell DT components are uploaded to the DTaaS platform (see Sect. 11.7 for further information on the DTaaS platform).

12.5.3 Summary and Future Work

The flex-cell case study provides a representative system for DT engineering in cooperative robotics, being composed of two robotic arms of different brands, with joints and grippers. It therefore gives an opportunity to explore the composition of DTs and models. The services for this case study are handled on its working space, i.e., the plate, mostly with a scope in robot positioning. There are multiple enablers for the modelling, communication and simulation aspects, such as the Robotics Toolbox and UniFMU for integrating the kinematic models of the robots, Maestro and RabbitMQFMU for running co-simulations, the DT Manager for administrating the interfaces and Unity and URDF for the visualisation components.

The most challenging aspect of this case study the integration of the models in a cooperative-synchronised setting. This is because there are multiple modelling approaches for robotic arms for different components, e.g., kinematics, dynamics, visualisation, programming. This makes it difficult to integrate all of them into a single DT if the scope is broad. Thus, it is better to narrow the scope to the applications that are needed, and hence choose the appropriate models for that scope. The current scope of the flex-cell is related to positioning, but it may be widened in the future to cover other relevant aspects.

The case study still evolves, mostly in terms of applications and services, such as collision detection, optimisation of assembly, and flexible and reconfigurable sequences. Additional models for other applications, such as payload estimation, are also possible evolution scenarios. The ongoing work of the flex-cell DT focuses on the generation of sequences for flexible manufacturing in a cooperative setting by integrating the skill-based programming approach [32].

12.6 The Research Vessel Gunnerus

The R/V Gunnerus, shown in Fig. 12.28, is a research vessel owned by the Norwegian University of Science and Technology (NTNU) that is used to support marine research and exploration. The vessel is equipped with the latest technology for a variety of research activities in biology, technology, geology, archaeology, oceanography and fisheries [33]. In addition to research, the ship is used for educational purposes and is an important platform for marine courses at all levels and disciplines including nautical operation and bio-engineering. The main dimensions of the vessel are given in Table 12.2. The vessel is arranged with a wet lab, dry lab and a computer lab in addition to a large aft deck. Accommodation comprises three double berth scientific personnel cabins and three single berth crew cabins. The large mess hall functions as a lecture room for 25 people.

Fig. 12.28: Starboard view of the R/V Gunnerus. © 2024 Fredrik Skoglund, reprinted with permission.

This case study makes use of the Gunnerus as the test-bed to demonstrate a sound conceptual framework that uses co-simulation as a fundamental technology *towards* realising a DT for ship manoeuvring. The current state of the study does not involve automatic data transmission back to the PT, thus making it a DS as discussed in Sect. 1.2.2 or "open-loop" case study. Based on this system, however, a lot of interesting research work has been implemented, including physics-informed data-driven approach for ship docking prediction [38], co-simulation-based pre-training of ship trajectory prediction [19], and incorporating approximate dynamics into a data-driven calibrator for ship manoeuvring [37].

Table 12.2: Main dimensions of the R/V Gunnerus.

Parameter	Value
Length overall (Loa)	36.25 m
Length between pp (Lpp)	33.90 m
Waterline length (Lwl)	29.90 m
Breadth middle (Bm)	9.60 m
Breadth extreme (B)	9.90 m
Depth mld. Main deck (Dm)	4.20 m
Draught, mld (dm)	2.70 m
Deadweight	165t

12.6.1 Fundamental Characteristics

This section presents the Gunnerus case study through the 14 fundamental DT characteristics introduced in Sect. 12.2.

12.6.1.1 C1: System-under-study

The PT in this case is the R/V Gunnerus. The environment it typically operates in is the sea off the west coast of Norway, with the ship's home port located in Trondheim. Ship operations in this area are affected by rapid variations in water depth and the potential obstruction of wind flow by high mountains in local areas. These effects introduce significant danger and make the physical operation and testing of the PT challenging. The agents include the ship and its crew, other vessels, the sea traffic centre and possibly the presence of a *Remote Operating Centre* (ROC).

The ship is regularly in operation, including navigation, Dynamic Propulsion (DP) operation, docking, and lifting, etc. It is fitted with a dynamic positioning system and a HiPap 500 unit to provide optimal conditions for remote operated vehicle operations and the positioning of any deployed equipment. The onboard diesel electric system has been specially designed for low hydro-acoustic noise levels to accommodate testing and development of hydro-acoustic equipment. In C7, we discuss the DT/DS with regard to ship steering. For docking and DP operations, the ship dynamics will be different from normal speed ship manoeuvring. Otherwise, the realisation of the DT/DS will be the same.

12.6.1.2 C2: Acting Components

In order to transmit data from the PT back to shore, a special-purpose PC has been installed onboard that is able to collect key information from the various systems onboard. As the ship operates offshore, data must be transmitted wirelessly over a cellular 4G/5G connection. Thus, Message Queuing Telemetry Transport (MQTT), a standard messaging protocol for the IoT, is used to transmit data onshore as it saves bandwidth by having a central broker relay data to potential clients onshore. This central broker needs to be reliable, so a virtual Linux server monitored and serviced by the NTNU IT department has been setup.

12.6.1.3 C3: Sensing Components

As a research vessel, there are many data acquisition systems installed onboard the Gunnerus, such as thrusters, power generators, navigational sensors and motion control systems, etc. The number of measurable sensor channels is high and for the purpose of motion prediction, we are interested in the data originating from GPS receiver, a Motion Reference Unit (MRU), compasses, a wind sensor and sensors reflecting orientation and rotational speed of thrusters. The sensor channels related to motion prediction are listed in Table 12.3. For all measurements in the data set, a sampling rate of 1 Hz is used.

Table 12.3: Data channels sensed onboard the R/V Gunnerus.

Signal	Channels	Unit
GPS	Latitude	ddmm.mmmm
	Longitude	ddmm.mmmm
	Surge velocity	knots
	Sway velocity	knots
	Course angle	deg
	Speed over ground	knots
MRU	Heading angle	deg
	Heading rate	deg/s
	Roll angle	deg
	Pitch angle	deg
	Heave displacement	m
	Roll rate	deg/s
	Pitch rate	deg/s
	Heave rate	m/s
Wind sensor	Wind direction	deg
	Wind speed	knots
Thruster	Port thruster rotational speed	%
	Port thruster angle	deg
	Starboard thruster rotational speed	%
	Starboard thruster angle	deg
	Tunnel thruster rotational speed	%

12.6.1.4 C4: Multiplicities

Currently there are two DTs implemented. One DT is for decision support for ship motion prediction and docking operations. The other DT is for system diagnostics for thruster failures detection and isolation for the ship in dynamically positioned operations for the same PT.

12.6.1.5 C5: Data Transmitted

Today's maritime engineering systems are often equipped with various sensors and operating in highly dynamic environments. All these data sources, including the status from ship, crane and engine, can be transferred via a 4G connection to an onshore control center to conduct operational planning. A PC is deployed on Gunnerus, from where the MQTT network protocol is used to publish data messages from different data sources as topics. Finally, the data is applied and analysed on the DT and could be transferred back to Gunnerus to enhance operational planning and decision-making. Relevant data is also stored to a database.

12.6.1.6 C6: Insights/Actions

Insights: The system can be used to visualise the current and historical state of the system in 2D and 3D using easily accessible web-services. Furthermore, the system can be used to provide on-board decision support for the crew and provide diagnostic services such as Remaining Useful Life (RUL) for serviceable equipment as in Fig. 12.29.

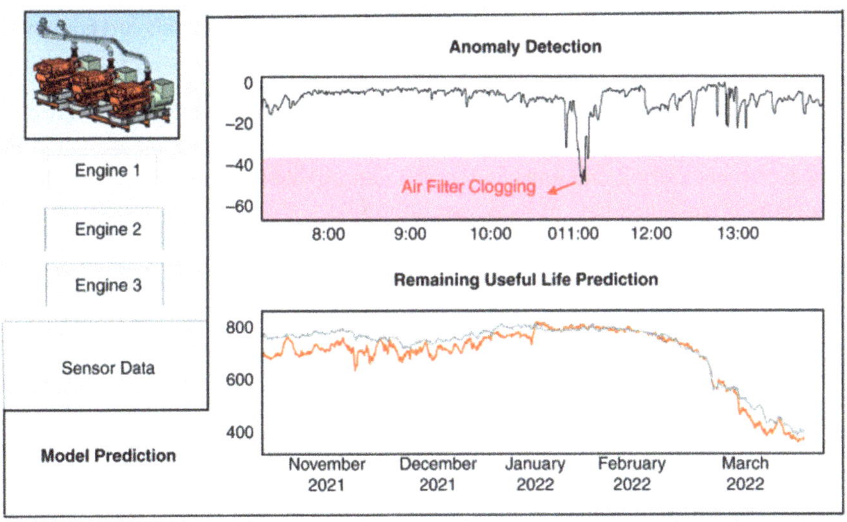

Fig. 12.29: Anomaly detection and fault prognostics of the vessel's engines. Figure taken from [39].

Actions: The PT/DT connection is currently running in an "open-loop" manner, where no automatic actions have been implemented yet.

12.6.1.7 C7: Services

Offshore vessels often perform in harsh environments with wind, waves, currents and low temperatures. These operations need to be carefully planned and rehearsed before execution. For this purpose, the DT/DS of the whole ship are used as a valuable tool for pre-operation and planning. It will allow the iterative design of operational procedures by identifying inconsistencies and blunders, exploring "what if" scenarios during onshore planning sessions based on historical methodological ocean records and raising awareness on critical phases. It can also be used in onboard "toolbox talks" just before the execution of an operation to rehearse the procedure, based on the current ship loading conditions and the short-term predictions of methodological ocean conditions.

As a short summary, the services that the DT/DS currently provides are:

1. Providing *what-if simulations* such that the user can enter in proposed conditions and examine the resulting behaviour (see Sect. 10.2).
2. Real-time *visualisation* for the stakeholders to understand the current state of the ship (see Sect. 8.4).
3. Providing extra sensory information about the current environment by comparing the simulation state with the real world state to *estimate the effect of environmental disturbances*.
4. Providing *anomaly detection* and *estimation of RUL* for the onboard equipment (see Sect. 9.4).

Different offshore operations have various features and requirements, and are quite demanding. The DT/DS currently provides valuable services for these operations. However, there are still ongoing challenges in improving the utility of these services, especially in the dimension of increased fidelity.

12.6.1.8 C8: Enablers

The case study relies on co-simulation frameworks such as the Open Simulation Platform (OSP) to coordinate the co-simulation of models [39]. A prognostics and health management system supports the RUL service, while a visualisation enabler supports the visualisation service.

12.6.1.9 C9: Models and Data

The following list briefly describes each of the FMUs used to create a DS of the Gunnerus.

Gunnerus: This model either contains historical sensor data for analytic purposes or access to live data through a MQTT connection. The data is sampled at 1Hz and includes information such as:

- Heading angle and percent-wise commanded Revolutions Per Minute (RPM) of the tunnel-thruster in the bow as well as the two azimuth thrusters in the aft.
- Longitude and latitude.
- Surge, sway, and heave.
- Yaw, pitch, and roll.
- Wind direction and speed.
- Positional and rotational velocities.

The model implements linear interpolation, which is convenient given the low sample rate of the sensors relative to the simulation, which runs at 20 Hz.

VesselModel: This model computes the vessel hydrodynamics such as the radiation forces, mass, and restoring forces as well as manoeuvring forces (resistance and cross flow drag as well as semi-empirical corrections). The equations of motions are solved by this model, summing up all the external forces acting on the vessel. The model was established during the SimVal [14] project, where the internal dynamics where calibrated based on standardised ship trails.

PMAzimuth: The hydrodynamic model of the azimuth thrusters without actuator/motor. Given a certain RPM command (issued by the *ThrusterDrive* FMU), location on the hull, azimuth angle, vessel speed, and the loss factor, the model will output the 3DoF (surge, sway, heave) force generated.

ThrusterDrive: A drive that converts force commands from the *SpeedController* into RPMs for the *PMAzimuth*.

PowerPlant: A marine power plant with two equally large gensets, including auxiliary load and circuit breakers [34].

SpeedCont: A controller used to regulate the force required by the *ThrusterDrives* so that the speed of the *VesselModel* and the vessel are aligned.

HeadingController: A controller that is used to align the heading of the *VesselModel* and the vessel by regulating the angle of the *PMAzimuths*.

VesselModelObserver: A simple model that computes the direction of travel and speed over ground of the *VesselModel* based on its current and previous position.

An overview of the models are also given in Table 12.4.

Additionally, CAD drawings and simplified 3D models as well as ample historical data in CSV format exists for the PT. This data is available for research internally in NTNU, but the data is proprietary and cannot be made public.

Table 12.4: FMUs involved in the DS for ship manoeuvring.

Component	Tool	Vendor	FMI version
HeadingController	FMI4j	NTNU	2.0
SpeedController	FMI4j	NTNU	2.0
Gunnerus	FMI4j	NTNU	2.0
VesselModelObserver	FMI4j	NTNU	2.0
ThrusterDrive	20sim	SINTEF Ocean	1.0
PowerPlant	20sim	SINTEF Ocean	1.0
VesselModel	VeSim	SINTEF Ocean	1.0
PMAzimuth	VeSim	Kongsberg Maritime	1.0

12.6.1.10 C10: Constellation

Fig. 12.30 shows the information flow between the models in the case study. As illustrated, the system is non-trivial with a total of 48 variable connections. Moreover, the VeSim and 20sim model instances cannot co-exist within the same process as other instances generated from the same tool due to shared library symbol conflicts. To overcome this challenge, execution of these must be split across processes.

12.6.1.11 C11: Time-Scale

The DS for the vessel operates at different rates:

- **Slower-than-real-time:** The anomaly alerts and dashboard information provided to the user may not be acted upon in real-time.
- **Real-time:** Dashboard updates.
- **Faster-than-real-time:** The simulator is able to simulate into the future.

As for the co-simulation time-step required to successfully run the simulation, 20Hz has been determined to be adequate as the system is fairly loosely coupled. In effect, this allows the system as a whole to run faster than real-time even on laptops with moderate specifications. The main bottleneck with regard to simulation speed is the fact that some of the models involved must communicate between processes due to implementation details related to shared library conflicts. The underlying issue is that a shared library can only be loaded once within a single process. If two models, or even two instances of the same model, needs to dynamically load the same shared library during runtime, then the code cannot be executed within one logical process.

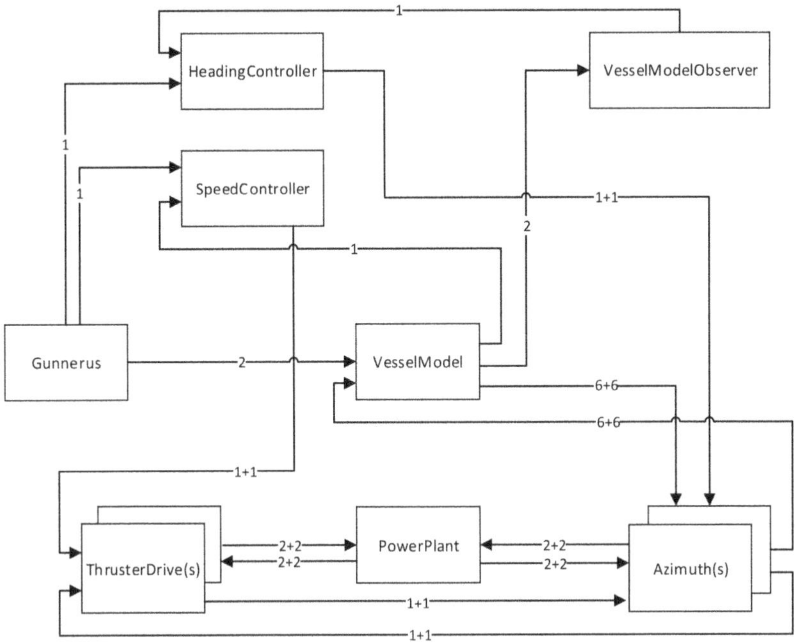

Fig. 12.30: Connection graph for the ship manoeuvring components for illustrative purposes. Numbers indicate the number of connected variables.

12.6.1.12 C12: Fidelity Considerations

The system requires high fidelity of the physics models, as the they would be used to predict anomalies and divergences between predicted and actual position of the ship. However, the complexity of the system could be decreased by removing certain behaviour models depending on the phenomena to be observed, i.e., removing the *ThrusterDrive* and *PowerPlant* models if one only wants to observe the hull characteristics.

12.6.1.13 C13: Life-cycle Stages

The DT for the ship is an open marine industrial infrastructure that serves multiple purposes. Firstly, it facilitates the overall system design by allowing system configuration and verification of operational performance. Additionally, it places significant emphasis on providing early warning capabilities, life cycle service support, and predicting system behaviour:

Design phase: in this phase, the ship could be modelled in a whole piece including all necessary important systems, so that the designer could find out if the design meets all technical requirements.

Manufacturing phase: all models and sub-models should be with high fidelity in order to make sure the single system with prefect performance. This phase should guarantee all components, systems running as expected. Depending on the complexity level, different models implementations contain different sets of properties describing the behaviour of the physical system.

Operation phase: during the operational phase, the DT becomes a system for integration, processing and analysis of the operational data. In this phase, many operational models in the DT ship could be light-weight and real-time so that ideally, the DT will provide decision support and predictions for a ship owner, subsystem manufacturers and safety training/improvement for crew and operators.

Maintenance and service: the DT tracks information on all parameters to define how each individual module and sub modules behaves over its entire useful life, including the initial design and further refinement, manufacturing related deviations, modifications, uncertainties, updates as well as sensor data from on-board systems, maintenance history and all available historical and fleet data obtained using data mining. All outcomes will be further used for ships that are either autonomous or remote-controlled for safety and reliability enhancement.

12.6.1.14 C14: Evolution

Table 12.5 details the reports on the development of this DT/DS.

Table 12.5: TwinShip articles and their topic.

Year	Topic	Citation
2019	Estimating useful life predictions	[3]
2020	Co-simulation framework to simulate Gunnerus case-study	[15]
2020	Gunnerus DT within a co-simulation environment	[16]
2021	Development of a ship trajectory predictor	[20]
2022	Improving trajectory prediction with a physics-based model	[22]
2023	Refining models between different ship size	[21]
2023	Enabling technologies for the Gunnerus DT	[39]

12.6.2 Summary and Future Work

This section has illustrated the research activity on a DT of the Gunnerus ship, especially focusing on co-simulation implementation. Key elements necessary for the success of this case-study include:

1. An onboard PC for collection and transmission of data.
2. A database with historic data for analytics.
3. High-fidelity models of the ship hull and internal systems, which is able to simulate the behaviour of the ship forward in time and perform what-if experiments.
4. Cellular 4G/5G coverage at sea as well as a reliable MQTT broker.
5. Real-time dashboard for information purposes.
6. 3D models for visualisation.
7. As some of the models involved depend on the same shared library, the simulation must be executed in part over multiple logical CPU processes. This can be solved by using a co-simulation orchestrator with distributed simulation facilities or using a wrapper solution like FMU-proxy [17] for the affected models.

Many challenges arise when trying to fully represent the physical counterpart by the twin system in a stochastic environment:

- **Developing high-fidelity models:** As seen in the Gunnerus twin ship case, the amount and quality of data is fundamental to establish high-fidelity models for the twin system, e.g., either for identification of system parameters, or for development of ML models.
- **Establishing a mechanism for model updating:** It is necessary to establish mechanisms to update models regularly, especially for these ML models. Thus, the computational cost of ML models, such as the training time for the model, must be relatively low, so that the model is able to adapt to operation/environmental changes.
- **Co-simulation:** Another challenge lies in co-simulation for DT implementation. Scalability must be considered when it comes to the industrial use of DTs. Insufficient synchronisation and tight coupling issues may arise if the system becomes complex and requires high computational effort. Domain-specific challenges in FMU implementation also need to be addressed.

The team at NTNU with our cooperated stakeholders will continue the work towards realising a full functional DT of the Gunnerus, gradually improving its accuracy. Continued development of the use-cases will provide meaningful on-board decision support for the crews of the Gunnerus.

References

1. Corke, P., Haviland, J.: Not your grandmother's toolbox–the Robotics Toolbox reinvented for Python. In: ICRA, pp. 11357–11363. IEEE (2021)
2. Ejersbo, H., Lausdahl, K., Frasheri, M., Esterle, L.: Dynamic Runtime Integration of New Models in Digital Twins. In: 2023 IEEE/ACM 18th Symposium on Software Engineering for Adaptive and Self-Managing Systems (SEAMS), pp. 44–55 (2023). DOI: 10.1109/SEAMS59076.2023.00016
3. Ellefsen, A.L., Bjørlykhaug, E., Æsøy, V., Ushakov, S., Zhang, H.: Remaining useful life predictions for turbofan engine degradation using semi-supervised deep architecture. Reliability Engineering & System Safety **183**, 240–251 (2019)
4. Feng, H., Gomes, C., Thule, C., Lausdahl, K., Iosifidis, A., Larsen, P.G.: Introduction to Digital Twin Engineering. In: 2021 Annual Modeling and Simulation Conference (ANNSIM), pp. 1–12. IEEE. DOI: 10.23919/ANNSIM52504.2021.9552135
5. Feng, H., Gomes, C., Thule, C., Lausdahl, K., Sandberg, M., Larsen, P.G.: The Incubator Case Study for Digital Twin Engineering. URL http://arxiv.org/abs/2102.10390
6. Foldager, F., Balling, O., Gamble, C., Larsen, P.G., Boel, M., Green, O.: Design Space Exploration in the Development of Agricultural Robots. In: AgEng conference. Wageningen, The Netherlands (2018)
7. Frasheri, M., Ejersbo, H., Thule, C., Esterle, L.: RMQFMU: Bridging the Real World with Co-simulation For Practitioners. In: H.D. Macedo, C. Thule, K. Pierce (eds.) Proceedings of the 19th International Overture Workshop. Overture (2021)
8. Frasheri, M., Ejersbo, H., Thule, C., Gomes, C., Kvistgaard, J.L., Larsen, P.G., Esterle, L.: Addressing time discrepancy between digital and physical twins. Robotics and Autonomous Systems **161**, 104347 (2023). DOI: https://doi.org/10.1016/j.robot.2022.104347
9. Frasheri, M., Lumer-Klabbers, G., Kvistgaard, J.L., Hausted, J.O., Thule, C., Esterle, L., Larsen, P.G.: Building safer robots: From simulation to hardware deployment. In: 2023 27th International Conference on Methods and Models in Automation and Robotics (MMAR), pp. 63–68 (2023). DOI: 10.1109/MMAR58394.2023.10242467
10. Frasheri, M., Thule, C., Macedo, H.D., Lausdahl, K., Larsen, P.G., Esterle, L.: Fault injecting co-simulations for safety. In: The 5th International Conference on System Reliability and Safety ICSRS'21 (2021)
11. Gil, S., Mikkelsen, P.H., Tola, D., Schou, C., Larsen, P.G.: A Modeling Approach for Composed Digital Twins in Cooperative Systems. In: 2023 IEEE 28th International Conference on Emerging Technologies and Factory Automation (ETFA), pp. 1–8. IEEE (2023). DOI: 10.1109/ETFA54631.2023.10275601
12. Gil, S., Oakes, B., Gomes, C., Frasheri, M., Larsen, P.G.: Towards a systematic reporting framework for digital twins: A cooperative robotics case study. SIMULATION (2024). Accepted for publication
13. Gil, S., Schou, C., Mikkelsen, P.H., Larsen, P.G.: Integrating Skills into Digital Twins in Cooperative Systems. In: 2024 IEEE/SICE International Symposium on System Integration (SII), pp. 1124–1131. IEEE (2024). DOI: 10.1109/SII58957.2024.10417610
14. Hassani, V., Ross, A., Selvik, Ø., Fathi, D., Sprenger, F., Berg, T.E.: Time domain simulation model for research vessel Gunnerus. In: ASME 2015 34th International Conference on Ocean, Offshore and Arctic Engineering. American Society of Mechanical Engineers Digital Collection (2015)
15. Hatledal, L.I., Chu, Y., Styve, A., Zhang, H.: Vico: An entity-component-system based co-simulation framework. Simulation Modelling Practice and Theory **108**, 102243 (2021). DOI: https://doi.org/10.1016/j.simpat.2020.102243
16. Hatledal, L.I., Skulstad, R., Li, G., Styve, A., Zhang, H.: Co-simulation as a Fundamental Technology for Twin Ships. Modeling, Identification and Control **41**(4), 297–311 (2020). DOI: 10.4173/mic.2020.4.2

17. Hatledal, L.I., Styve, A., Hovland, G., Zhang, H.: A language and platform independent co-simulation framework based on the functional mock-up interface. IEEE Access **7**, 109328–109339 (2019). DOI: 10.1109/ACCESS.2019.2933275
18. Hausted, J.O., Kvistgaard, J.L., Lumer-Klabbers, G.: Towards a Digital Twin for an Autonomous Robot. Aarhus University, BSc thesis (2020)
19. Kanazawa, M., Hatledal, L.I., Li, G., Zhang, H.: Co-simulation-Based Pre-training of a Ship Trajectory Predictor. In: International Conference on Software Engineering and Formal Methods, pp. 173–188. Springer (2021)
20. Kanazawa, M., Skulstad, R., Li, G., Hatledal, L.I., Zhang, H.: A multiple-output hybrid ship trajectory predictor with consideration for future command assumption. IEEE Sensors Journal **21**(23), 27124–27135 (2021). DOI: 10.1109/JSEN.2021.3119069
21. Kanazawa, M., Wang, T., Ichinose, Y., Skulstad, R., Li, G., Zhang, H.: Bridging similar ships' dynamics for safeguarding the system identification of maneuvering models. Ocean Engineering **280**, 114874 (2023). DOI: https://doi.org/10.1016/j.oceaneng.2023.114874
22. Kanazawa, M., Wang, T., Skulstad, R., Li, G., Zhang, H.: Knowledge and data in cooperative modeling: Case studies on ship trajectory prediction. Ocean Engineering **266**, 112998 (2022). DOI: https://doi.org/10.1016/j.oceaneng.2022.112998
23. Legaard, C.M., Tola, D., Schranz, T., Macedo, H.D., Larsen, P.G.: A Universal Mechanism for Implementing Functional Mock-up Units. In: G. Wagner, F. Werner, T.I. Ören, F.D. Rango (eds.) Proceedings of the 11th International Conference on Simulation and Modeling Methodologies, Technologies and Applications, SIMULTECH 2021, Online Streaming, July 7-9, 2021, pp. 121–129. SCITEPRESS (2021). DOI: 10.5220/0010577601210129
24. Lehner, D., Gil, S., Mikkelsen, P.H., Larsen, P.G., Wimmer, M.: An architectural extension for digital twin platforms to leverage behavioral models. In: 2023 IEEE 19th International Conference on Automation Science and Engineering (CASE), pp. 1–8 (2023). DOI: 10.1109/CASE56687.2023.10260417
25. Liu, M., Fang, S., Dong, H., Xu, C.: Review of digital twin about concepts, technologies, and industrial applications. Journal of Manufacturing Systems **58**, 346–361 (2021). DOI: https://doi.org/10.1016/j.jmsy.2020.06.017. Digital Twin towards Smart Manufacturing and Industry 4.0
26. Lumer-Klabbers, G., Hausted, J.O., Kvistgaard, J.L., Macedo, H.D., Frasheri, M., Larsen, P.G.: Towards a digital twin framework for autonomous robots. In: SESS: The 5th IEEE International Workshop on Software Engineering for Smart Systems. COMPSAC 2021, IEEE (2021)
27. Madsen, E., Tola, D., Hansen, C., Gomes, C., Larsen, P.G.: AURT: A Tool for Dynamics Calibration of Robot Manipulators. In: 2022 IEEE/SICE International Symposium on System Integration (SII), pp. 190–195 (2022). DOI: 10.1109/SII52469.2022.9708769
28. Oakes, B., Parsai, A., Van Mierlo, S., Demeyer, S., Denil, J., De Meulenaere, P., Vangheluwe, H.: Improving Digital Twin Experience Reports. In: Proceedings of the 9th International Conference on Model-Driven Engineering and Software Development, pp. 179–190. SCITEPRESS - Science and Technology Publications. DOI: 10.5220/0010236101790190
29. Oakes, B.J., Gomes, C., Larsen, P.G., Denil, J., Deantoni, J., Cambeiro, J., Fitzgerald, J.: Examining model qualities and their impact on digital twins. In: 2023 Annual Modeling and Simulation Conference (ANNSIM), pp. 220–232. IEEE (2023)
30. Oakes, B.J., Parsai, A., Meyers, B., David, I., Van Mierlo, S., Demeyer, S., Denil, J., De Meulenaere, P., Vangheluwe, H.: A Digital Twin Description Framework and its Mapping to Asset Administration Shell. In: Model-Driven Engineering and Software Development, vol. 1708, pp. 1–24. Springer Nature Switzerland. DOI: 10.1007/978-3-031-38821-7_1
31. Oakes, B.J., et al.: Structuring and accessing knowledge for historical and streaming digital twins. In: 1st Workshop on Ontology-Driven Conceptual Modeling of Digital Twins, pp. 1–13
32. Schou, C., Andersen, R.S., Chrysostomou, D., Bøgh, S., Madsen, O.: Skill-based instruction of collaborative robots in industrial settings. Robotics and Computer-Integrated Manufacturing **53**(June 2016), 72–80 (2018). DOI: 10.1016/j.rcim.2018.03.008

33. Skjetne, R., Sørensen, M.E., Breivik, M., Værnø, S.A., Brodtkorb, A.H., Sørensen, A.J., Kjerstad, Ø.K., Calabrò, V., Vinje, B.O.: AMOS DP research cruise 2016: Academic full-scale testing of experimental dynamic positioning control algorithms onboard R/V Gunnerus. In: International Conference on Offshore Mechanics and Arctic Engineering, vol. 57632, p. V001T01A080. American Society of Mechanical Engineers (2017)
34. Skjong, S., Pedersen, E.: A real-time simulator framework for marine power plants with weak power grids. Mechatronics **47**, 24–36 (2017)
35. Talasila, P., Gomes, C., Mikkelsen, P.H., Arboleda, S.G., Kamburjan, E., Larsen, P.G.: Digital Twin as a Service (DTaaS): A Platform for Digital Twin Developers and Users. In: 2023 IEEE Smart World Congress (SWC), pp. 1–8 (2023). DOI: 10.1109/SWC57546.2023.10448890
36. Van Mierlo, S., Oakes, B.J., Van Acker, B., Eslampanah, R., Denil, J., Vangheluwe, H.: Exploring validity frames in practice. In: Proceedings of the First International Conference, ICSMM 2020, Bergen, Norway, June 25–26, 2020, pp. 131–148. Springer, Cham. DOI: 10.1007/978-3-030-58167-1_10
37. Wang, T., Li, G., Hatledal, L.I., Skulstad, R., Æsøy, V., Zhang, H.: Incorporating approximate dynamics into data-driven calibrator: A representative model for ship maneuvering prediction. IEEE Transactions on Industrial Informatics **18**(3), 1781–1789 (2021)
38. Wang, T., Skulstad, R., Kanazawa, M., Li, G., Æsøy, V., Zhang, H.: Physics-informed data-driven approach for ship docking prediction. In: 2022 IEEE International Conference on Real-time Computing and Robotics (RCAR), pp. 111–117. IEEE (2022)
39. Zhang, H., Li, G., Hatledal, L.I., Chu, Y., Ellefsen, A., Han, P., Major, P., Skulstad, R., Wang, T., Hildre, H.P.: A Digital Twin of the Research Vessel Gunnerus for Lifecycle Services: Outlining Key Technologies. IEEE Robotics and Automation Magazine **30**(3), 6–19 (2023). DOI: 10.1109/MRA.2022.3217745

Part V
Advanced Topics

This final part covers advanced or research-based topics that relate to our ability to deliver dependable cyber-physical systems enabled by Digital Twins (DTs). Given the ability of a DT to exercise some control over a Physical Twin (PT) and the capacity to develop DTs that act with a measure of autonomy, it is important to consider the verification of key safety and security properties. Chapter 13 considers the balance between the benefits of DTs and the risks that come from networked communications between the real and virtual worlds. Chap. 14 takes us forward to the stage where the DT is able to reconfigure the PT autonomously. Finally, Chap. 15 rounds off the book by identifying a range of possible directions for future research and innovation addressing challenges in DT dependability, formal verification, platforms and novel applications.

Chapter 13
Security and Privacy-related Issues in a Digital Twin Context

Tomas Kulik, Zahra Kazemi and Peter Gorm Larsen

Abstract Digital Twin (DT)-enabled systems depend on communications and networking, and so face a range of threats to their security. This chapter reviews approaches to managing or mitigating these threats. Approaches that are already well established in the software industry, techniques based on formal methods and finally, control theory approaches using features such as state estimation ensure that the developers of DT-enabled systems are faced with a plethora of options. The chapter also presents a standardised notation for communication of security features between engineers working within the DT area.

13.1 Introduction

This chapter introduces security-related issues relevant to Digital Twins (DTs). The data exchanged between the PT and its DT is typically communicated close to real-time, and in some cases can carry critical signals. Thus, the need for cyber security assurance within the systems employing the DT technology becomes important. Recently, an increase in attack surfaces within the industrial control systems has been observed [27], largely due to the additional connectivity that is being added to these systems in order to satisfy increasing requirements for data availability. Utilising a DT further extends the attack surface because the models underlying the DS/DT might also become attack vectors.

It is important to consider the nature of the DT itself. In order to provide a good representation of the PT it needs to be provided with a multitude of signals from the PT. Furthermore the DT can often be used as a reconfiguration tool for the PT, sending a signal to the PT on how to reconfigure itself. To this extent the DT is often seen as a solution to the security problem rather than as an increasing attack vector [15]. While a DT can often act as a security component, a question of what happens if the DT itself is compromised needs to be asked [19]. This chapter attempts

J. Fitzgerald et al. (eds.), *The Engineering of Digital Twins*,
https://doi.org/10.1007/978-3-031-66719-0_13

to answer the question of how we can trust a DT within critical systems. It does this by introducing example attacks within a DT environment, considering the impacts of these attacks and discussing the protections against such attacks.

A final consideration within a DT enabled systems is sharing of the data from the DT between multiple types of users with different requirements. For example, some users require read access to a DT while others need write access. It is often necessary to ensure that different entities can access different parts of a DT, requiring strong isolation in order to prevent disclosure of sensitive information. This becomes an important challenge in cases where a DT is hosted by a service provider separate from the system and the DT designers as well as the system operators. Protection of different stakeholders' privacy is also important, especially in systems where user interactions with the DT critical, for example in healthcare applications [11, 24].

Security and privacy in DT ecosystems have not been studied for long as an integrated discipline, but fall into the areas of security of digital communication and control systems. Currently the trends are emerging where DT-based solutions merit specific consideration [3]. It could be argued that, with increasing complexity of DT-enabled systems, security topics are expected to become one of the most important design aspects of these systems. Different kinds of model checking technologies are relevant in this context [18].

This chapter starts with an overview of the general security architecture for DTs in Sect. 13.2. This is followed in Sect. 13.3 with approaches that can be used for mitigating different security and privacy attacks. Section 13.4 considers Intellectual Property (IP) protection approaches before Sect. 13.5 finally considers security concerns in the real world.

13.2 DT Security Architecture

In order to discuss the security of DT-enabled systems one must consider their architecture. Often the PT carries out actions that impact the physical environment, while the DT is executed remotely. The PT and its DT are then interconnected using one or more communication protocols, depending on the complexity of the system. A simple architectural model for the incubator is presented in Fig. 13.1. Even a simple system like this can expose multiple DT services potentially running on several independent computers, requiring a network connection.

The presented architecture uses the RabbitMQ message broker and the Advanced Message Queuing Protocol (AMQP) protocol flowing from the DT to the PT. In order to secure this communication several questions have to be asked. What is the nature of the data transmitted? How could the PT deal with communication delays? Does DT communicate any commands back to the PT? For example, if it is only essential to protect the data flowing from DT to the PT, data encryption might be considered sufficient. The question then becomes about the responsibility of carrying out this

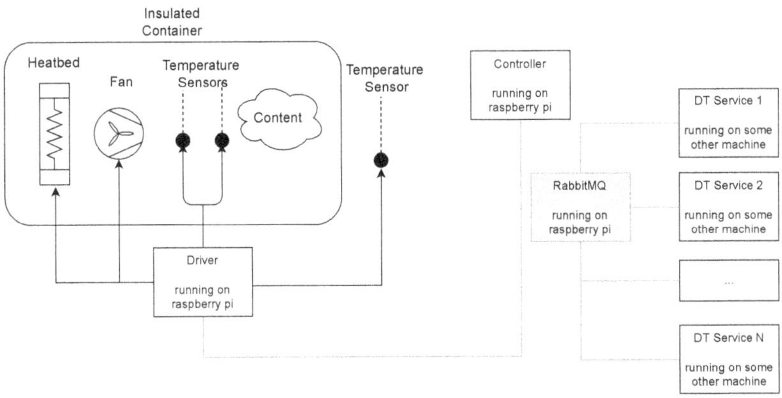

Fig. 13.1: Diagram of the communication among different DT services of the incubator.

encryption, as the DT can encrypt the data and provide it over an open channel, or the AMQP could be setup with network encryption using Transport Layer Security (TLS), which would furthermore let the AMQP client know it is connecting to the expected broker. If only a specific PT can provide data to a DT's services, as well as only specific DT services having access to this data, a mutual authentication using TLS can authenticate the PT and the DT service AMQP clients as well as provide a network encryption [38]. The decision on use of different security mechanisms often has to be taken considering the impact on data availability and potential latency it would introduce.

In this chapter we consider five cyber attacks applicable to DT-enabled systems. Four of them, the *man-in-the-middle*, *command injection*, *denial of service and data modification* are well known from communication and control systems, while the final attack, *model malware*, is known from the area of industrial automation. These attacks present a good base for discussion of cyber security within DTs, especially since they can have a significant impact on the sensing capabilities of DTs as presented in Chap. 7. It has to be noted that these attacks do not present an exhaustive list of possible cyber attacks. The security mechanisms described in this chapter are applicable to a wide variety of potential cyber attacks. In order to provide a full overview the attacks are described in detail as:

Man-in-the-middle attack: An attack in which a malicious actor joins a network channel used to transport data between the PT and the DT. The malicious actor is not seeking to actively disturb the operations of the DT-enabled system but rather obtain data from it in order to extract potential value from this data as shown in Fig. 13.2. The value obtained by the malicious attacker can be utilised for learning more about the DT-enabled system and subsequently using this knowledge for further attacks or simply directly obtaining secret information from the data transmitted.

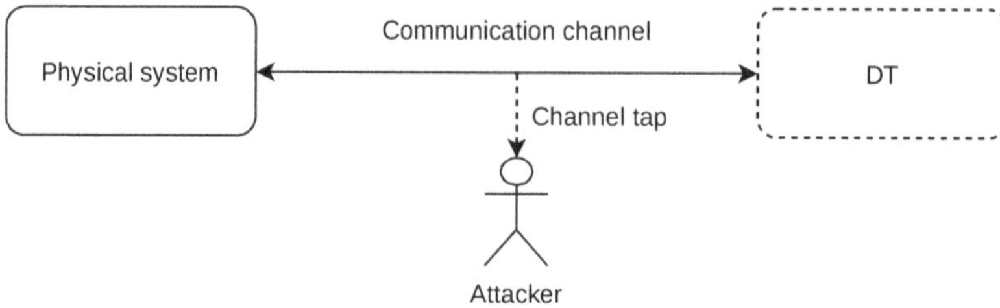

Fig. 13.2: Man-in-the-middle attack within a DT-enabled system

Command injection attack: An attack in which a malicious actor joins a network channel used to transport commands from the DT to the PT and issues malicious commands. This attack is aimed directly at disturbing operations of the PT in order to damage its intended operation. The positioning of the attacker and the DT-enabled system components is shown in Fig. 13.3. Attacks of this type are known from the field of industrial control systems [37] where, if undetected, they could result in the destruction of the physical equipment that the control system is operating. In order to be able to inject the commands, the attacker needs to have knowledge about the DT-enabled system and formatting of the command data packets. This could for example be obtained by utilising the man-in-the-middle attack prior to attacking it using a command injection attack.

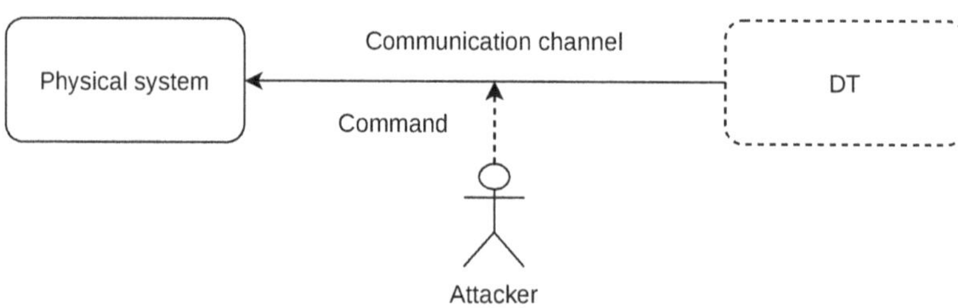

Fig. 13.3: Command injection attack within a DT-enabled system

Denial of Service (DoS) attack: An attack in which a malicious actor attempts to disturb the DT-enabled system by either delaying or altogether preventing the data transfer either from the PT to the DT or from the DT to the PT. Both options are explored in Fig. 13.4. Depending on the nature of the DT-enabled system this can have critical impact as the DT can lose overview of the state of the PT and hence mislead a potential operator. The PT might not receive important parameter updates or commands from the DT leading to potential damage of the system or the system might simply shut down due to the unavailability of the necessary data. In order to execute this type of attack, the attacker needs to flood the communication channel with messages to such an extent that it overwhelms the processing capabilities of the affected communication endpoints.

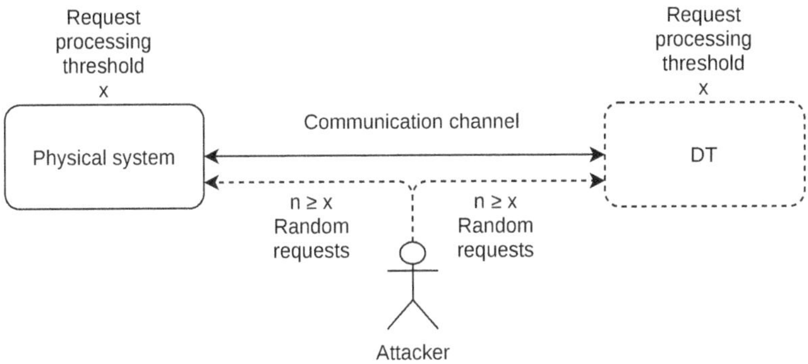

Fig. 13.4: Denial of Service attacks within a DT-enabled system

Data modification attack: An attack in which a malicious actor attempts to modify the data transferred from the PT. This attack is aimed on providing false information about the state of the PT to the operator of the DT-enabled system as well as to the DT. If the DT acts as a supervisory controller in the loop, modified data might cause wrong regulation input and subsequent damage to the PT. The original data flow is often intercepted by the man-in-the-middle attack and then modified by the attacker [49] as shown in Fig. 13.5. Depending on the nature of the DT-enabled system, small modifications to the values would lead to incorrect regulation if left undetected for a longer time period.

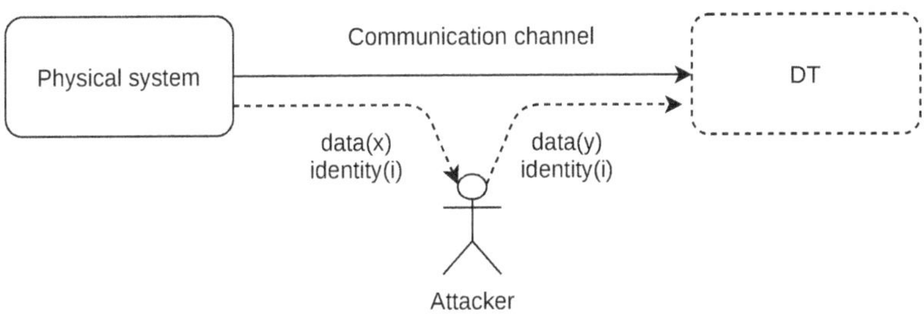

Fig. 13.5: Data modification attack within a DT-enabled system

Model malware attack: An attack in which a DT is targeted directly. The attacker aims to inject malware to a DT model, in order to impose malicious behaviour on a DT. In order to execute the attack, the attacker either introduces malware directly to the content of the model by gaining access to the repository or indirectly by either introducing malware to some of the libraries that DT is based upon or even if a DT is capable of side loading executable code, by injecting it directly as presented within Fig. 13.6. In order to succeed the attacker must have prior knowledge of the DT and its execution. The infected model can then either mislead the system operator or even issue commands to the system causing operational problems. Similar attacks are known from the area of industrial automation, specifically control loop programs executed by Programmable Logic Controllers (PLCs) [58].

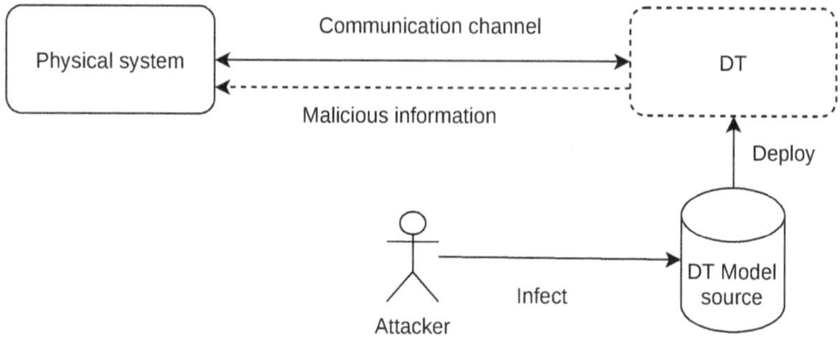

Fig. 13.6: Model malware attack aimed at the DT within a DT-enabled system

Security Attacks on the Incubator

Consider how these attacks may impact the DT-enabled incubator system. Considering the *Man-in-the-middle* attack, the attacker might dump data packets transmitted over the AMQP protocol and gather information about the system by observing the exchanged data. With the *Command injection* attack, the attacker can send a command to the PT, for example to have the heater on continuously, leading to overheating of the system. With the *denial of service* attack, the attacker can simply prevent the communication between the incubator and its DT, leaving the DT without knowledge of the system state. In case that the DT acts as a primary regulator, this would practically stop the regulation of the active elements within the incubator. The *Data modification* attack would allow the attacker to change the content of the value provided by the incubator to the DT, where the regulation provided by the DT would not correspond to the real situation of the incubator and would lead to too low or high temperature. Finally, the *Model malware* attack would affect the DT services running on separate machines, considering for example the DT service responsible for the heater once again the malware in the model might simply offset the target temperature and hence keeping it too low or too high. The attacker would simply provide an innocent library for example logging library that if permitted by the programming stack might gain access to global and local variables within different scopes [44].

These are only some of the attacks that could be anticipated by systems engineers for the overall DT-enabled system. It is the aim of this chapter to show how to mitigate these attacks as well as provide a toolkit for understanding how other types of attack could be mitigated. The different attacks presented share several properties, and utilisation of access control technologies and encryption provides good starting points for a secure DT-enabled system.

13.3 Approaches to a DT Security and Privacy

Having identified some possible attacks, we now look at how to defend against them. It has to be noted that DT engineering is an evolving field and so it could be expected that tooling specifically for security analysis of DTs will become available. An interesting approach is the use of frameworks based on the game theory, where a non-cooperative game model is used to assess behaviour of different nodes within the system, including DTs [12]. This chapter considers several different approaches to security and privacy challenges: *Standard* approaches, which are widely used within the software industry not necessarily specifically aimed at DTs, *Formal Methods based* approaches, that utilise formal (logical and mathematical) analysis to discover and address security challenges, and finally *Control theory based* approaches utilising control theory and state estimation to address the potential security challenges.

13.3.1 Standard approaches to cyber security in a DT context

The standard approaches are those that have been widely used in the software industry. This chapter considers the security by design [8, 17] approach, security reviews and risk assessments in this group. In addition, concepts such as encryption and authentication play a significant role within a secure DT ecosystem. Finally, all of the approaches need to consider operational trade-offs between flexibility and security.

The **Security by Design** approach is built upon a principle that the foundation of security starts within the design of the system. This specifically means that the design process is not only concerned with the features and operations of the DT but from the very start of the design process considers how to make sure that the features of the DT are secure. One way to carry out this process is using the Confidentiality, Integrity and Availability (CIA) framework, taking into account the basic attributes of *Confidentiality, Integrity* and *Availability* of the information within the DT enabled systems. The attributes are complementary, as indicated in Fig. 13.7. In the case of the DT-enabled system, further attributes that need to be considered include *Safety* and *Accountability*. Security by Design employs security patterns (often based on best practices) to satisfy the CIA and other attributes. Within the area of a DT enabled system, the considerations could be:

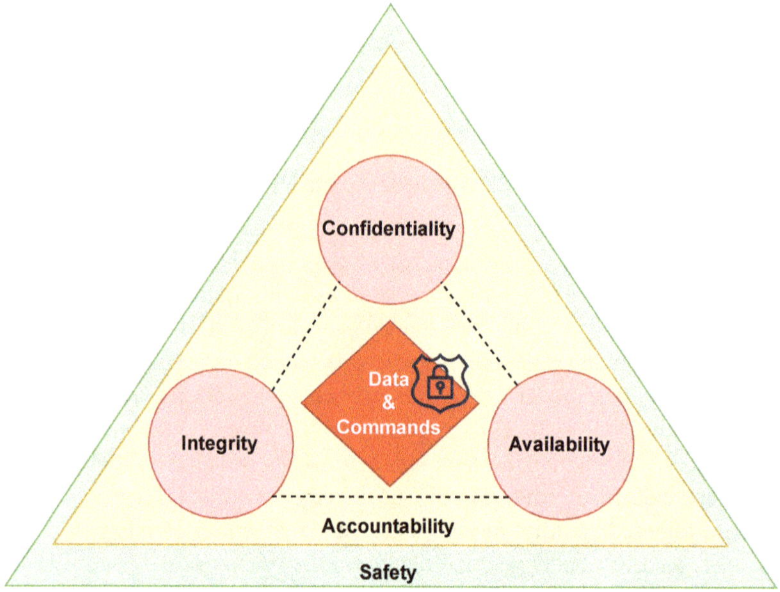

Fig. 13.7: The enhanced Confidentiality, Integrity and Availability framework

Confidentiality making sure the information transmitted on the channel stays secret. From the best practices, an approach with channel encryption or data encryption could be used, with the specific encryption algorithm adapted to the DT use case. However, a foundational requirement for secrecy of exchanged information shall be considered within the DT-enabled system as even data which seemingly does not need to be considered secret would provide a lot of information out the system to a potential attacker.

Integrity ensuring that the information exchanged within the system is accurate and consistent, i.e., the information carries the true value within an expected range of values. To achieve this, a so called semantic integrity it is necessary for the DT to have a mechanism that can establish what the correct range of values is. The strategies for information integrity could be input validation, error detection and cryptographic signing of data blocks carrying the information providing tampering protection.

Availability making sure the information is available within the system even under degraded conditions (for example, if the system is under attack). The availability requirements again depend on the use case for the DT-enabled system, however generally the information needs to be available between the PT and the DT in order for the DT to operate correctly. Different strategies for ensuring information availability are use of redundancy and avoiding single points of failure, for example a DT can consist of multiple redundant processes where these processes use independent communication channels.

Safety ensuring the system remains in a safe operation state or shut downs safely in case of a disturbance. The safety requirements again depend on the use case of the system, however prevention of the unsafe condition of the PT is often the priority. In case of a DT-enabled system, the DT itself could be used to detect unsafe conditions. In order for this, the CIA attributes must be satisfied. One approach to increasing PT safety is utilisation of what-if analysis performed on the DT [40], i.e. executing potential commands against the DT with a current system state and only pushing these commands to the PT if the DT does not execute unsafe behaviour.

Accountability making sure that the process or an operator that has executed a specific command or series of commands can be identified and the identifying information is readily available. This is especially important in complex environments where a DT might consist of multiple processes with the ability to issue commands to the PT. One strategy for providing accountability is wide use of logging of operations and collecting the logs to a central system (e.g., inside the DT).

One standard approach to security is use of an island mode, or an isolation of the system from external systems. This approach has been used for a long time within the area of industrial control systems. Currently, industrial control systems are becoming more interconnected which increases their utilisation but also poses additional security risks [53]. The isolation approach could also be applied to DT-enabled systems, where both the DT and the PT would be in close proximity, connected on a local network without any connectivity to wider external networks, which would however in several ways decrease the utility of the DT. For example remote monitoring or model updates would not be possible. This is why isolation is regarded as an option to secure the system, but to be used sparingly. A possible use is a fallback mode where, in case of a cyber attack (or even an increased threat of a cyber attack), the connectivity to a remote network is deactivated and the DT-enabled system operates in an island mode.

The possible utilisation of island mode leads to an important consideration of trade-offs between flexibility and security. The security options that the system designers face need to be tailored to the system at hand. Several trade-off considerations are:

1. Remote monitoring versus local monitoring only
2. Remote DT updates versus local updating only
3. Remote commanding versus local commanding only
4. Direct operations versus network analysis

One of the most important points for a system designer is to decide whether the security requirements are strong enough to utilise a DT. In addition, the direct operations (not analysing the network packets) could lead to lower latency within the system. This can be important for close-to-real-time systems.

The concepts often used to satisfy the cyber security requirements are *encryption*, *authentication* and *authorisation*.

The concept of encryption provides secrecy to the data within the system. This could be for data in transit, data at rest (in a data store), data loaded to the system memory and in some cases even data that is being manipulated [1]. The basic concept considers cleartext data d, cryptographic key (secret) k and a cryptographic function f. The cipher-text c (encrypted data) is created as $c = f(d, k)$ and then d can be obtained back as $d = f^{-1}(c, k)$, in this case a symmetric cryptography has been used where a single key is used for both encryption and decryption and it needs to be transported securely between the entities that are using it. Another option is use of asymmetric cryptography that utilises key pairs as public and private key. In this case a public key pk and a private key sk are forming a key pair (pk, sk). The public key does not contain secret information and can be freely shared. The data is then encrypted as $c = f(d, pk)$ and can be decrypted as $d = f^{-1}(c, sk)$. It is important that sk remains protected. It is beyond the scope of this book to discuss individual encryption functions and advanced mechanisms, but in the case of a DT-enabled system it is important to determine how to use encryption to protect sensitive data while ensuring that the encryption overhead does not interfere with the operation of the system.

The concept of authentication is the ability to prove the identity of a specific principal. This could for example be a computing service or a user. The authentication practice ensures that only entities possessing a specific piece of information can be allowed to interact with the system. For example, this could be a password, a cryptographic token or some other secret. The DT-enabled system then knows the entities that are allowed to interact with it and checks their secret before allowing interaction. An example is a *basic* authentication known from use with HTTP, where a username and password (credential) is simply sent as clear text. The DT-enabled system then receives the credential and checks it against a database of known users. This database typically stores the username and cryptographic hash of a password. The system computes the hash of the received password, and if it matches the user is allowed to interact with the system. This is just a simple example as it is possible to utilise many different authentication mechanisms utilising different types of credentials. The Simple Authentication and Security Layer (SASL) provides evolving definitions for multitude of mechanisms and protocols [21] and is often used within computing systems. The DT-enabled system designers need to consider authentication mechanisms in order to ensure that only the DT services with valid authentication can interact with the physical system.

Finally, the concept of authorisation builds on top of authentication by adding granularity, stating which entities have access to what resources. This can be accomplished by use of Access Control Lists (ACLs) where a specific resource such as a module within the physical system can only be accessed by a specific module within the DT. In this case ACL defines the necessary access permissions, very often in categories or as read and write permission from and to the resource. For example an entity x wants to access the resource R, the access control list would define: $x \longrightarrow rR$ or $x \longrightarrow wR$ or $x \longrightarrow rwR$ for x is allowed to read R, x is allowed to write R and x is allowed to read and write R. This is a basic example that works on a *deny by default* principle, i.e., if no ACL rule is defined, x is denied all operations on R. This does not need to be the case as ACLs can also be used to specify deny rules. The system needs to be able to store and enforce ACLs. This leads to the need to manage the ACLs on a system level, often on individual components carrying the resources. Another option is utilisation of Role Based Access Control (RBAC), which allows for centralised management of roles and assignment of component principals (users) to specific groups. As RBAC is often managed centrally, it is suitable for larger fast evolving systems [4]. These are the two most utilised approaches that span from networked systems to operating systems. The DT-enabled system designers need to consider the scale of the system and its evolution when choosing the authorisation strategy as well as consider use of authorisation frameworks such as for example OAuth 2.0 [6] (this was also used in Sect. 11.9).

During the design process of a DT-enabled system it is important to undertake *Rrisk Assessment* and *Security Review*. Risk assessment is a process that identifies security risks to a DT-enabled system, which then are to be mitigated. This process usually follows a specific model for identification of security threats. One commonly used model (thanks to the support of modelling tools) is the STRIDE model covering threats from areas of Spoofing, Tampering, Repudiation, Information disclosure, Denial of Service and Elevation of Privilege [29]. The security review is a process that is used to review the proposed mitigations and determine if the security the provide is sufficient by identifying the residual risks and deciding if these are to be accepted. If the risks are too high to be accepted, specific actions are taken, such as redesign or mitigation improvements. The system can only be built after a successful security review. This process then starts over for new features added to the system. The full process is shown in Fig. 13.8. It is important that the system features are built with minimum known vulnerabilities and accepted risks are not of a critical nature.

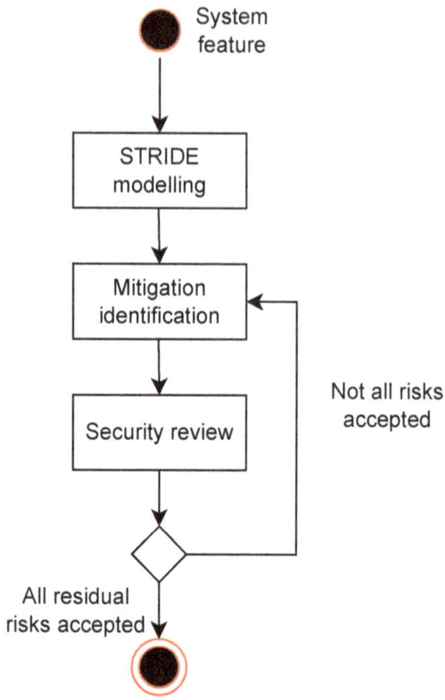

Fig. 13.8: Security process in the design stage

13.3.2 Formal Methods-Based Approaches to Cyber Security for DTs

In some cases the DT-enabled system is of a critical nature and designers must provide formal (in the sense of being highly mathematically rigorous) guarantees of certain attributes of the system [56]. Formal methods have been used to analyse security aspects of different types of systems and provide security guarantees on multiple levels of abstraction [31]. An example is model checking, a type of formal analysis where a state space of a formal model is explored for violation of given properties also expressed in a specific formal notation. Another approach is the use of theorem proving, or expressing the system as a set of theorems and proving their validity. Finally, an approach sometimes used is guided formal validation, where a formal model of the system is created and then many scenarios are executed on the model level (this was already introduced in Sect. 5.6 and in Sect. 9.2).

In all cases, the system is formally specified and then subsequently analysed. The guarantees provided by this approach are considered strong because of the rigour of the mathematical assessment. It needs to be noted that the use of formal methods presents its own challenges, for example a learning curve for extra set of techniques as well as assurance that the implemented system follows the model used for formal analysis.

To demonstrate the approach, let us consider a simple high level model of a DT model update for the incubator. This example is based on a model checking and formal specification system called Temporal Logic of Actions (TLA+) [35]. Consider a system requirement that the DT model is considered potentially malicious if it does not pass a series of checks. This could be checking the validity of a cryptographic signature, checking the validity of the integrity of the model and checking the expected size of the model. This can be expressed more compactly for a model m as: $is\_malicious(m) = \neg\ sigValid(m) \vee \neg\ integValid(m) \vee size(m) > limit$ where the *limit* is an arbitrary size limit for a valid model. Furthermore let us consider a system with a simple set of state transitions in order to transport the model to the DT. This set of transitions is shown in figure Fig. 13.9. The set of transitions can then be encoded in the formalism known as TLA+, utilising an algorithm modelling language PlusCal [36]. The property checking that the received model is not malicious can then be noted as $is\_valid(m) = \neg is\_malicious(m)$. In order for the model checker to be able to search for this property it needs to be expressed in a Linear Temporal Logic (LTL) formalism (also used for monitoring purposes in Chap. 9). In summary, several components need to be used together to carry out the formal analysis, these are:

1. Formal model
2. Property to verify
3. TLC model checker

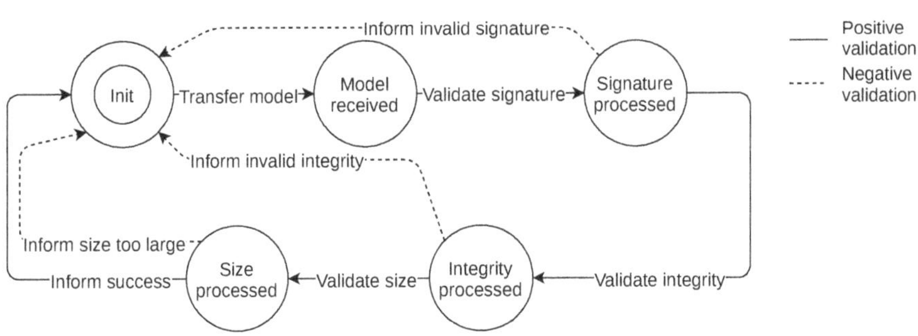

Fig. 13.9: Model delivery state transitions

Model checking suffers from the problem of state space explosion, which refers to the fact that the number of states in a system grows exponentially with the number of variables in the system [10] (recall Sect. 5.6.3). Consequently, it is regarded as being feasible for small systems or abstract models of a complicated system, although research and innovation in model checking is ongoing.

The formal model represents the behaviour of the system. The system model created in PlusCal is shown in Listing 13.1. The model covers the functionality of receiving the model and validating it against the specified conditions.

Listing 13.1: Processing of a DT model update (PlusCal)

```
fair process dt = DT
begin
DTBegin:
    skip;
    DTReceiveModel:
        if ModelHasValidSignature(dtModel) then
            DTModelValidSignature:
                \* Perform integrity check
                if ModelIsGenuine(dtModel) then
                    \* Check for size
                    DTModelGenuine:
                        if ModelFitsStorage(dtModel) then
                            DTModelSpaceAvailable:
                                \* Store model
                                assert dtStorage = NO_MODEL;
                                dtModelStatus := MODEL_STATUS_OK;
                                dtStorage := dtModel;
                        else
                            DTModelSpaceNotAvailable:
                                \* Security log: wrong size
                                dtModelStatus :=
                                      MODEL_STATUS_INSUFFICIENT_STORAGE;
                         end if;
                else
                    DTModelNotGenuine:
                        \* Security log: not genuine
                        dtModelStatus := MODEL_STATUS_NOT_GENUINE;
                        skip;
                end if;
        else
            DTModelInvalidSignature:
                \* Security log: invalid signature
                dtModelStatus := MODEL_STATUS_INVALID_SIGNATURE;
                skip;
        end if;
    DTHandlePendingRequests:
            goto DTBegin;
end process;
```

The property to verify is the property the model of the DT-enabled system is analysed against. This property can be expressed in LTL is as in Listing 13.2. The property simply states that it is expected that the model is always valid.

Listing 13.2: TLA property related to the model validation

```
DTModelAlwaysValid ≜
  □(ModelValid(dtInstalledModel) ∧
    (dtStorage ≠ NO_MODEL ⇒ ModelValid(dtStorage)))
```

The TLC model checker is used to build the entire state space of the system given by the model and tries to find a path in which the property does not hold. It can then provide a counterexample showing at which state of the system model the property does not hold.

These methods make it possible to explore the security properties of the system deeply. While the example only assesses very high-level functionality, it can be broken down and expressed in more detail, depending on the needs of the system designer. Often formal methods are used alongside standard approaches. For example the output of the risk assessment could be used to determine the specific security properties that need to be analysed by use of formal methods. In this way the approaches are complementary.

It may be necessary to specify system models, potential attacks and mitigations even without the need for formal analysis. This can be accomplished if the system design is to be communicated among a wide variety of personnel working on different aspects of the system. An example approach is the use of semi-formal notation that provides a specific way of describing a system or a problem while not necessarily utilising rigorous mathematics. In the area of DTs a specific notation has been proposed to communicate about their security aspects [32]. The notation is based on context diagrams from the Problem Frames notation, and is meant to provide a clear way of communicating between DT engineers, minimising (in many cases complementing) the use of natural language that can lead to misunderstandings.

The context diagrams are composed of domains that describe the key aspects and participants, while not describing the system architecture. In fact, they only contain one machine domain representing "the software" or "the computer". In our context diagrams it is always called the *Digital Twin*. Some domains can be controlled like the *CPS* (whose behaviour in general is predictable and therefore called *causal* and marked with a C). It is the objective of the DT to re-configure or augment the CPS. The behaviour of the *Attacker*, whose domain is called *biddable* and marked with a B, is not predictable in causal terms. Another relevant kind of domain is called *lexical*, representing some form of data. Domains that are *designed* are marked with one vertical bar on the left and machine domains with two vertical bars. Domains that interact in some way are connected by edges that are annotated with the phenomena they share. Phenomena are abstractions of concepts from the real world that can be observed or measured (for a more thorough discussion see [22, 23]). In order to emphasise the role of the network, we draw all edges connecting to it by "double bars". Attacks are then described in oval dashed ellipses using natural language and reference to points of the system they interact with. An example of this is a simple man-in-the-middle attack as defined in Sect. 13.2 and is shown in Fig. 13.10, where it can be compared with the informal diagrams first presented in this chapter.

The notation is being used actively within the area of DT security and is further being expanded upon [42]. As with any discipline in digital security, clear communication of vulnerabilities, attack and mitigations is one of the cornerstones of secure systems.

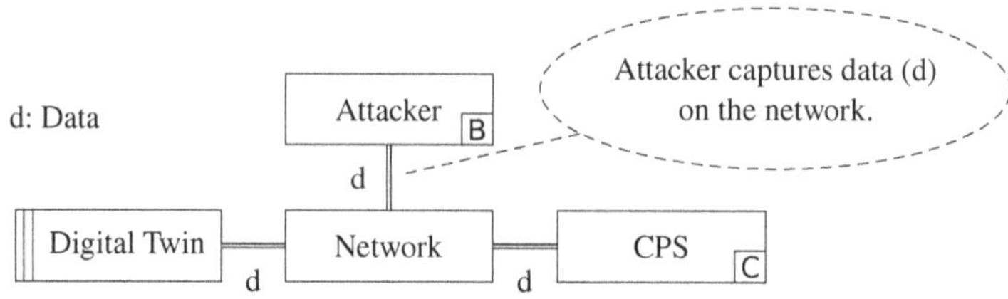

Fig. 13.10: Context diagram for man-in-the-middle attack.

13.3.3 Attack mitigations

This section will focus on mitigation of cyber attacks against a DT-enabled system. It will explore standard mitigations but also describe use of formal methods against a DT-enabled system at runtime. Several techniques have been described previously, namely use of encryption and authentication. Let us consider how these techniques could be utilised in mitigating the attacks defined in Sect. 13.2. The mitigations for most of the attacks are presented informally while an example of semi-formal mitigation definition is presented for the *Man-in-the-middle* attack.

The *Man-in-the-middle* attack can be mitigated using channel or data encryption. The difference between these two approaches is that for channel encryption a tunnel between the communication is created that encrypts all of the traffic flowing through that channel. This could for example be use of the TLS protocol. Data encryption on the other hand encrypts the data packets prior to sending them and then transfers them over an open channel, this could be used on a hardware not suitable for use of the TLS, for example in an IoT setting [50] or on network types not supporting channel encryption. An example of channel encryption is shown in Fig. 13.11. Note that the attacker is only capable of capturing the encrypted communication, given sufficiently strong encryption and key establishment protocol, the decryption of data would be beyond practical means of the attacker. This mitigation is expressed in a semi-formal notation presented in Sect. 13.3.2 as shown in Fig. 13.12. The mitigation, while simple, provides a powerful protection for the system.

The *Command injection* attack can be mitigated by enforcing authentication for any entity issuing commands. This means that the system would reject any command sent from an anonymous source. While this approach provides only limited protection against insider attacks (a potential attacker is among entities that can authenticate), it is an important starting point for securing of the system against malicious attacks. The authentication requires that the entity issuing the command submits a credential that is then evaluated; only a command with correct credentials will be executed. This approach is shown in Fig. 13.13. It is important to note that the credential needs to be protected from a potential attacker, it needs to be stored securely and encrypted during transport, for example by use of a channel encryption. In order to increase the level of protection, the system might consider authorisation to further

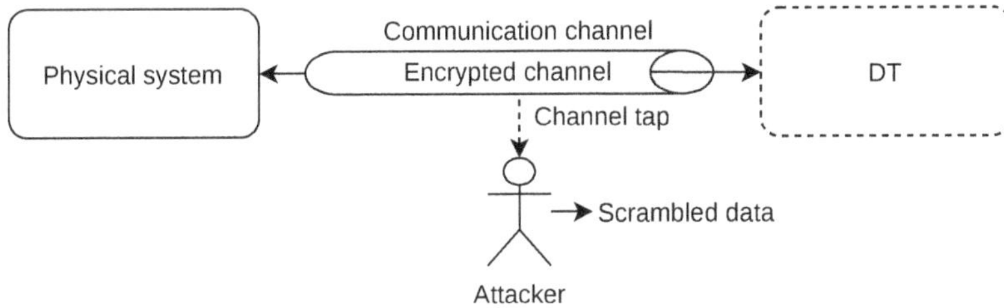

Fig. 13.11: Channel encryption mitigating Man-in-the-middle attack within a DT-enabled system

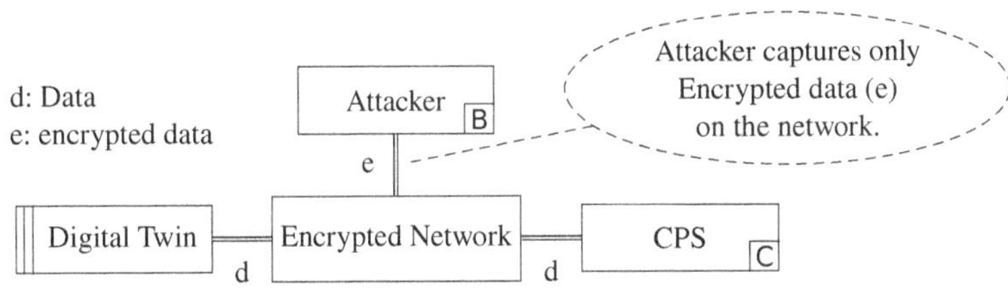

Fig. 13.12: Context diagram for man-in-the-middle attack mitigation.

limit which authenticated entities can issue what types of commands. The choice of the appropriate approach is dependent on the system complexity and the ability of the system operator to manage the authentication (i.e., keep a database of communication entities) or potential authorisation (by managing ACLs or RBAC).

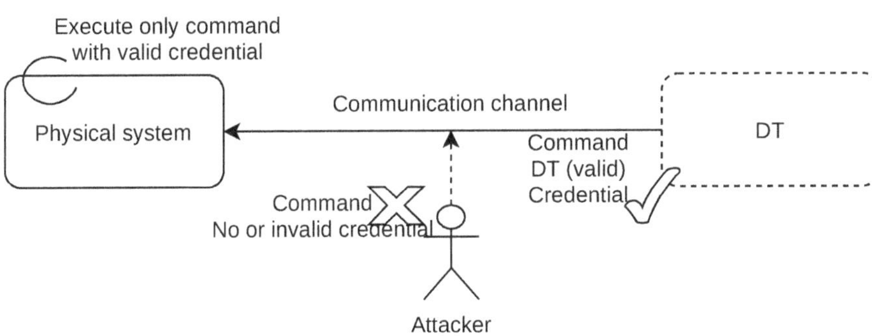

Fig. 13.13: Authentication mitigating the command injection attack within a DT-enabled system

The *Denial of Service* attack often provides a challenge for successful mitigation. One way of mitigating DoS attacks is by using request gateways [20] that can filter out traffic identified as potential DoS attack. This requires extra hardware and such a mitigation can introduce latency, limiting its viability for some systems. This approach is shown in Fig. 13.14. The PT itself and/or the DT shall have mitigation measures implemented such as refusing traffic from a specific source if the number of data packets exceeds a specific threshold. This approach prevents the computing entities from attempting to process all of the data and become overloaded. The filtering algorithm, however, needs to be implemented directly within the system what might lead to false positives. Potential improvement within a DT-enabled system is utilisation of a DT for traffic monitoring, where the DT could issue a command to the PT to drop specific traffic if a DoS attack is suspected. As in other aspects of security, the chosen approach needs to take into account the goal of the DT-enabled system.

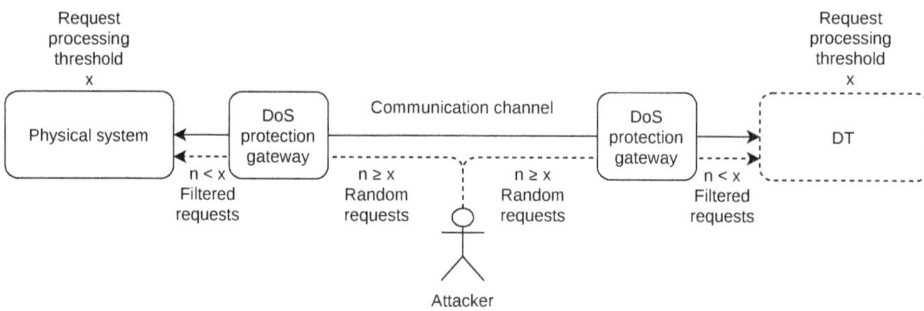

Fig. 13.14: Denial of Service mitigation using a network gateway

The *Data modification* attack can be mitigated using cryptographic means such as using signatures created from data packets computed from the identity of the system components and the data value itself. This permits the creation of a signature that cannot be simply recreated by the attacker, and any change within the data would not correspond to the signature. The mechanism should be understood as computing a hash of the value and then signing it by the private key of the entity providing this data. This approach is shown in Fig. 13.15. The entity receiving the messages needs to trust the public keys of the entity providing the messages. It is important that the legitimate private keys are protected from possible attackers as if an attacker gains an access to a private key, it is possible to modify the data, re-sign and it and the data would be accepted by the system as genuine. The protection of private keys is a wide topic beyond the scope of this book. However, approaches such as Trusted Platform Modules or software-based key storage systems should be utilised.

The *Model malware* attack can be mitigated by model integrity checks and the setup of authorisation for model contributions. This is especially important when third party libraries are used. These should be scanned for vulnerabilities and reviewed before being integrated into the DT model. The necessary depth of reviews and scanning depends on the criticality of the system. The approach is shown in Fig. 13.16. All the changes to the model should also be traceable to individual de-

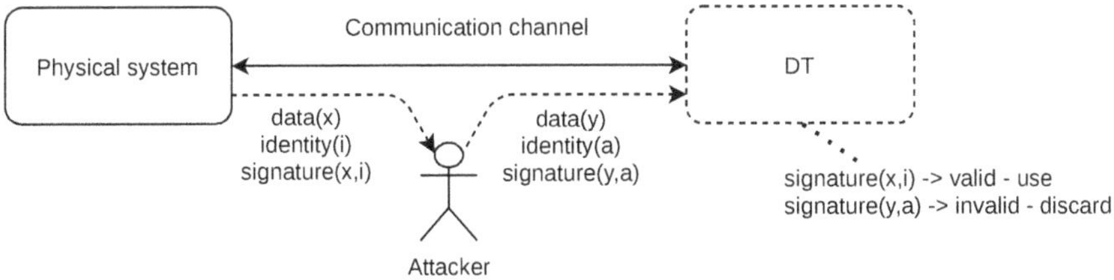

Fig. 13.15: Mitigation of data modification attack

velopers and any updates to the third party libraries should trigger the scanning and review process. This mitigation can be extended to protect against insider attacks by attaching multiple reviewers and automated scanning to the process. Furthermore, formal methods can be utilised at runtime to continuously check the system security properties are not being violated.

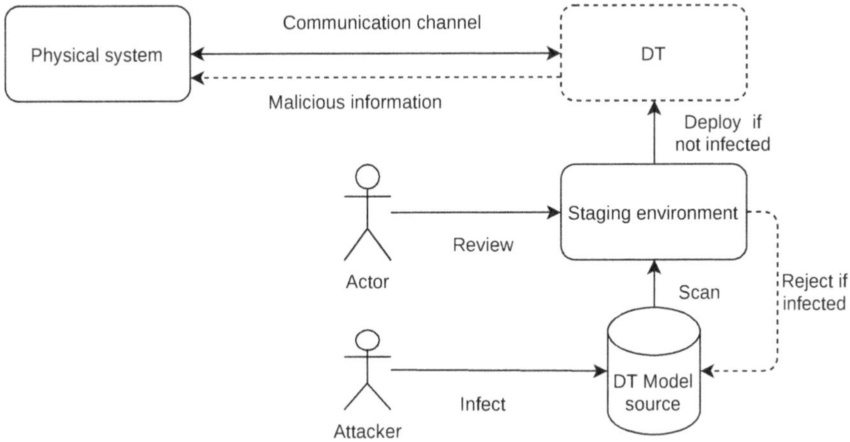

Fig. 13.16: Mitigation of model malware attack via scanning and review process

The approach of using a formal model in the loop to provide continuous assurance that specific security properties hold has been applied in industrial control systems [52] where a formalism known as Vienna Development Method (VDM) [13] has been used (see Sect. 5.6). This approach could be carried out in two ways. In the first option, the formal model understands the networking protocol between the system components and ensures that all the communication conforms to the expected protocol. In the second option, the model includes internal workings of the system and as such can monitor properties of the actual system component functionality rather than just the networking protocol. Both approaches can be combined to provide coverage of security properties affecting the network as well as the functionality.

This approach can be considered as creation of a **formal security DT** for the system, hence extending the system with an extra DT service specifically aimed at monitoring the security properties and providing formal guarantees for secure execution. The approach utilised within the industrial control systems uses a formal model m that expresses protocol and execution scenarios s_i where each scenario is bound to a security contract expressed as pre-condition *pre*, post-condition *post* and global invariant *inv*. This binding then creates a security property p_i.

During the system execution the security critical scenarios, matching the scenarios formally expressed within the model are evaluated as:

$$holds(p_i) = (m.s_i.pre \wedge m.s_i.post \wedge m.inv)$$

while $holds(p_i)$ is satisfied the system continues its execution as normal, however if the property p_i is violated the operator is informed and potential automated mitigation measures could be taken, such as disconnecting an affected component or switching to a fully isolated mode. This approach, while valuable, presents an additional learning curve for system designers and as such is more suited to security-critical functionality.

13.3.4 Attack Detection in DTs

Cyber attack detection is the process of identifying and recognising malicious activities or unauthorised actions in the CPS and the DT. The goal of the cyber attack detection mechanisms is to discover and respond to security breaches promptly in different layers of the CPS including the DT, communication channels, sensors, actuators, etc., and identify malicious activities before they can compromise the integrity of a DT and cause significant damage. Attack detection is a critical component of a robust DT cybersecurity strategy in addition to prevention and mitigation measures like firewalls, encryption and access controls that are discussed further in Sect. 13.3.3. Rapid identification and response to cyber attacks has the following advantages:

Early intervention: Fast detection and swift response to a cyber attack significantly reduces the potential impact on safety and performance [14]. It also prevents attackers from escalating their actions to cause severe impact.

Reduced damage: Identification and neutralisation of cyber attacks reduces the extent of damage such as service interruptions, financial losses, safety risks, etc [43].

Protecting critical infrastructure: In industry sectors such as transportation, energy, agriculture, or health care, cyber attacks have a great impact on public safety and national security [51] and thus, detecting cyber attacks is very important to protect these vital sectors.

Preserving and enhancing the DT integrity: Attack detectors reveal the security weaknesses of the DT and this knowledge can be used to cover security gaps and to strengthen DTs' security posture and integrity [19].

Protection strategies such as those discussed above can be devised to reduce the probability of experiencing successful attacks. However, if such attacks are successful, they can have expensive or catastrophic consequences for the PT. Given the risks, effective detectors should be devised and embedded within the DT to identify security breaches in the DT infrastructure.

13.3.4.1 Design considerations for DT attack detectors

The design consideration of a DT attack detector is briefly discussed in the following.

High accuracy: The accuracy of the DT attack detectors should be high to minimise false positive alarms which may endanger DTs security [19]. This is important to avoid unnecessary disruption and downtime, and reduce the burden on cybersecurity teams. False negatives should also be minimised to ensure the reliability of the detector and DT as a whole. The detector should also be robust against disturbances and noise that could negatively affect the accuracy [46].

Fast detection: The detection speed should be high so that cyber attacks can be detected soon after they are launched, so as to minimise the disruption and time needed to respond and limit damage.

Low complexity: The attack detector mechanism should not be complex so as to maintain its responsiveness and speed and also to avoid imposing additional burdens on DT computations [25].

Low impact on nominal performance: The attack detector should be designed to have a minimal impact on the performance of the DT and ensure smooth operations [46]. As will be discussed in Sect. 13.3.4.3, some detection methods against replay attacks such as Linear–Quadratic–Gaussian (LQG)-based techniques [41] are based on slight perturbation or violation of nominal system performance, which will cause non-optimal operation.

Regular updates and calibration: The nature of a DT requires it to be continuously updated based on the conditions of the PT, which creates possibilities for emerging threats [30]. Detectors should be regularly updated and/or calibrated (either manually or automatically) to ensure they remain effective over time.

Configurability: The attack detector should be configurable to adapt to the unique requirements of the DTs in different applications, systems and processes [25].

Scalability: The attack detector should be scalable to handle large volumes of data and traffic in dynamic DT environments [25].

It should be noted that the exact requirements and design considerations of the DT attack detectors depend on the type of DT Sect. 3.6, the application area and the corresponding security requirements. Attack detectors should be strategically placed at key points within the DT to ensure comprehensive coverage. Examples may include the DT application layer (API and software interfaces, etc.), communication channels (e.g., for detecting man-in-the-middle attacks), databases (e.g., to identify unauthorised access to data), middleware, etc.

13.3.4.2 Motivational Examples Showing Significance of Attack Detection

Cyber attacks, if they remain undetected, can lead to catastrophic failures in DTs. Several past cyber attack incidents have been reported in the literature emphasising the critical need for effective detection mechanisms. For example, on August 2003, a cyber attack was launched against power system controllers that led to an electric power outage in the Midwest United States and Ontario, Canada, resulting in an estimated economic loss of about $6 billion [28]. A cyber attack was also launched against electricity distribution companies in Ukraine in December 2015, which impacted almost 225,000 customers [7]. A large-scale attack on SCADA systems was also reported in 2010 which directly exploited targeted control loops of centrifuges in uranium-enrichment plants in Iran [34]. The plant was first compromised to record its information silently and use it to manipulate sensing and actuating data, destabilise and destroy the controlled processes. During the attack, the supervisory controllers were provided with false data, to mislead plant operators into thinking that processes are running in normal conditions. This type of attack is termed a *replay cyber attack*.

These examples show that designing effective attack detectors is critical to ensuring DT integrity, availability and confidentiality. It is therefore appropriate to consider attack detection mechanisms in DTs from a control engineering perspective.

13.3.4.3 Attack Detection Strategies

Different attack types and scenarios have been reviewed in Sect. 13.2. The variety of attack types and scenarios means that special and dedicated attack detectors may need to be used for the detection of different attack types against DTs. For example, a detector for the identification of a False Data Injection (FDI) attack may not be effective for the detection of a replay attack. This subsection studies different attack detection methods but multi-attack detection (i.e., detecting multiple attacks with one detector) is beyond the scope. Likewise, we clarify that the methods reviewed for detecting attacks may be different from tools for the identification of faults which are unintentional incidents and rather deal with technical issues, software bugs, hardware failures, or misconfigurations.

In the following we consider attack detector design for three types of cyber attack applicable to DT-enabled systems introduced in Sect. 13.2, including command injection, DoS and data modification. To this end, we first need to formulate the attack models and their impact on the system parts including DT and PT. Fig. 13.17 shows the positioning of the DT, the PT and the attacker. The PT is formed of the physical plant, related sensors and actuators. In Fig. 13.17, u_p and y_p are the input and output of the PT, respectively. Likewise, u is the command signal calculated by the controller and transmitted to the PT and y is the PT output received by the DT-enabled system. It should be noted that the input signal received by the PT (u_p) is equal to the command signal calculated by the controller in the DT (u) under normal conditions without any cyber attacks manipulating the command signal. An attack detector is placed in the DT-enabled system and can read the command signal (u) and signal received from the PT output (y). Signal transmission between the PT and the DT parts are accomplished through communication channels. Three main types of attacks against communication channels illustrated in Fig. 13.3 to 13.5 can be represented as switches that can disconnect the data transmission in case of DoS attacks, attack signals a_u injected to command signal u in case of command injection and attack signals (a_y) injected to the data transferred from the PT (y_p) in case of data modification. The command injection and data modification attacks are considered as two different types of attacks in the FDI attack category. In the following we discuss the design process of FDI attack detectors in the general case.

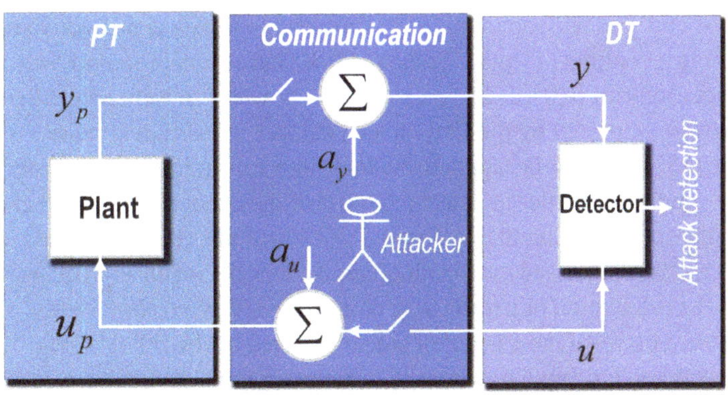

Fig. 13.17: Block diagram of a DT-enabled system in the presence of cyber attacks.

According to Fig. 13.17, the physical plant in the PT can be modelled using a state space representation to describe the system's dynamic behaviour by defining its state variables, inputs, outputs and the relationships among them. Considering the system matrix $\mathbf{F} \in \mathbb{R}^{n\times n}$, the input matrix $\mathbf{B} \in \mathbb{R}^{n\times p}$, the output matrix $\mathbf{C} \in \mathbb{R}^{m\times n}$ and the attack matrices $\mathbf{H} \in \mathbb{R}^{n\times q}$ and $\mathbf{D} \in \mathbb{R}^{m\times q}$, the state space representation of the physical plant can be written as follows:

$$\begin{aligned} \mathbf{x}(k+1) &= \mathbf{Fx}(k) + \mathbf{Bu}(k) + \mathbf{Ha}(k) + \mathbf{w}(k) \\ \mathbf{y}(k) &= \mathbf{Cx}(k) + \mathbf{Da}(k) + \mathbf{v}(k) \end{aligned} \tag{13.1}$$

where $\mathbf{x}(k) \in \mathbb{R}^n$ is the state vector. The plant is assumed to be subjected to process and measurement noises $\mathbf{w}(k) \in \mathbb{R}^n$ and $\mathbf{v}(k) \in \mathbb{R}^m$, respectively. Vector $\mathbf{a}(k) \in \mathbb{R}^q$ also models a specific cyber attack and can be composed of a_u and a_y (see Fig. 13.17) as $\mathbf{a}(k) = \begin{bmatrix} a_u \\ a_y \end{bmatrix}$.

FDI Attack Detection

The FDI attack against physical plant can be modelled considering $\mathbf{a}(k) \neq 0$ in Eq. (13.1). Three different FDI cases can be considered [47]:

1. command injection ($a_u \neq 0, a_y = 0$), in which the attack injects false data into the command signal u;
2. data modification ($a_y \neq 0, a_u = 0$), in which false data is injected into the PT output vector y_p;
3. coordinated FDI ($a_y \neq 0, a_u \neq 0$) in which both u and y_p are subjected to the attack.

Such attacks can influence the process states, control and measurements of the sensors.

The FDI attack detection can be done by determining the value of the attack vector, i.e., to decide whether $\mathbf{a}(k) \neq 0$ or $\mathbf{a}(k) = 0$. This helps operators to determine the attack type and affected communication links and to subsequently activate an alarm and possibly a contingency plan. Some methods have been proposed to estimate the attack vector entries $a(0), a(1), \ldots, a(k)$, which will precisely reveal the nodes or data that were impacted by the FDI attack [47]. A recursive generalised likelihood-ratio test [55] can also be applied on real-time transmission data from physical plant to/from the controller to detect the onset of the cyber attack. Another approach based on the so-called Cumulative Sum (CUSUM) test has been discussed in [5]. The CUSUM test is categorised into residue-based attack detection approaches which are based on the calculation of a residual vector. The residual vector (denoted as $\mathbf{r}(k)$) is the difference between the PT output signal received by the DT ($\mathbf{y}$ in Fig. 13.17) and its estimated value using a dynamic state estimation mechanism such as a Kalman filter. The Kalman filter utilises the same system, input and output matrices of $\mathbf{F}$, $\mathbf{B}$ and $\mathbf{C}$, respectively, as the physical plant model (defined in Eq. (13.1)) and estimates the system state vector ($\mathbf{x}(k)$ in Eq. (13.1)) as follows:

$$\begin{aligned} \hat{\mathbf{x}}(k+1) &= \mathbf{F}\hat{\mathbf{x}}(k) + \mathbf{B}\mathbf{u}(k) + \mathbf{K}\mathbf{r}(k) \\ \mathbf{r}(k) &= \mathbf{y}(k) - \mathbf{C}\hat{\mathbf{x}}(k) \end{aligned} \tag{13.2}$$

where $\mathbf{K}$ is the Kalman gain that can be tuned with the data from non-attack conditions and $\hat{\mathbf{x}}(k) \in \mathbb{R}^n$ is the estimated state vector. In the presence of the attack vector ($\mathbf{a}(k) \neq 0$), the estimated values by the dynamic state estimator Eq. (13.2) deviates from the actual values and therefore the value of the residual vector $\mathbf{r}(k)$ increases. In each step, the residual vector $\mathbf{r}(k)$ is calculated using Eq. (13.2).

The attack detection is fulfilled based on determining a proper detection criterion. For instance, in case of the χ^2 detector [16], the following detection criterion is defined:

$$\begin{cases} |\mathbf{r}(k)|^2 < \tau & \text{no alarm} \\ |\mathbf{r}(k)|^2 \geq \tau & \text{alarm} \end{cases} \tag{13.3}$$

Despite their simplicity, the residual-based detection tests may be challenged to detect some attack types such as undetectable and stealthy attacks. For instance, for the χ^2 detector, deterministic stealthy attacks can be designed by attackers to bypass bad data detectors satisfying the following constraint for $k \geq 0$ [47]:

$$|\mathbf{r}(k; y, x, u, a)|^2 \leq \tau + \sigma \tag{13.4}$$

where $\mathbf{r}(k; y, x, u, a)$ is the residual signal under the attack condition. By choosing $\sigma = 0$, the attack will not generate an alarm signal for the χ^2 detector.

The foregoing problem has been addressed in several studies in the literature involving techniques based on Kullback–Leibler divergence [9], interleaved hop-by-hop authentication [57] and generalised likelihood ratio [39] and most recently methods based on state observers and estimators.

Replay attack detection

In replay attacks, delayed versions of the real sensor measurements (y_p) are transmitted to the controllers in the DT, and at the same time, the command signal ($\mathbf{u}$) is manipulated to degrade the DT-enabled system's performance. An example of a replay attack has been discussed in [41] which collects online sensory measurements between the sensors and the controllers, and changes the correct measurements to the values that were previously recorded. Assuming the replay attack starts at time $k = k_0$ for T steps, one can write:

$$y_a(k) = \mathbf{y}(k - T), \quad 0 \leq k \leq T - 1, \tag{13.5}$$

After this step, the cyber attack tries to change the control actions or control commands transmitted from the controller by injecting an external signal $u_a(k)$ for $0 \leq k \leq T - 1$. Replay attacks are most dangerous when the system is at steady-state conditions when the measurement values correspond to well-regulated system states [47].

It is difficult to detect replay attacks with statistical test methods that are based on a sequence of measurements. To improve the detectability of such attacks, a basic technique is to change the controller output purposefully and randomly to make the system deviate slightly from the steady-state conditions. Such variations are not reflected in the injected or fake measurements applied by a replay attack which will reveal the occurrence of the attack.

As a typical example, one can assume a system described by Eq. (13.1) with $\mathbf{a}(k) \neq 0$. Assume the original controller is based on LQG control objective minimising the quadratic cost and the optimised control command denoted as $\mathbf{u}^*$ is obtained. The replay attacker collects the system responses to the control command $\mathbf{u}^*$. Once the attacker injects false data to the system, it changes the actual responses to the recorded values to remain undetectable from residual-based detectors such as ones discussed in Eqs. (13.2) to (13.4). To assure replay attack detectability, the control signal can be redesigned as follows:

$$\mathbf{u}(k) = \mathbf{u}^*(k) + \Delta\mathbf{u}(k). \tag{13.6}$$

where $\Delta\mathbf{u}$ is considered as an authentication signal independent of $\mathbf{u}^*$ and is derived from an i.i.d. Gaussian distribution with zero mean. The system response to the redesigned control signal Eq. (13.6) deviates from the values recorded and injected into the system by the attacker and thus the attack will be detected. While this approach improves detectability, the control performance will need to be sacrificed to some extent since the control signals are no longer fully optimised in the LQG sense [41].

DoS Attack Detection

DoS attacks usually have severe impacts on the operation of the system since they will cause temporary or continuous interruptions in the feedback control loops which may destabilise a complete plant, system, or process [47]. In the control sense, DoSs can be modelled by introducing switches as shown in Fig. 13.17. Whenever the DT-enabled system is under attack conditions, the corresponding switch will transit to the open state. The attack can be represented as:

$$u_p(k) = \gamma(k)\mathbf{u}(k), \quad \mathbf{y}(k) = \eta(k)y_p(k) \tag{13.7}$$

where u_p is the command signal applied to the PT and y_p is the PT output. The binary variables $\gamma(k)$ and $\eta(k)$ represent DoS attacks, such that $\gamma(k) = 0$ if the control command communication is interrupted by an attack and $\gamma(k) = 1$ otherwise, and $\eta(k)$ is defined in a similar way. Detection methods for DoSs are diverse and the choice of method depends on several parameters such as anomaly type, data types and sensitivity, complexity constraints and the required safety levels that must be assured [45]. The existing techniques can be classified into Signature-based, Anomaly-based and Hybrid detection as described in [2]. Likewise, Artificial Intelligence-based methods have become popular in the detection of DoS attacks [48]. A full review of DoS attack detection methods can be found in [26].

13.4 Intellectual Property Protection

Protection of IP is an important aspect of any commercial system. In the DT context the IP regarding the inner workings of the system might be inferred from a DT without the need to reverse engineer the physical system itself. This leads to the need to protect the DT model. It is of course important to manage access to the repositories where the model source is being stored, but precautions also need to be taken when the model is deployed within an active DT. Technical means such as obfuscation and encryption at rest can be applied at relatively low cost. Access to the system shall be monitored for all users, and a clear licence needs to be issued for use of the model. An interesting aspect of a DT-enabled system is that a system could be represented by a multitude of DTs that need to be interconnected.

Let us consider a DT of an engine, where the mechanical parts have a DT model representing them, fuel injection control has its own DT model and the oil filtration subsystem has its own DT model, see Fig. 13.18. These different models would often be provided by different enterprises that are sub-suppliers of the engine builder. While the engine builder would be able to access the state of the entire engine, deep diagnosis of the injection profile could for example only be visible to the supplier of the injection subsystem. This means that clear isolation of data access and model access needs to be available within the system. Similarly, the oil filtration device manufacturer might be interested in the performance of the oil filtration. However they should not learn about inner workings of the fuel injection subsystem. This is partially achievable by use of authorisation frameworks, where a set of authorisation rules guides specific user access only to the parts that should be available to those users. To this end, the DT can implement ACLs, where engine builder users would get overall engine read privileges that would include the sub-supplier subsystems on a high level. The fuel injection sub-system manufacturer would then have a diagnostic (detailed) read privilege within the fuel injection DT providing data that are usable to infer the inner workings of the subsystem.

The platform on which the DT is composed also needs to prevent IP disclosure between different parties. Specifically, the hardware, operating system and a potential hypervisor need to provide functionality for component isolation. On the operating system level this can be achieved by process isolation techniques, where different processes executed on top of the operating system forbid access to each other resources, most often the memory allocated by the process. The techniques could further be enforced by using cryptographic hardware such as trusted platform modules or in specific cases (if supported by CPU) per thread memory encryption. Recently, isolation techniques became available even for processes requiring fast domain switching, such as creation of network facing cryptographic sessions [54], though research within this area is rapidly advancing. In cases where fast switching is not necessary, a hypervisor (a virtual machine monitor and executor) can provide its own isolation, which is often easier to achieve as it can allocate a specific memory area for the virtualised context it is executing and hence can encrypt this entire

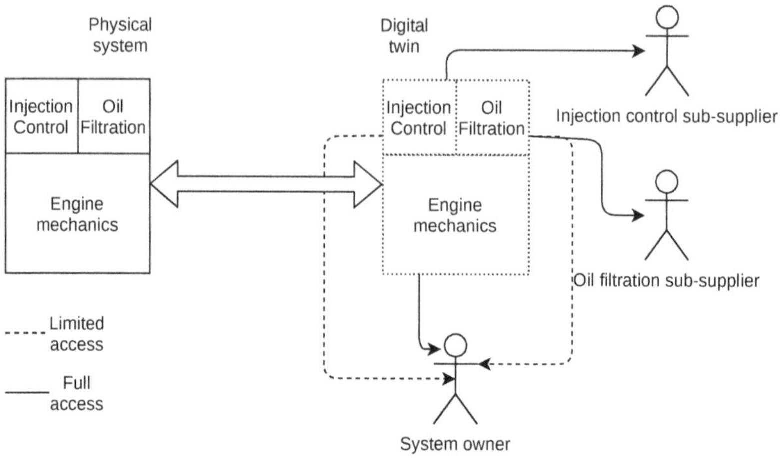

Fig. 13.18: Composite system with corresponding DT and multiple users

memory space under a specific key (rather than dynamically creating keys for every resource of every thread within the memory). Similar measures are often utilised by hypervisors for data-at-rest protection as well as network components especially within multi-tenancy environments such as cloud based platforms [33]. If there is a need to be stricter than software-based isolation within cloud platforms, it is often possible to utilise single tenant hosting, where the hosting hardware is assigned to a single tenant, what however often comes at a higher financial cost.

The primary driver for the decision on where and how to deploy a composite DT is the consideration of trust between the different entities as well as mission criticality of the DT.

13.5 Security in the Real World

An important aspect of any secure system is the use of correct security approaches to meet the security goal for the system. It is never utilisation of just one technique or approach; different security techniques need to work together in order to provide the overall security. Let us consider the composite DT once more. Systems of this type would need to consider secure network, secure user access, isolation of components and correct execution. This means that approaches such as network and data encryption, use of authentication and authorisation for users, isolation techniques such as use of hypervisors, security analysis by use of formal methods, and detection and mitigation mechanisms to prevent performance degradation in the presence of attacks would be used at the same time. This leads to more detailed decisions, such as how to encrypt the network traffic, what authorisation model should be used, what level of isolation is needed, whether the use of formal analysis is warranted within the specific system, and how to deal with complexity of implementation and execution of Kalman filter-based detectors in real world applications where assumptions of system model, process and measurement noises are not met. There is not a single answer as it depends on the system under consideration. For example, restarting a device could be a simple attack mitigation strategy for an IT system, but interrupting an autonomous vehicle or an infrastructure system could be impossible in most practical situations. However, it is important that the system designer poses these questions and considers security as a guiding principle within the system design.

This chapter has presented some of the basic concepts applicable within secure development, but only at an introductory level. The use of encryption, for example, comes with caveats in cases where potential real-time deadlines are to be considered. It is also important that the system designers have a good understanding of primitives of encryption and authorisation in order to make the correct decisions. There are often considerations such as speed of execution, user friendliness, reconfiguration flexibility and many more that can clash with the security design. It is therefore important to focus on the risk assessment and decide on the baseline for the security of the system. The baseline dictates the minimum amount of security mechanisms for the system to be considered production secure. In case of a DT-enabled system, specific consideration needs to be given to the DT itself and its communication with the physical system, however tried and trusted approaches to secure control and computing as presented here could be utilised in the DT paradigm.

It must be said that only a system that does nothing can be considered truly secure. In the real world it is important to have systems secured to meet their designated goals. A final note for a secure system is the need not only to design and deploy the system securely, but also continuously review the threat landscape and update the system with new mitigations against vulnerabilities that were not known at a design time. As such the risk assessment and threat analysis are activities that shall be maintained during the lifetime of the system.

References

1. Acar, A., Aksu, H., Uluagac, A.S., Conti, M.: A Survey on Homomorphic Encryption Schemes: Theory and Implementation. ACM Comput. Surv. **51**(4) (2018). DOI: 10.1145/3214303
2. Agarwal, B., Mittal, N.: Hybrid approach for detection of anomaly network traffic using data mining techniques. Procedia Technology **6**, 996–1003 (2012)
3. Alcaraz, C., Lopez, J.: Digital twin: A comprehensive survey of security threats. IEEE Communications Surveys & Tutorials **24**(3), 1475–1503 (2022). DOI: 10.1109/COMST.2022.3171465
4. Alramadhan, M., Sha, K.: An overview of access control mechanisms for Internet of Things. In: 2017 26th International Conference on Computer Communication and Networks (ICCCN), pp. 1–6 (2017). DOI: 10.1109/ICCCN.2017.8038503

5. Basseville, M., Nikiforov, I.V., et al.: Detection of abrupt changes: theory and application, vol. 104. prentice Hall Englewood Cliffs (1993)
6. Bihis, C.: Mastering OAuth 2.0. Packt Publishing Ltd (2015)
7. Case, D.U.: Analysis of the cyber attack on the Ukrainian power grid. Electricity Information Sharing and Analysis Center (E-ISAC) **388**(1-29), 3 (2016)
8. Casola, V., De Benedictis, A., Rak, M., Villano, U.: A novel Security-by-Design methodology: Modeling and assessing security by SLAs with a quantitative approach. Journal of Systems and Software **163**, 110537 (2020). DOI: `https://doi.org/10.1016/j.jss.2020.110537`
9. Chaojun, G., Jirutitijaroen, P., Motani, M.: Detecting false data injection attacks in AC state estimation. IEEE Transactions on Smart Grid **6**(5), 2476–2483 (2015)
10. Clarke, E.M., Klieber, W., Novácek, M., Zuliani, P.: Model checking and the state explosion problem. In: B. Meyer, M. Nordio (eds.) Tools for Practical Software Verification, LASER, International Summer School 2011, Elba Island, Italy, Revised Tutorial Lectures, *Lecture Notes in Computer Science*, vol. 7682, pp. 1–30. Springer (2011)
11. De Maeyer, C., Markopoulos, P.: Future outlook on the materialisation, expectations and implementation of digital twins in healthcare. In: 34th British HCI Conference 34, pp. 180–191 (2021)
12. Feng, H., Chen, D., Lv, H., Lv, Z.: Game theory in network security for digital twins in industry. Digital Communications and Networks (2023). DOI: `https://doi.org/10.1016/j.dcan.2023.01.004`
13. Fitzgerald, J.S., Larsen, P.G., Verhoef, M.: Vienna Development Method. Wiley Encyclopedia of Computer Science and Engineering (2008). Edited by Benjamin Wah, John Wiley & Sons, Inc.
14. Gandi, V.P., Jatla, N.S.L., Sadhineni, G., Geddamuri, S., Chaitanya, G.K., Velmurugan, A.K.: A comparative study of AI algorithms for anomaly-based intrusion detection. In: 2023 7th International Conference on Computing Methodologies and Communication (ICCMC), pp. 530–534 (2023). DOI: `10.1109/ICCMC56507.2023.10084186`
15. Gehrmann, C., Gunnarsson, M.: A digital twin based industrial automation and control system security architecture. IEEE Transactions on Industrial Informatics **16**(1), 669–680 (2020). DOI: `10.1109/TII.2019.2938885`
16. Giraldo, J., Urbina, D., Cardenas, A., Valente, J., Faisal, M., Ruths, J., Tippenhauer, N.O., Sandberg, H., Candell, R.: A survey of physics-based attack detection in cyber-physical systems. ACM Computing Surveys (CSUR) **51**(4), 1–36 (2018)
17. Guggenmos, F., Häckel, B., Ollig, P., Stahl, B.: Security First, Security by Design, or Security Pragmatism – Strategic Roles of IT Security in Digitalization Projects. Computers & Security **118**, 102747 (2022). DOI: `https://doi.org/10.1016/j.cose.2022.102747`
18. Hensel, C., Junges, S., Katoen, J.P., Quatmann, T., Volk, M.: The probabilistic model checker Storm. International Journal on Software Tools for Technology Transfer **24**(4), 589–610 (2022). DOI: `10.1007/s10009-021-00633-z`
19. Holmes, D., Papathanasaki, M., Maglaras, L., Ferrag, M.A., Nepal, S., Janicke, H.: Digital twins and cyber security – solution or challenge? In: 2021 6th South-East Europe Design Automation, Computer Engineering, Computer Networks and Social Media Conference (SEEDA-CECNSM), pp. 1–8 (2021). DOI: `10.1109/SEEDA-CECNSM53056.2021.9566277`
20. Huang, X., Du, X., Song, B.: An effective DDoS defense scheme for SDN. In: 2017 IEEE International Conference on Communications (ICC), pp. 1–6 (2017). DOI: `10.1109/ICC.2017.7997187`
21. Iqbal, Z., Mehmood, A., Ghafoor, A., Ahmed, H.F., Shibli, A.: Authenticated service interaction protocol for multi-agent system. In: 2007 International Symposium on High Capacity Optical Networks and Enabling Technologies, pp. 1–5 (2007). DOI: `10.1109/HONET.2007.4600276`
22. Jackson, M.: Software Requirements and Specification: a lexicon of practice, principles and prejudices. Addison-Wesley (1995)

23. Jackson, M.: Problem Frames: Analysing and Structuring Software Development Problems. Addison-Wesley, New York (2001)
24. Jeske, S.J.: Digital twins in healthcare. Paris Lodron University of Salzburg and Aalborg University in Copenhagen (2020)
25. Kandasamy, N.K., Venugopalan, S., Wong, T.K., Leu, N.J.: An electric power digital twin for cyber security testing, research and education. Computers and Electrical Engineering **101**, 108061 (2022). DOI: https://doi.org/10.1016/j.compeleceng.2022.108061
26. Kaur, P., Kumar, M., Bhandari, A.: A review of detection approaches for distributed denial of service attacks. Systems Science & Control Engineering **5**(1), 301–320 (2017)
27. Kayan, H., Nunes, M., Rana, O., Burnap, P., Perera, C.: Cybersecurity of industrial cyber-physical systems: A review. ACM Comput. Surv. **54**(11s) (2022). DOI: 10.1145/3510410. URL https://doi.org/10.1145/3510410
28. Kazemi, Z., Safavi, A.A., Setoodeh, P.: Efficient resilient dynamic co-estimation framework for cyber-physical systems under sensor attacks. IET Control Theory & Applications **14**(20), 3526–3536 (2020)
29. Khan, R., McLaughlin, K., Laverty, D., Sezer, S.: Stride-based threat modeling for cyber-physical systems. In: 2017 IEEE PES Innovative Smart Grid Technologies Conference Europe (ISGT-Europe), pp. 1–6 (2017). DOI: 10.1109/ISGTEurope.2017.8260283
30. Klein, M., Carpentier, T., Jeanclaude, E., Kassab, R., Varelas, K., de Bruijn, N., Barbaresco, F., Briheche, Y., Semet, Y., Aligne, F.: AI-augmented multi function radar engineering with digital twin: Towards proactivity. In: 2020 IEEE Radar Conference (RadarConf20), pp. 1–6 (2020). DOI: 10.1109/RadarConf2043947.2020.9266455
31. Kulik, T., Dongol, B., Larsen, P.G., Macedo, H.D., Schneider, S., Tran-Jørgensen, P.W.V., Woodcock, J.: A Survey of Practical Formal Methods for Security. Formal Aspects of Computing (2022). DOI: 10.1145/3522582
32. Kulik, T., Gomes, C., Macedo, H.D., Hallerstede, S., Larsen, P.G.: Towards secure digital twins. In: T. Margaria, B. Steffen (eds.) ISoLA 2022, *Lecture Notes in Computer Science*, vol. 13704. Springer (2022). DOI: 10.1007/978-3-031-19762-8_11
33. Kumar, V., Rathore, R.S.: Security issues with virtualization in cloud computing. In: 2018 International Conference on Advances in Computing, Communication Control and Networking (ICACCCN), pp. 487–491 (2018). DOI: 10.1109/ICACCCN.2018.8748405
34. Kushner, D.: The real story of Stuxnet. IEEE Spectrum **50**(3), 48–53 (2013)
35. Lamport, L.: Specifying systems: the TLA+ language and tools for hardware and software engineers. Addison-Wesley Longman Publishing Co., Inc. (2002)
36. Lamport, L.: The PlusCal algorithm language. In: International Colloquium on Theoretical Aspects of Computing, pp. 36–60. Springer (2009)
37. Langner, R.: Stuxnet: Dissecting a cyberwarfare weapon. IEEE Security & Privacy **9**(3), 49–51 (2011). DOI: 10.1109/MSP.2011.67
38. Leines-Vite, L., Arriaga, J.C.P., Limón, X.: Confidentiality and Integrity Mechanisms for Microservices Communication. CoRR (2021). URL https://arxiv.org/abs/2111.01218
39. Li, S., Yılmaz, Y., Wang, X.: Quickest detection of false data injection attack in wide-area smart grids. IEEE Transactions on Smart Grid **6**(6), 2725–2735 (2014)
40. Löcklin, A., Müller, M., Jung, T., Jazdi, N., White, D., Weyrich, M.: Digital twin for verification and validation of industrial automation systems – a survey. In: 2020 25th IEEE International Conference on Emerging Technologies and Factory Automation (ETFA), vol. 1, pp. 851–858 (2020). DOI: 10.1109/ETFA46521.2020.9212051
41. Mo, Y., Sinopoli, B.: Secure control against replay attacks. In: 2009 47th annual Allerton conference on communication, control, and computing (Allerton), pp. 911–918. IEEE (2009)
42. Naseri, F., Kazemi, Z., Larsen, P.G., Arefi, M.M., Schaltz, E.: Cyber-physical cloud battery management systems: Review of security aspects. Batteries **9**(7) (2023). DOI: 10.3390/batteries9070382

43. Negrete-Pincetic, M., Yoshida, F., Gross, G.: Towards quantifying the impacts of cyber attacks in the competitive electricity market environment. In: 2009 IEEE Bucharest PowerTech, pp. 1–8 (2009). DOI: 10.1109/PTC.2009.5282237
44. Pfretzschner, B., Othmane, L.b.: Dependency-based attacks on node.js. In: 2016 IEEE Cybersecurity Development (SecDev), pp. 66–66 (2016). DOI: 10.1109/SecDev.2016.023
45. Raut, A.S., Singh, K.R.: Anomaly based intrusion detection-a review. International Journal on Network Security **5**(3), 7 (2014)
46. Saad, A., Faddel, S., Youssef, T., Mohammed, O.A.: On the implementation of IoT-based digital twin for networked microgrids resiliency against cyber attacks. IEEE Transactions on Smart Grid **11**(6), 5138–5150 (2020). DOI: 10.1109/TSG.2020.3000958
47. Sandberg, H., Gupta, V., Johansson, K.H.: Secure networked control systems. Annual Review of Control, Robotics, and Autonomous Systems **5**(1), 445–464 (2022). DOI: 10.1146/annurev-control-072921-075953
48. Singh, N., Hans, A., Kumar, K., Birdi, M.P.S.: Comprehensive study of various techniques for detecting DDoS attacks in cloud environment. International Journal of Grid and Distributed Computing **8**(3), 119–126 (2015)
49. Sridhar, S., Manimaran, G.: Data integrity attacks and their impacts on SCADA control system. In: IEEE PES General Meeting, pp. 1–6 (2010). DOI: 10.1109/PES.2010.5590115
50. Thakor, V.A., Razzaque, M.A., Khandaker, M.R.A.: Lightweight Cryptography Algorithms for Resource-Constrained IoT Devices: A Review, Comparison and Research Opportunities. IEEE Access **9**, 28177–28193 (2021). DOI: 10.1109/ACCESS.2021.3052867
51. Thakur, K., Ali, M.L., Jiang, N., Qiu, M.: Impact of Cyber-Attacks on Critical Infrastructure. In: 2016 IEEE 2nd International Conference on Big Data Security on Cloud (BigDataSecurity), IEEE International Conference on High Performance and Smart Computing (HPSC), and IEEE International Conference on Intelligent Data and Security (IDS), pp. 183–186 (2016). DOI: 10.1109/BigDataSecurity-HPSC-IDS.2016.22
52. Ubys, L., Vancea, V.N., Kulik, T., Larsen, P.G., Boudjadar, J., Aranha, D.F.: Formal model in-the-loop for secure industrial control networks. In: 18th International Conference on Formal Aspects of Component Software. Oslo (virtual) (2022)
53. Upadhyay, D., Sampalli, S.: Scada (supervisory control and data acquisition) systems: Vulnerability assessment and security recommendations. Computers & Security **89**, 101666 (2020). DOI: https://doi.org/10.1016/j.cose.2019.101666
54. Vahldiek-Oberwagner, A., Elnikety, E., Duarte, N.O., Sammler, M., Druschel, P., Garg, D.: ERIM: Secure, efficient in-process isolation with protection keys (MPK). In: 28th USENIX Security Symposium (USENIX Security 19), pp. 1221–1238 (2019)
55. Willsky, A., Jones, H.: A generalized likelihood ratio approach to the detection and estimation of jumps in linear systems. IEEE Transactions on Automatic control **21**(1), 108–112 (1976)
56. Woodcock, J., Larsen, P.G., Bicarregui, J., Fitzgerald, J.: Formal Methods: Practice and Experience. ACM Computing Surveys **41**(4), 1–36 (2009). DOI: http://doi.acm.org/10.1145/1592434.1592436
57. Zhu, S., Setia, S., Jajodia, S., Ning, P.: An interleaved hop-by-hop authentication scheme for filtering of injected false data in sensor networks. In: IEEE Symposium on Security and Privacy, 2004. Proceedings. 2004, pp. 259–271. IEEE (2004)
58. Zonouz, S., Rrushi, J., McLaughlin, S.: Detecting industrial control malware using automated plc code analytics. IEEE Security & Privacy **12**(6), 40–47 (2014)

Chapter 14
Autonomous Reconfiguration Enabled by Digital Twins

Lukas Esterle, Mirgita Frasheri and Peter Gorm Larsen

Abstract Digital Twins (DTs) integrate models with data gathered from Physical Twin (PT) counterparts, adding value through enablers and services that allow informed decision-making about the reconfiguration of the PT, either autonomously, or by human stakeholders. We consider the potential of DTs for improving the capabilities and performance of autonomous systems through two lenses. First, we outline how different agent architectures can integrate DT models to enhance decision-making and reconfiguration. Second, we consider how DT technology can further support self-* properties within autonomous (multi-)agent systems. We identify open challenges that we argue need to be tackled in order to improve the utilisation of DT technology for autonomously reconfiguring systems at runtime.

14.1 Introduction

The important difference between a DT and a Digital Shadow (DS) as discussed in Sect. 1.2.2 is the DT's ability to support the reconfiguration of the Physical Twin (PT) with a degree of autonomy. The DT thus allows the PT to operate in dynamic environments and respond to at least some changes in the environment. In this chapter, we aim to shed light on requirements and approaches to "closing the loop" by allowing the DT to support the reconfiguration process. This provides a basis for autonomous and self-adapting Cyber-Physical Systems (CPSs) while having the advantage of a DT that is able to marshal a set of models that offer different levels of granularity and performance.

Self-adaptation procedures can be very intricate, rely heavily on domain knowledge and are application-specific (e.g., a self-adaptation procedure for our running example of the tempeh incubator is very different to the self-adaptation loop for a robot manipulator). Incidentally, the people who know best how to reconfigure the CPS are not necessarily the ones who can best implement those automated proce-

J. Fitzgerald et al. (eds.), *The Engineering of Digital Twins*,
https://doi.org/10.1007/978-3-031-66719-0_14

dures. Therefore these CPSs have to rely on self-reconfiguration approaches utilising DTs.

In this chapter, we first describe autonomous CPSs, their architectures and the incorporation of DTs in general terms in Sect. 14.2. Thereafter, Sect. 14.3 details the self-* properties that are core to autonomous CPSs before focussing on self-awareness, a capability required by autonomous CPSs to make decisions on their abilities and how they can be reconfigured. In order to implement self-* properties, it is necessary to consider how we describe the goals that direct the decision-making about a reconfiguration, and this is discussed in Sect. 14.4. Afterwards, Sect. 14.5 discusses the capabilities required of autonomous agents to collaborate with each other, even in the absence of human operators. The components required to exploit DT technology to facilitate collaboration are highlighted. Then Sect. 14.6 discusses safety issues stemming from autonomous CPSs and how they might be addressed using DTs. Finally, a roadmap is provided in Sect. 14.7, indicating required research to bring our vision of autonomous DT-enabled systems to fruition.

14.2 Autonomous Systems and DTs

Autonomous CPSs are governed by autonomous software *agents*, able to respond to changing conditions, allowing higher flexibility in their deployment. An agent is generally defined as a computer system that is able to sense and act, through its sensors and actuators, in the environment in which it is placed, as well as carry out actions autonomously [39, 26]. This definition can be expanded to capture intelligent agents, provided that there is a capability for flexible autonomous actions [39]. Flexibility targets three main aspects of behaviour of agents such as: (i) social ability, i.e., being able to interact with others, (ii) reactivity to external stimuli and (iii) pro-activity, or in other words being able to decide autonomously on a course of action. Wooldridge argues that a balance between reactivity and pro-activity is crucial in intelligent agents, ensuring neither blind execution of procedures, nor blind reactivity impeding the agent from actually completing any goals [40]. Similarly for CPSs, sensing mechanisms are required to identify changes in the environment and actuators to perform actions on the environment. Dedicated decision-making mechanisms allow the CPS to determine which action to perform in a given situation [38].

We differentiate between *reactive* and *deliberative* agent architectures for autonomous systems. In the following, both types of agent are explored and expanded by introducing a DT. This means the autonomous CPS has a model of itself available that is constantly updated based on sensory input about the environment and the action it takes, and is able to provide feedback to the DT-enabled system.

We also differentiate between *loosely* and *tightly* coupled DT-enabled systems. A tightly coupled DT-enabled system requires the DT to operate and utilise all its capabilities. Specifically, for autonomous DT-enabled systems this would mean that all decisions are being made by the DT itself and imposed on the physical counterpart. In contrast, a *loosely* coupled DT-enabled system can operate without access to the DT and decisions can be made through other local mechanisms. This becomes relevant, when we consider the DT to be external to the system. A *loosely* coupled DT-enabled system will remain operational while a *tightly* coupled DT-enabled system will become unoperational. Loosely coupled DT-enabled systems are able to operate continuously but require synchronisation between the DT and the physical system when a re-connection is established. In the following, we will consider *loosely* coupled DT-enabled systems, replicating their decision-making approaches within the DT and the local DT-enabled system.

Reactive agent architectures operate on direct sensory inputs and immediate reactions to them. In contrast to other types of architectures, reactive agents focus on quick, reflexive responses to the current state of the environment. These architectures are designed to perform specific tasks efficiently without the need for complex planning or long-term memory.

Reactive agent architectures rely on a stimulus-response mapping to select the right actions in each situation. The agent receives sensory inputs from the environment and maps them directly to appropriate actions. This mapping is typically implemented through a set of predefined rules or behaviours which are often organised in a hierarchical or layered structure. During development, the goal is to define rules or behaviours allowing a response for each specific situation or stimulus. At runtime, the agent iterates through the possible actions based on the current sensory input and the environmental response. One of the most prominent examples of a reactive architecture is Brooks's Subsumption Architecture [10]. Fig. 1(a) illustrates a generic version of Brooks's idea. Here, lower-level actions (e.g., Action C) can be overridden by more fine-grained actions in the higher levels (e.g., Action A). With a dedicated DT involved in the reactive architecture illustrated in Fig. 1(b), the DT itself allows actions to be reviewed before generating an output, if the environment and the respective results of the actions are modelled correctly. In practice, without the DT actions would need to be executed for the CPS to understand whether the action has no result and the next action needs to be utilised. However, with the DT involved, the CPS can already exploit this information to determine if another action should be utilised. However, to remain a reactive architecture, the DT is required to model the response of the environment.

One of the key advantages of reactive agent architectures is that, due to their simplicity, they are able to work in their ability to operate in real-time, dynamic and resource-constrained CPS environments. However, since such agent architectures lack internal representations or models of the environment, they cannot reason or plan in a given situation, but only react to immediate stimuli. This means they do not consider the broader context or high-level goals. As a result, these architectures may not be suitable for complex tasks that involve strategic thinking, learning from experience, or maintaining persistent states.

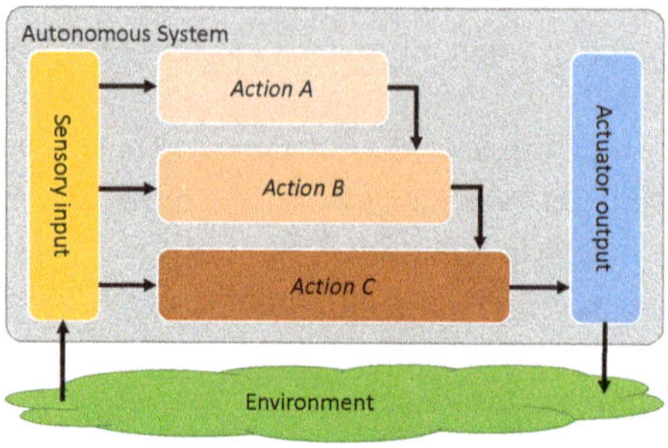

(a) Traditional subsumption architecture

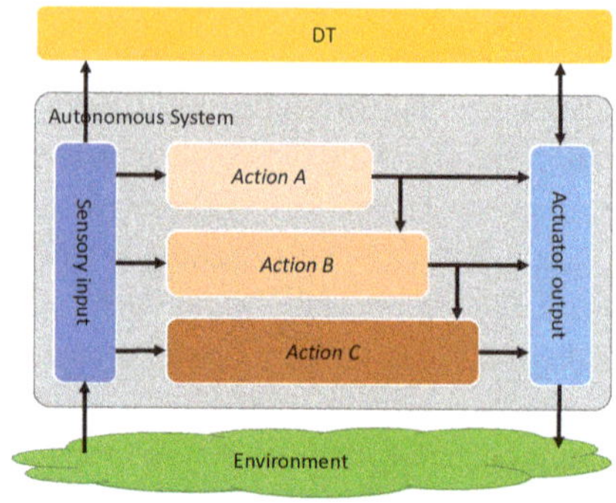

(b) DT-enabled Subsumption architecture

Fig. 14.1: Brooks's Subsumption Architecture [10] extended with a DT. Lower level Actions can be can *subsumed* by higher level behaviour.

Deliberative agent architectures, in contrast to reactive agent architectures, focus on reasoning, planning and decision-making based on internal representations or models of the environment. This means they prioritise a more strategic and goal-oriented approach. Generally, deliberative agents consider various possible actions and their impact on long-term objectives before making a decision.

At the heart of deliberative agent architectures is a model of the environment. This model represents the agent's understanding of the environment, including its own current state, the available actions the agent can perform and deliver the expected / potential outcomes. Depending on its granularity, this model may include explicit representations of objects, relationships between itself and other objects and PTs, various rules and available goals. The agent utilises this model to simulate possible future scenarios, evaluate different courses of action and make informed decisions based on desired outcomes. This is where DTs can play a crucial role when combining autonomous agent architectures with autonomous reconfiguration of PTs.

Deliberative agents utilise algorithms for planning and reasoning in order to create sequences of actions, enabling them to achieve their goals [33, 3, 36, 35]. Proper understanding of their own state and the state of their environment as well as the impact of their actions is required. This can be achieved by introducing domain knowledge through the DT paradigm.

There are various architectures for deliberative agents. One example is the Belief-Desire-Intention (BDI) model [9], which attributes three internal states to each agent: (i) beliefs, representing the knowledge about the environment the agent is embedded, (ii) desires, the goals and preferences each agent has and (iii) intentions, a subset of desires consisting of the immediate plans and sequence of actions to achieve the goals. The overall architecture is illustrated in Fig. 14.2, extended with a DT. Here the Belief Revision Function (BRF) updates the beliefs based on experiences gained through the sensors. In this arrangement, the BRF can additionally interact with the digital copy of the PT, and therefore reduces the degree to which actions have to be executed on the environment. Beliefs feed into the function to generate options. The DT-enabled system replicates the BDI architecture within the DT in order to extend the functionality. Sensory input is not only given to the DT but specifically to the BRF. Similarly, the Actuators not only affect the real world, but also inform the DT. The assumption is, that the generated beliefs, desires and intentions are deterministic. If they are not, a synchronisation between beliefs, desires and intentions is required. The generated goals might or might not be achievable for the agent. Again, the function of generating options can interact with the DT to refine the options more accurately. The filter function ensures that only feasible desires become intentions. Reconfiguring the agents physical presence (i.e., moving a robot arm or traversing through the environment) is based on a local reasoning approach.

Another architecture is MAPE-K, introduced by Kephart and Chess [28]. At its core, it utilises an autonomic manager consisting of the Monitor, the Analyser, the Planner, the Executer and the Knowledge base, as shown in Fig. 14.3. The elements of the architecture are as follows:

- *The Monitor* collects data about the CPS's internal state and relevant external or environmental factors. Information is directly received through the sensors of the system as well as monitoring the CPS's performance and required resources.
- *The Analyser* processes the information from the Monitor. Utilising machine learning approaches allows the CPS to create meaningful insights such as patterns, trends, anomalies or critical events. Furthermore, it can be utilised to perform diagnostics or predictions.
- *The Planner* creates appropriate plans and strategies for the CPS in order to achieve predefined goals or optimisations. It prepares necessary reconfiguration actions in order to improve the performance of the CPS towards its goals. The planning process may involve decision-making algorithms, optimisation techniques or rule-based approaches to generate an executable plan.
- *The Executer* implements and carries out the reconfiguration of the CPS based on the previously generated plans. The Executer interacts with the CPS's actuators or effectors to make the required adjustments to the CPS's configuration or behaviour.
- Finally, *the Knowledge base* contains all information and knowledge of the CPS. This ranges from domain-specific knowledge, models, policies and rules to historical data and information created by the Monitor. The Knowledge base provides the information required by the various components within the MAPE-K architecture.

When introducing a DT into the MAPE-K loop, the autonomic manager needs to be expanded and the knowledge base generated within the loop needs to be tightly coupled with the DT model as illustrated in Fig. 14.3. Feng et al. [21] have detailed the inclusion of a DT in the MAPE-K loop. As with other architectures, if knowledge

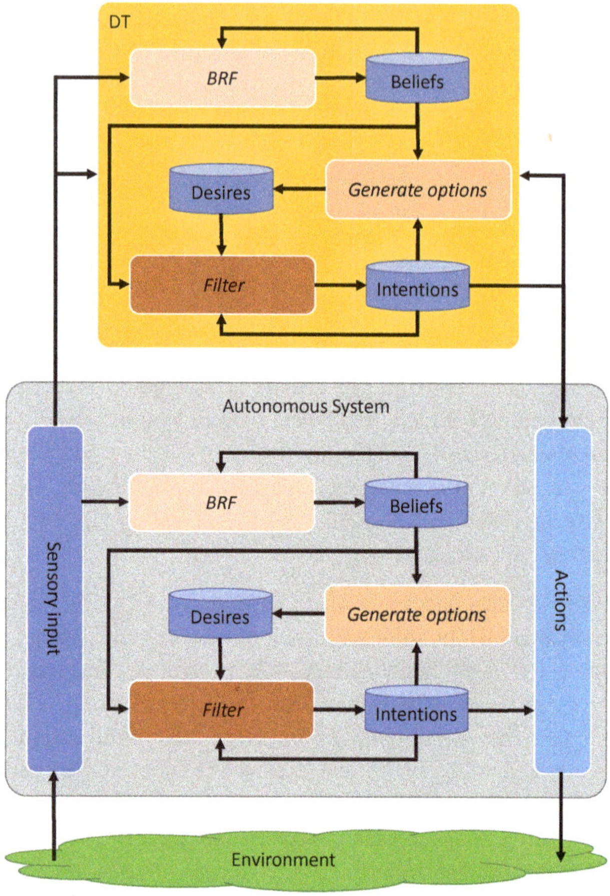

Fig. 14.2: Belief-Desire-Intention architecture for Autonomous Systems with an extension for DT following Bratman [9]. Beliefs represent knowledge about the environment shaped by BRF. Desires are goals of the CPS generated by the Generate options function and filtered by the Filter based on feasibility to create Intentions.

cannot be generated deterministically, a synchronisation between the PT and DT is required on this specific aspect.

We also consider the architecture for computational self-aware deliberative agents following [12, 20, 11, 29]. Computational self-awareness considers four fundamental levels of self-awareness and a meta-self-aware level:

- *Stimulus awareness* considers all stimuli and can differentiate between them. This is the most basic form of awareness.
- *Time awareness* allows to distinguish stimuli over time, enabling the CPS to consider patterns in the environment.

- *Interaction awareness* enables the CPS to identify different sources of stimuli and how other CPSs interact with it.
- *Goal awareness* is the last level, enabling the CPS to distinguish between its goals, allows it to adapt, select from among them and synthesise them.

In addition, *meta-self-awareness* gives the CPS an understanding of its awareness. Furthermore, the self-aware CPS architecture has a self-expression component, coordinating actions on the environment. As with all deliberative agents, models are at the core of the architecture, enabling the CPS to keep models about themselves and their environment. When incorporating a DT into the self-aware architecture, the self-aware mechanisms and algorithms are replicated in the DT. This means, similar to the MAPE-K loop, that the CPS needs to identify when the first-principles models need to be updated and calibrated with new information and learned models during runtime. With missing ground truth and in the absence of human operators, this can become very challenging even with simple CPSs [32]. Additionally, when goals are being updated within the DT-enabled PT, this needs to be reflected in the DT. In a similar way, self-expressive behaviour needs to be synchronised between the PT and the DT.

Introducing autonomous reconfiguration abilities into a DT-enabled system requires adequate models of the reconfigurable system itself and of the environment in which it operates, and the quality of the knowledge represented in those models needs to be known or maintained. As examples consider: (i) a change made by the

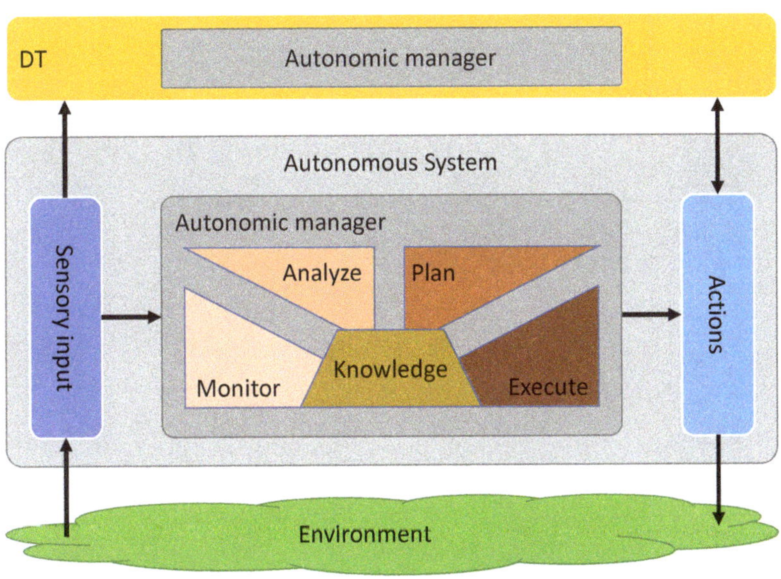

Fig. 14.3: MAPE-K architecture for Autonomous CPSs following [28] with a DT extension.

PT's actuators need to be reflected in the DT's models, (ii) DT-enabled system needs to take account of the bounds on the accuracy of data. However, through proper models of itself and the environment, so-called self-* properties can be realised.

With the proposed architectures shown in this chapter, DT-enabled systems have to ensure they can synchronise their information and state accordingly in order to profit from their DT. Additionally, explicit and secure communication channels are required. The issue of where architecturally a DT is deployed (fog, cloud, etc., as discussed in Chap. 11) clearly plays a central role in determining the quality of the system's autonomous actions.

14.3 Self-* properties

Self-* properties are core elements of autonomous CPSs and relevant for any autonomous reconfiguration. Different self-* properties do have dependencies on each

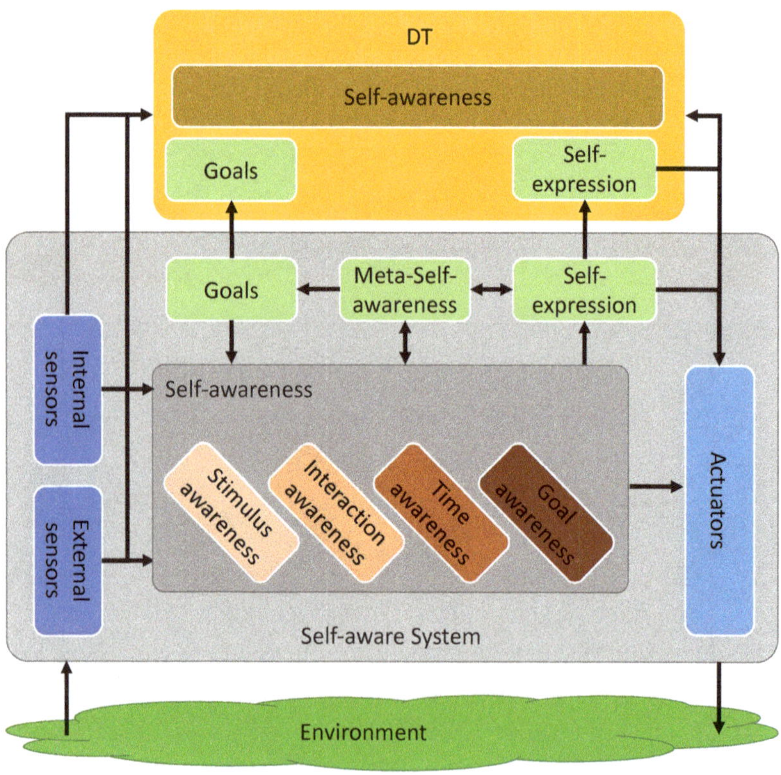

Fig. 14.4: Computational self-awareness architecture for Autonomous CPSs [11]

other in order to enable useful reconfiguration in autonomous CPSs. For example, self-adaptation requires a level of self-monitoring. In the following different self-* properties are discussed.

Self-monitoring: CPSs able to observe and assess their own behaviour, performance and state are considered self-monitoring. The process involves collecting, analysing and processing of data about the CPS's own components as well as interactions with the environment. Through insights into its own operation, identified patterns, anomalies or deviations from the expected behaviour, the CPS can detect issues or areas for improvement. This makes self-monitoring a crucial capability for self-maintenance, self-improvement and optimisation, self-healing, self-organisation with others and computational self-awareness. A Digital Shadow (DS) as discussed in Sect. 1.2.2 brings together data and models to deliver a monitoring capability. This can be used to increase the level of self-monitoring in the DT-enabled system, as it can inform external stakeholders through visualisation capabilities of the kind outlined in Chap. 8.

Self-adaptation: CPSs able to adapt and change their actions, behaviour, configurations or even physical structure in response to changing conditions in the environment or the requirements, are considered to have self-adaptive capabilities. Self-adaptation builds heavily on self-monitoring, i.e. observing and understanding ones own capabilities, performance and effects on the environment. In many cases, self-adaptive behaviour will synthesise new actions and behaviours and analyse its performance either in internal simulations or directly in the physical environment. Note that, while simulations are resource intensive, they avoid risks of potential damage to the CPS through harmful actions. Self-adaptation enables systems to dynamically change their configuration, allocate resources, adjust parameters or change their operational modes. By adapting their behaviour at runtime, CPSs can maintain optimal functioning, handle uncertainties and respond effectively to evolving circumstances. Self-adaptation is crucial for other self-* properties such as self-optimisation or self-maintenance, among others.

Given the enablers and services that we discussed in earlier chapters, we would argue that a DT that has some decision-making ability and an actuation link to the PT is able to increase the extent of self-adaptation at the level of the DT-enabled system. Such a decision-making capability might come from the implementation of either a reactive or deliberative approach in the DT, exploiting the data, models, enablers and services such as what-if analysis (see Sect. 10.2), that the DT offers.

Self-healing: This property enables autonomous CPSs to recover from faults, errors or failures without the need for external intervention such as a human operator [24]. Self-healing mechanisms vary in capability: from simply applying alternative actions, behaviours or modes of operations to more complex techniques aiming to identify root causes and respective recovery operations to ensure a continuous functional state. These, more complex, mechanisms can involve error detection algorithms, fault-tolerant design, redundancy or proactive measures to prevent failures [34]. This means that self-healing requires an approach for fault diagnosis. Self-healing CPSs aim to minimise downtime, maintain continuous operation and recover from disruptions, ensuring the CPS's reliability and availability, contributing to the overall resilience of the CPS. DTs can support self-healing activities through monitoring (Chap. 9), and fault diagnosis (Sect. 10.3) as well as taking appropriate recovery actions.

Self-maintenance: CPSs able to ensure their availability, performance and reliability proactively are considered self-maintaining [2, 31]. To maintain the CPS's operational health, preventive measures and corrective actions are applied to avoid degradation of the CPS's performance. To achieve this, these CPSs need to continuously self-monitor their performance, resources and activities. Applying approaches of self-maintenance and proactively identifying and addressing potential problems reduces downtime, extends the CPS's lifetime and potentially optimises performance under changing conditions. Self-maintenance is, in contrast to self-healing, a pro-active approach that aims to avoid failures and reduce downtime. Through the DT paradigm, predictive maintenance can be applied, allowing the exploration of various adaptation and self-maintenance strategies. The link between self-maintenance and predictive maintenance, discussed in Sect. 10.4, is natural, as in both cases techniques and data are needed to make predictions on when things will go wrong in order to apply either preventive or corrective measures. Thus, predictive maintenance can be seen as part of self-maintenance for a given CPS.

Self-protection: As discussed in Chap. 13, CPSs operating in a shared environment are prone to threats and malicious activities, and DTs arguably further increase the attack surface. CPSs with an ability to safeguard themselves against potential vulnerabilities and threats are considered self-protecting [23]. For this, the CPS is required to have proactive measures and mechanisms implemented to detect, prevent and mitigate risks that may compromise its security, integrity or availability. This can include authentication, access control, encryption, intrusion detection and firewalls. DTs can be utilised to find best defensive measures against unauthorised access, data breaches or malicious attacks. This ensures the provision of confidentiality, integrity, as well as availability of resources and information.

Self-optimisation: While continuous operation is provided by self-healing, self-maintenance, self-protection and self-adaptation, the property of self-optimisation goes one step further and aims at improving performance, efficiency and effectiveness of the CPS. This is achieved by runtime exploration and exploitation of adaptive parameters in combination with continuous monitoring and analysis. In static or known environments, self-optimisation requires identification of the state and matching with optimal parameters. In highly dynamic environments, the CPS might also be required to predict the future states in order to perform adaptation in a timely manner. As discussed in Sect. 10.2, the DT can be used in both cases, (i) to explore optimal parameters for all given configurations of the CPS and the environment by performing simulations and (ii) by making predictive simulations of the DT to identify potential future states and apply self-adaptive activities pro-actively. By autonomously adapting and fine-tuning its operation, a self-optimising CPS can continually enhance its performance and adapt to evolving requirements. Exploration of better parameters can be performed within a single, established DT. However, recently, Ejersbo et al. [14] have proposed an approach that allows CPSs to replace entire models and change their internal structure at runtime. This can be utilised for exploration and optimisation of behaviour in co-simulations.

Self-management: In contrast to self-adaptation, CPSs with self-management abilities are also able to regulate and control their own behaviour and resources [13]. Self-management is often used as an umbrella term as it covers self-configuration, self-optimisation, self-healing and self-protection capabilities. By taking responsibility for tasks that would traditionally require human intervention, self-management allows CPSs to operate efficiently, adapt to changing conditions and maintain desired performance levels. In contrast to simple self-adaptation, the CPS cannot only dynamically adjust the its own behaviour, but also regulate and control itself as a whole. Therefore, self-adaption is only a specific aspect of self-management, making adaptations of individual aspects of the CPSs part of the overall self-management process [25]. Specifically, self-management in CPSs can be concrete implementation of a DT when the management aspects are applied immediately.

Self-organisation: When multiple autonomous CPSs collaborate, structure, patterns and order may emerge from their concurrent operation in the absence of external or centralised controller. This means, the interaction of individual CPSs gives rise to collective behaviour and organisational structures. As a result, the system of CPSs appears to be controlled even though it is only governed by individual rules. These rules might even differ from system to system. In the same way that self-organisation is found in biological systems [27] and social networks [4], it has also been utilised to coordinate multiple autonomous agents towards a common goal [40]. Utilising a DT in this setting can enable individuals to improve their individual actions towards mutual or personal goals. Therefore the self-optimisation is not necessarily only for the personal gain but may also benefit the greater good of the social community of other systems. It is tempting to draw a connection between this possibility and the potential for self-organising fleets of DT-enabled systems that function effectively as systems-of-systems (see Sect. 11.11), although realising this remains a research goal.

Self-awareness: Computational self-awareness enables autonomous CPS to understand their own internal state, behaviour, capabilities and interactions. Self-monitoring plays a crucial role in such CPSs, allowing it to analyse and interpret data about itself, including information about its resources, performance and interactions with the environment [30]. Furthermore, it is argued that, through computational self-awareness, an autonomous CPS can reason about its own limitations, strengths or goals. This enables informed decision making about adaptation processes [8]. Expanding the concept in more detail towards the environment of the CPS, concepts of social awareness [5, 7] and networked self-awareness [16, 17] enable the autonomous CPS to consider their interactions with other CPSs in more detail, allowing for better performance in their interaction. Simple forms of self-awareness can be achieved with DSs as discussed in Sect. 1.2.2. Full computational self-awareness, as discussed in [11], also allowing to apply actions based on self-aware knowledge, will require a DT as a basis of the CPS.

14.4 Goals

In order to implement any of the self-* properties and perform meaningful autonomous reconfiguration, a CPS either needs to have an explicit goal or an awareness and understanding of goals. In contrast to properties of a CPS, goals describe states and properties that are not currently reached. Goals, whether defined and described explicitly or not, indicate and guide actions in order to achieve them. These actions might affect internal properties as well as external aspects and environmental states, including other CPSs.

Having goals: Having a simple goal function enables the CPS to adjust, reconfigure, self-optimise its actions in order to achieve this goal more efficiently and effectively. Utilising a DT, enables the exploration of optimal adaptations for the physical counterpart.

Knowing goals: With additional knowledge, the autonomous agent cannot only *have* goals but also reason about them. This means the agent has additional information related to the goals such as requirements from the environment or resources. With this the agent can select the most appropriate goals during runtime in the event that prerequisites for goals change. For example, a resource, that heavily affects the efficiency of achieving a goal has been removed from the environment. With this an alternative goal yields higher expected performance utility and is therefore selected by the agent [15]. Utilising a DT enables us to explore these differences in performance utility in order to maximise the overall performance of the autonomous CPS.

Dividing goals: Reconfiguration of CPSs may not only entail changing parameters for actions or even replacing one action with another, but also includes adapting goals themselves while keeping the underlying goal function. For example, let us consider the construction of a complex object, requiring various parts that need to be prepared beforehand. In this context, if some of the tools are sporadically provided externally, the CPS would not need to tool them and in doing so would waste valuable resources. Utilising a DT would allow the exploration of potential sub-goals and split up the original goal into individual steps with its own goals. With *knowing these sub-goals* the CPS could choose which goals are most profitable for their performance.

New goals: The last step, when it comes to dealing with goals, enables a CPS not only to work with existing goals, but also generate new goals based on observations. This might initially seem counter-intuitive. However, consider a CPS that would be idle and not generate any revenue. To overcome this, the DT could be utilised to explore potential new revenue streams defined by new goals. More so, a CPS could identify tasks and underlying goals it could perform in order to support another CPS, leading to better performance of the other CPS and an overall better utilisation of resources across all CPSs. This kind of collaboration is discussed further below.

14.5 Collaboration between Systems

When CPSs operate in the environment, they may interact with other CPSs. Integrating these other CPSs with one's own actions can lead to emergent and unintended behaviour and results [37]. Introducing more reliability and control in such integration processes, leading to collaboration between individual CPSs, is needed and desired. Utilising DT-enabled systems provide additional benefits when enabling CPSs to collaborate reliably [18]. In previous work, several steps have been outlined to achieve this:

1. *Detection:* A CPS that has the potential to interact with others in a (potentially mutually) beneficial way, needs first to be able to detect the other systems whether they are highly controlled or relatively autonomous. Additionally, the environment as a whole should be considered, as it may be affecting the chosen actions as well as interfering with the reconfiguration towards specific goals [6].
2. *Refinement:* Knowing that there is a *something else*, is not enough. In addition, the system also needs to be able to understand what actions do affect it and how. An accurate representation of the causes might not be relevant. Consider a situation in which a resource required by an agent has been displaced. The fact that the resource is displaced might require a reconfiguration. However, the cause of the displacement may or may not be relevant, dependent on whether one wants to simply use the displaced resource or alter the displacement in itself.
3. *Utilisation:* Finally, systems have to utilise their potential collaboration with other systems in order to achieve their individual goals more efficiently. This might require individual reconfiguration to ensure the best integration of the actions of others within the set of actions required to reach the goal.
4. *Repetition:* For best results, steps 2 and 3 need to be repeated in regular intervals. This ensures that changes to the configuration in one system do not interfere with the actions of another system. In addition, parallel reconfiguration may lead to a perpetual effect that requires continuous reconfiguration in order to perform well. With this, systems may spend more time on reconfiguration than actually performing well towards their goal.

To achieve these steps, a DT-enabled architecture has been proposed for collaboration and integration of new systems [18]. Fig. 14.5 illustrates this architecture. At its core, there is the DT manager, aiming to generate, modify, handle, but also destroy DTs of other systems locally. These DTs can then be utilised with its own DT model in the *Planner* to achieve most useful actions. While it would be preferred if all DT models are complete and shared by all CPS willingly, in reality DT models will be limited by the requirements of the CPSs they reflect. If now another CPS wants to use the DT, the relevant model needs to be adapted and adjusted to the requirements of the new CPS. Additionally, since the DT model does not affect the actual CPS it represents anymore, it is downgraded to a DS.

Specifically, the following components are proposed:

Creator: By performing a DT creation, a CPS can generate a DT from another CPS. This can be achieved by either applying first-principles observations to the model or by exploring the available DT models from the database and utilising the Mapper to find the best match for the observations.

Mapper: By performing a DT mapping, a CPS can identify similarities, differences or overlaps between models and facilitate the exchange of information or knowledge. It enables the integration of heterogeneous models, the conversion of models from one representation to another and the synchronisation of data and semantics across diverse modelling languages or systems. DT mapping plays a crucial role in areas such as data integration, model-driven engineering, ontology alignment and semantic interoperability.

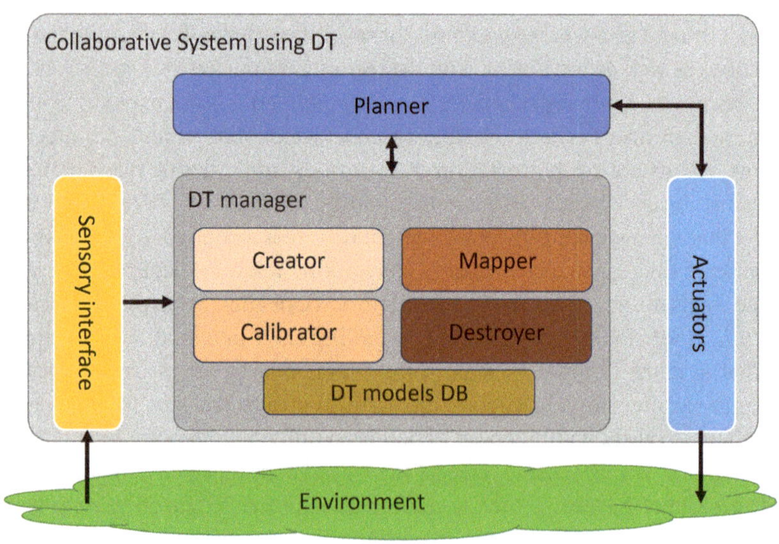

Fig. 14.5: Components required for a collaborative autonomous CPS using DT [18]

Calibrator: By performing a DT calibration, a CPS can modify the DT itself, making it more accurate to the actions and tasks performed. It allows the integration of new information into the DT at runtime while the actual CPS is being observed (see Sect. 6.2). A challenge arises as observation might not be accurate or obstructed and reliable calibration becomes difficult.

Destroyer: By performing a DT destruction, a CPS can free resources by not further utilising a DT of another CPS. In such a situation, the DT needs to be destroyed and removed from memory. Before doing so, the component needs to decide whether the underlying model should be stored in the DT models database for future reference. This should only be done if it is not a comparable DT model already. Again, the Mapper can be used to make such a decision.

14.6 Safety and uncertainty in reconfiguration

The interplay of DT-enabled systems and their changing environment lead to rapidly unfolding situations that may not have been foreseen when the individual DT-enabled system elements were designed and created. Nevertheless, CPSs operating along with other systems, including humans or other animals, have to ensure safety even in unexpected situations. Due to the emergent nature of the CPSs, not all safety measures, i.e., reconfigurations to ensure safety, can be defined at design-time, requiring autonomous approaches to establish them at run-time. This requires the CPS to identify similar situations and map those towards the required safety measures.

The monitoring of requirements itself is dependent on the architecture. In MAPE-K, this would be achieved in the Planner with information from the knowledge base while BDI agents would take those requirements into consideration during the filtering process. Almeaibed et al. [1] have discussed safety framework incorporating a DT explicitly for autonomous vehicular systems.

However, measurements of a CPS might be prone to uncertainty, either be due to insufficient knowledge and experience (epistemic uncertainty) or due to the inherent randomness of the environment itself (aleatoric uncertainty) [19]. While we often need to estimate the degree of uncertainty (i.e., uncertainty quantification), we are also interested in *uncertainty classification*, i.e., deciding whether an experienced randomness is epistemic or aleatoric. Approaches to identify and classify various uncertainties can apply what-if simulations as discussed in Sect. 10.2 utilising fault injection [22] and model-swapping [14] to explore the affects of various uncertainties.

14.7 Roadmap

While initial steps have been made towards meaningful reconfiguration utilising DTs, there are still several aspects that remain to be resolved in order to improve autonomous reconfiguration for CPSs. These aspects include, but are not limited to:

Mapping: At this point, mapping individual models against each other can only be done based on their natural language description. There are no dedicated techniques that allow us to compare the semantic levels of two arbitrary models. In order to achieve this, machines are required to build up meaning from the execution or observation of models directly. Even harder is the problem of generalisation from one model to another. A model that seems to be similar to another model by 95%, may still have a vastly different goal and purpose.

Uncertainty classification: Deciding whether an observed uncertainty is aleatoric or epistemic is inherently hard without a fixed point of reference. Only by having comparable and reliable information can estimations about the source of uncertainty be made.

Observation-based calibration: The task of generating and calibrating models based on observations faces a dilemma around deciding the nature of the uncertainty. For example, when a robotic arm is observed to be jittering, it is not possible to know if this jitter is due to some inaccuracies in the driving motor or if the robot is deliberately jittering due to some reasons only known to the robot itself. Even without the issue of external uncertainty classification, calibration of models based on observation requires complete information about the actions of the system and its effect on the environment. Overall, the external calibration represents a magnitude harder problem than direct calibration which has already been identified as an ongoing research direction in itself.

Micro/Macro models: Particularly when operating in a self-organised manner and in collaboration with numerous other systems, models of individual systems might result in a level of fidelity that is higher than is required for the goals of interaction with the collective. In such situations, it is required to abstract away information from individual models and generate collective models, representing the overall system-of-systems. Conversely, it is necessary to refine collective models down to individual ones when required. The former seems to be straight-forward, requiring the combination of relevant information from individuals within a single model. The latter might be impossible without explicit information exchange from the CPS represented by the model.

References

1. Almeaibed, S., Al-Rubaye, S., Tsourdos, A., Avdelidis, N.: Digital Twin Analysis to Promote Safety and Security in Autonomous Vehicles. IEEE Communications Standards Magazine,

5(1), 40–46 (2021). DOI: 10.1109/MCOMSTD.011.2100004

2. Asami, S., Talagala, N., Patterson, D.A.: Designing a self-maintaining storage system. In: Proceedings of the Symposium on Mass Storage Systems, pp. 222–233. IEEE (1999)
3. Badran, O., Mekhilef, S., Mokhlis, H., Dahalan, W.: Optimal reconfiguration of distribution system connected with distributed generations: A review of different methodologies. Renewable and Sustainable Energy Reviews **73**, 854–867 (2017)
4. Barabási, A.L.: Linked: The New Science of Networks. Perseus Pub. (2003)
5. Barnes, C.M., Ekárt, A., Lewis, P.R.: Social action in socially situated agents. In: Proceedings of the International Conference on Self-Adaptive and Self-Organizing Systems, pp. 97–106. IEEE (2019)
6. Barnes, C.M., Esterle, L., Brown, J.N.: "When you Believe in Things that you don't Understand": the Effect of Cross-Generational Habits on Self-Improving System Integration. In: 2019 IEEE 4th International Workshops on Foundations and Applications of Self* Systems (FAS* W), pp. 28–31. IEEE (2019)
7. Bellman, K., Botev, J., Hildmann, H., Lewis, P.R., Marsh, S., Pitt, J., Scholtes, I., Tomforde, S.: Socially-sensitive systems design: Exploring social potential. IEEE Technology and Society Magazine **36**(3), 72–80 (2017)
8. Bellman, K., Landauer, C., Dutt, N., Esterle, L., Herkersdorf, A., Jantsch, A., Taherinejad, N., Lewis, P.R., Platzner, M., Tammemäe, K.: Self-aware cyber-physical systems. ACM Transactions on Cyber-Physical Systems **4**(4), 1–26 (2020)
9. Bratman, M.: Intention, Plans, and Practical Reason. Cambridge, MA: Harvard University Press, Cambridge (1987)
10. Brooks, R.: A robust layered control system for a mobile robot. IEEE Journal on Robotics and Automation **2**(1), 14–23 (1986)
11. Chen, T., Faniyi, F., Bahsoon, R., Lewis, P.R., Yao, X., Minku, L.L., Esterle, L.: The handbook of engineering self-aware and self-expressive systems. arXiv preprint arXiv:1409.1793 (2014)
12. Diaconescu, A., Bellman, K.L., Esterle, L., Giese, H., Götz, S., Lewis, P., Zisman, A.: Architectures for collective self-aware computing systems. In: Self-Aware Computing Systems, pp. 191–235. Springer (2017)
13. Diao, Y., Hellerstein, J.L., Parekh, S., Griffith, R., Kaiser, G., Phung, D.: Self-managing systems: A control theory foundation. In: Proceedings of the International Conference and Workshops on the Engineering of Computer-Based Systems, pp. 441–448. IEEE (2005)
14. Ejersbo, H., Lausdahl, K., Frasheri, M., Esterle, L.: Dynamic Runtime Integration of New Models in Digital Twins. In: 2023 IEEE/ACM 18th Symposium on Software Engineering for Adaptive and Self-Managing Systems (SEAMS), pp. 44–55 (2023). DOI: 10.1109/SEAMS59076.2023.00016
15. Esterle, L.: Goal-aware team affiliation in collectives of autonomous robots. Proceedings of the International Conference on Self-Adaptive and Self-Organizing Systems, pp. 90–99 (2018). DOI: 10.1109/SASO.2018.00020
16. Esterle, L., Brown, J.N.: Levels of networked self-awareness. Proceedings of the International Workshops on Foundations and Applications of Self* Systems (FAS*W), pp. 237–238 (2018). DOI: 10.1109/FAS-W.2018.00054
17. Esterle, L., Brown, J.N.A.: I Think Therefore You Are: Models for Interaction in Collectives of Self-Aware Cyber-Physical Systems **4**(4) (2020). DOI: 10.1145/3375403
18. Esterle, L., Gomes, C., Frasheri, M., Ejersbo, H., Tomforde, S., Larsen, P.G.: Digital twins for collaboration and self-integration. In: 2021 IEEE International Conference on Autonomic Computing and Self-Organizing Systems Companion (ACSOS-C). IEEE (2021). DOI: 10.1109/ACSOS-C52956.2021.00040
19. Esterle, L., Porter, B., Woodcock, J.: Verification and uncertainties in self-integrating system. In: 2021 IEEE International Conference on Autonomic Computing and Self-Organizing Systems Companion (ACSOS-C). IEEE (2021). DOI: 10.1109/ACSOS-C52956.2021.00050

20. Esterle, L., Rinner, B.: An Architecture for Self-Aware IoT Applications. In: Proceedings of the IEEE International Conference on Acoustics, Speech and Signal Processing, pp. 6588–6592 (2018). DOI: `10.1109/ICASSP.2018.8462005`
21. Feng, H., Gomes, C., Gil, S., Mikkelsen, P.H., Tola, D., Sandberg, M., Larsen, P.G.: Integration of the MAPE-K Loop in Digital Twins. IEEE (2022). DOI: `doi:10.23919/ANNSIM55834.2022.9859489`
22. Frasheri, M., Thule, C., Macedo, H.D., Lausdahl, K., Larsen, P.G., Esterle, L.: Fault injecting co-simulations for safety. In: The 5th International Conference on System Reliability and Safety ICSRS'21 (2021)
23. Frincke, D., Wespi, A., Zamboni, D.: From intrusion detection to self-protection. Computer Networks **51**(5), 1233–1238 (2007). DOI: `j.comnet.2006.10.004`
24. Ghosh, D., Sharman, R., Rao, H.R., Upadhyaya, S.: Self-healing systems—survey and synthesis. Decision support systems **42**(4), 2164–2185 (2007)
25. Hinchey, M.G., Sterritt, R.: Self-managing software. Computer **39**(2), 107–109 (2006)
26. Jennings, N.R., Sycara, K., Wooldridge, M.: A roadmap of agent research and development. Autonomous Agents and Multi-agent Systems **1**(1), 7–38 (1998)
27. Kauffman, S.: The Origins of Order: Self-organization and Selection in Evolution. The Origins of Order: Self-organization and Selection in Evolution. Oxford University Press (1993)
28. Kephart, J.O., Chess, D.M.: The vision of autonomic computing. Computer **36**(1), 41–50 (2003)
29. Kephart, J.O., Diaconescu, A., Giese, H., Robertsson, A., Abdelzaher, T., Lewis, P., Filieri, A., Esterle, L., Frey, S.: Self-adaptation in collective self-aware computing systems. In: Self-Aware Computing Systems, pp. 401–435. Springer (2017)
30. Kounev, S., Lewis, P., Bellman, K.L., Bencomo, N., Camara, J., Diaconescu, A., Esterle, L., Geihs, K., Giese, H., Götz, S., et al.: The notion of self-aware computing. In: Self-Aware Computing Systems, pp. 3–16. Springer (2017)
31. Mamei, M., Zambonelli, F.: Self-maintaining overlay data structures for autonomic distributed computing. In: Proceedings of the International Conference on Autonomic Computing, pp. 376–377. IEEE (2005)
32. Nelson, P.R., Bellman, K.L., Landauer, C.: Self-modeling - a practical example of why it's hard. In: 2022 IEEE International Conference on Autonomic Computing and Self-Organizing Systems Companion (ACSOS-C), pp. 88–94 (2022). DOI: `10.1109/ACSOSC56246.2022.00039`
33. Piciarelli, C., Esterle, L., Khan, A., Rinner, B., Foresti, G.L.: Dynamic reconfiguration in camera networks: A short survey. IEEE Transactions on Circuits and Systems for Video Technology **26**(5), 965–977 (2015)
34. Psaier, H., Dustdar, S.: A survey on self-healing systems: approaches and systems. Computing **91**, 43–73 (2011)
35. Thalamy, P., Piranda, B., Bourgeois, J.: A survey of autonomous self-reconfiguration methods for robot-based programmable matter. Robotics and Autonomous Systems **120**, 103242 (2019)
36. Tipaldi, M., Glielmo, L.: A survey on model-based mission planning and execution for autonomous spacecraft. IEEE Systems Journal **12**(4), 3893–3905 (2017)
37. Tola, D., Madsen, E., Gomes, C., Esterle, L., Schlette, C., Hansen, C., Larsen, P.G.: Towards easy robot system integration: Challenges and future directions. In: 2022 IEEE/SICE International Symposium on System Integration (SII), pp. 77–82 (2022). DOI: `10.1109/SII52469.2022.9708846`
38. Watson, D.P., Scheidt, D.H.: Autonomous systems. Johns Hopkins APL technical digest **26**(4), 368–376 (2005)
39. Wooldridge, M.: Intelligent agents: The key concepts. In: ECCAI Advanced Course on Artificial Intelligence, pp. 3–43. Springer (2001)
40. Wooldridge, M.: An introduction to multiagent systems. John Wiley & Sons (2009)

Chapter 15
Future Directions and Challenges

Peter Gorm Larsen, John Fitzgerald, Cláudio Gomes, Jim Woodcock, Stylianos Basagiannis, Alessandro Ulisse, Lukas Esterle, Daniel Enrique Lucani Rötter, Simon Thrane Hansen and Bentley James Oakes

Abstract Although digital twins have been around for decades, there are many areas of research that need to be pursued in order to enable trustworthy predictions. The existing challenges and the future directions to take can be organised in many different ways. In this chapter, the foundational issues that we still think require clarification comes first. This is followed by research challenges in the different platforms needed to be able to produce digital twins in a cost-efficient manner. Afterwards the research changes for enabling autonomy of digital twins (uncertainty in many areas may hinder an accurate estimation of the state). Finally, research in composition of digital twins is considered and this really goes into system of systems considerations where an eco-system of stakeholders needs to agree about the means of composition.

15.1 Introduction

Although the concept of the Digital Twin (DT) is very well established, the rapid growth in interest in the technology is remarkable. Although many tools are emerging that enable the construction of DT-enabled systems, and the wave of new applications is almost overwhelming, we seem hardly to have scratched the surface of this technology. The potential of this combination of models, data, computation and communication is enormous and yet to be realised. In this final chapter, we identify directions for research and innovation towards the goal of a well-founded and established "discipline" of DT engineering that helps us to realise more of the benefits in the future and deliver DT-enabled systems on which people and businesses can rely.

There are many potential directions for research and innovation on DTs. It is striking that established organisations like the International Council on Systems Engineering (INCOSE) have visions for the future of digital transformations that include DTs [34]. In general, developing a DT research and innovation community

J. Fitzgerald et al. (eds.), *The Engineering of Digital Twins*,
https://doi.org/10.1007/978-3-031-66719-0_15

with a joint research agenda that cuts across application domains is perhaps needed to outcomes that are focussed on achieving such a vision. After the process of developing the content of this book, we propose four strands for future work:

Foundations: This includes the development of a basis for placing well-understood levels of reliance on the services offered by DTs. Key to this is mastering the uncertainties that arise from elements of DT-enabled systems and their environments, so that a DT's predictions are fit for purpose and can safely and securely be used to inform decisions that affects DTs' corresponding Physical Twins (PTs). We discuss directions for future work here in Sect. 15.2.

Platform: It can be costly and challenging to develop and maintain a DT, and so there is a line of valuable work to be done in making the engineering of DTs for applications both more affordable and sustainable. Goals here include the automation of DT operation through offering platforms and tools that have potentially wide user bases. We consider this further in Sect. 15.3.

Autonomy: Can DTs present a route towards dependable autonomy for Cyber-Physical Systems (CPSs)? As shown in Chap. 14, there are decisions to be made in either trusting DTs to influence the physical world directly or in augmenting existing autonomy. Research and innovation here also involves ways to provide a "learned cognitive capability" inside the DT by deductive and inductive means. We discuss this further in Sect. 15.4.

Composition: The work on a platform for DTs as a Service discussed in Chap. 11 leads towards the possibility of the composition of DTs and by extension the composition of their DT-enabled systems. Such compositions range from simple fleets of near-identical and non-interfering products to more heterogeneous Systems of Systems (SoSs) in which reliance comes to be placed on the qualities of the emergent collaboration. We consider this subject further in Sect. 15.5.

We provide a short overview of novel and perhaps controversial applications of DTs beyond the area of engineering CPSs in Sect. 15.6 and end the book with concluding remarks in Sect. 15.7.

15.2 Firm Foundations for Digital Twin Engineering

Although DTs arise in many existing fields (see Sect. 1.3.3), there is still a lack of formal foundations that justify placing appropriate reliance on the outcome of some DT services. Below, we argue for work on the overall limits of the predictability of DTs (Sect. 15.2.1). This is closely linked to the need to quantify and cope with uncertainties at different levels (Sect. 15.2.2), and to support verify properties of DTs (Sect. 15.2.3). As discussed in Chap. 13, security-related properties are a particular concern in the context of DTs, and so we discuss this further in Sect. 15.2.4, alongside safety in Sect. 15.2.5. Finally, Sect. 15.2.6 considers our reliance on good calibration and the challenge of calibrating multiple parameters simultaneously.

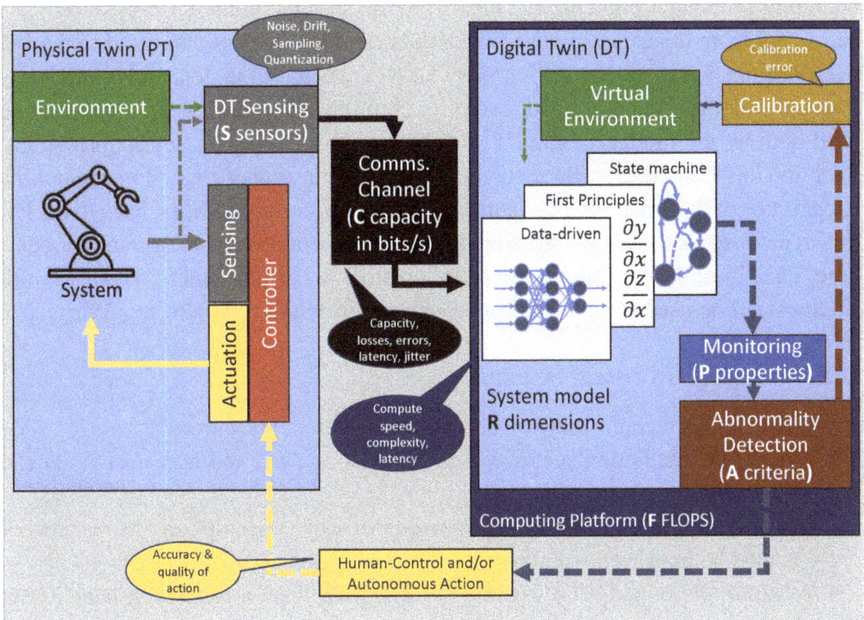

Fig. 15.1: Aspects affecting the prediction limitations from a DT.

15.2.1 Understanding the Limits of Predictions

To be able to rely on the predictions made by a DT, we must understand the limits on its accuracy. These limits include the high dimensionality and scale of the DT's different variables and stages of its operation, from data capture and communication to modelling and predictions. Preliminary work has proposed examining key qualities of models, namely relevance, verifiability, substitutability and faithfulness [51]. Earlier work in has focused on the model validity frame, that is, the set of inputs for which the models holds its relevance, verifiability and substitutability [18, 63]. The DT services that depend on models will naturally depend on, and in some sense propagate, these qualities, but it is currently unclear how they are affected by the algorithms employed in these services.

By understanding the limits on model-based analyses in the DT context, we can answer questions such as "What is the prediction accuracy of a DT with a model with complexity order **R** and a set of **P** properties testing **A** abnormality criteria using information from **S** sensors sharing a communication channel of capacity **C** and a computational resource of **F** flops?". An overview of the connections between these different aspects is shown in Fig. 15.1. From left to right, we have the PT operating in its environment. The data from the sensors on the PT may have noise and suffer from other inaccuracies. That data needs to be communicated to the DT over a channel that can have limitations in its capacity, time delay and potentially lose information. When the data comes over to the DT on the right, it needs to be used both by the models of the PT and the monitors that check for abnormalities. Such models may need to be reduced to work within timing requirements and the capabilities of the underlying computing platform. In communicating back to the PT, the predictions' accuracy will be affected by the uncertainties of the different aspects here. The "formula" for this has not been scientifically established for DTs. Network calculus can be used for making upper- and lower-bound estimates to performance estimates [29].

Further questions arise from this:

- What are more appropriate models of computational resources?
- What is the prediction accuracy time horizon of a DT and how does it depend on the evolution of the PT?
- Can the over-approximations or over-conservative assumptions from network calculus be improved with control of the uncertainties?
- What are the bounds and feasible operating regions on prediction latency and accuracy?

Various effects characterise the limits of prediction accuracy. Naturally, when we wish to act on the predictions we need to communicate to the actuators, and here too there may be uncertainties in the form of inertia, accuracy and failure rates. If we think of this metric as an error in prediction, we could visualise it as a result of four sources: sensing, communication, modelling and monitoring.

15.2.2 Uncertainty: Quantification and Propagation

Extensive work has been done on traditional uncertainty configuration and propagation for CPS models. There is a need to continue the research in this area, but at the DT level, to provide valuable input for the challenges mentioned in Sect. 15.2.1. Different application domains have different ways to represent and propagate uncertainty. This includes reachability analysis [21], Monte-Carlo simulation [68], stochastic differential equations [49] and sensitivity analysis [14]). The novelty of this challenge is how to represent that uncertainty in the models used by the DT, how this propagates through the DT layers and whether that propagation affects the operation of those layers themselves. A DT must be able to cope with substantial changes to the environment of the PT [68]. DT-enabled systems must continue to make decisions in the presence of uncertainty [16]. To do this, it must clearly distinguish between aleatoric and epistemic uncertainty as explained in Chap. 14. The propagation of the uncertainties of different aspects is essential to understanding how to determine the tolerances that can be allowed. The actions to be taken if such tolerances are exceeded will depend on the specific DT-enabled system in particular in relation to the criticality to the environment for the PT.

There is a need to work on the challenge of modelling and characterising the effects of sources of uncertainty that are not considered in the existing state of the art from a theoretical perspective, but which are known to be critical to the prediction accuracy and, thus, operation of DTs from a practical standpoint. For example, a physical model that incorporates friction effects usually considers it a known constant. This could be measured beforehand for controlled conditions with some reasonable measurement error. However, a DT aims to uncover the value of this constant from prior measurements (model calibration) and direct interactions with the environment. Thus, it will be able to adapt to changes in the environment not envisaged when the PT was originally engineered. This is related to the notion of "validity frames" indicating when the DT is able to provide acceptable predictions [2].

15.2.3 Towards Verified Digital Twins

Formal methods provide mathematical specifications, designs and realisations that can be analysed using simulation (see Sect. 5.6). Formal methods systematise and introduce rigour into all the phases of the development of software artefacts to be used in a DT context. The usual workflow is as follows:

- Write a specification of the PT from a system perspective.
- Write the desirable properties the PT should have and capture the assumptions that are necessary for its environment.
- Directly "test" the specification to check the properties. This testing may involve conventional validation techniques or more rigorous verification with model checkers and theorem provers.
- If this finds a problem, fix it without building the wrong PT.
- Once the system passes the test against the properties, start realising it while being confident that it will satisfy the needs.
- If necessary, the documentation for this process can be used for the certification of the PT.

If we deal with a DT-enabled system, we can go further. A DT enables the development of a self-adaptive CPS that reconfigures its behaviour in response to novel environments. These self-adaptations must be formally verified against the properties at runtime to avoid expensive re-certification of the reconfigured CPS [69]. A formal definition of DTs would enable the following:

- Develop a digital twinning research community that clarifies and agrees on a universally accepted concept.
- Continue to develop the CPS foundations that underpin DT engineering. Formalise these foundations in agreed tools, such as Isabelle [56] and Coq [8] on the discrete side and incorporate differential equations using tools such as KeYmaera X [58]. In the case of probabilistic models it makes sense to use model checkers such as Storm [17].
- Build user communities for application domains to increase industrial adoption and commercialisation.

The formalisation of DTs will enable use cases. For example, probabilistic DTs will be augmented with Markov and Bayesian models, particularly probabilistic risk models. They will be used for online risk assessment to support operational decisions.

15.2.4 Protection against Security and Privacy Attacks

We may need to add sensors and actuators to the PT so that the DT can perform appropriately. As we indicated in Chap. 13, malicious stakeholders can misuse these features [3, 65]. So, we must consider how to protect the DT-enabled system from attacks [40]. The objective is to increase security with the availability of the DT. How we do this will depend on the specific circumstances, but patterns of potential mitigation strategies need to be explored [39]. If we operate a Digital Twin as a Service (DTaaS) platform including many DTs, it will be essential to guarantee that each of their data spaces is separated and protected against intruders. As we discuss further in Sect. 15.5, organisations will be very eager to protect their Intellectual Property (IP).

Further research directions for secure DTs include the following:

- Developing a taxonomy of security properties for DTs across different application domains.
- Identifying DT vulnerabilities and how they can be exploited.
- Developing and specialising methods and tools for verifying security in DTs.
- Compiling a catalogue of DT attack models and mitigation strategies.
- Developing low-overhead security protocols for bidirectional communication between DTs and PTs.
- Agreeing on community-based, domain-specific case studies and benchmarks.

15.2.5 Synthesising Safe Scenarios

It can be advantageous to request the PT to perform specific actions that can be used to calibrate the models in a DT (e.g., determining the friction affecting a movement). However, before carrying out such experiments, it is necessary to ensure that the PT is in a state where it is possible (and not harmful) to carry out such an experiment. We imagine having a fundamental exploration layer inside the DT which can synthesise possible scenarios to be conducted to clarify something the DT thinks has changed. This is comparable to what is known as *design of experiments* for which a methodology was proposed by Fisher [30].

An exploration layer in the DT would require information about the PT and its interaction with the environment to create models. During the operation of the PT, it may have to devise physical experiments that the PT can conduct, the results of which will enable adjustments of the models (in particular of the environment). An example is an agricultural robot driving in a straight line. This movement does not give enough information about the mass distribution of the robot. So, the exploration layer needs to command the robot to experiment. It tells it to turn through 90° at a certain speed. It can then estimate the mass distribution from the experiment's results. Naturally, the exploration layer must know that this action is safe and will not interfere with the normal operation of the PT. There is extensive research in system identification [35] and anomaly detection [12] that tackles this problem. Scenarios can also provide predictive analytics for accidents by identifying poor or unsafe procedures in domain-specific workplaces. However, all of this research needs to be raised to the DT-enabled systems level and it needs to be possible to determine when it is safe to carry out such "experiments".

15.2.6 Mutual Calibration

Following a PT for extended periods will be necessary to track the PT and estimate the PT's state constantly. The uncertainties mentioned above mean that the perception of the PT's current state and the PT's real state will not always be fully aligned, and in addition, there may be a time offset inside the DT. The DT layers use multiple models for different aspects of the system. Many of these may overlap. For example, the PT model contains the plant updated by the PT tracking and state estimation layer [27]. When the plant model parameters are updated, the PT model must also receive those updates because other layers depend on it. The PT tracking and state estimation layer will use different state estimators (as in interacting multiple model filters [46]) that rely on overlapping models. Therefore, we need a way to synchronise them so they do not interfere.

When a DT detects some discrepancy, its root cause may either be some abnormality or the need to perform a (re-)calibration since uncertainties have accumulated to the extent that deviation from prediction is now too great. If it is possible to carry out designs of experiments like those introduced in Sect. 15.2.5, this may be a fast way to get back in sync and reestablish the correct estimation of the current state of the PT. The fact that a DT contains multiple models complicates such calibrations further. In the long term, the vision is that it will be possible to shift autonomously between these models (see Sect. 1.4.2).

15.3 Digital Twin Platforms

Although DT frameworks and platforms already exist, there is scope for increasing automation in the engineering of DTs to reduce the development time and risk, as well as maintenance and operational costs. A visionary goal might be to reach the level of automation achieved for computer-based languages with scanners and parsers. Compiler construction has become one of the most mature examples of domain engineering. A similar standardisation of frameworks and tools may allow the DT community to move away from hand-built DTs and to focus efforts on more interesting avenues of research and development.

In this section, we first consider the potential for automating the production of DTs using frameworks and platforms (Sect. 15.3.1). Such improved automation entails using modelling languages that could be used to specify DTs and the services we require from them (Sect. 15.3.2). For these to be effective we need to have confidence that concepts are shared consistently, for which ontologies and knowledge graphs are critical in providing a common semantic backbone (Sect. 15.3.3).

As we implement and deploy more capable but computationally demanding DTs, there is a growing need to improve the ways in which work, especially simulation, should be distributed among the computing resources available (Sect. 15.3.4), and to ensure scalable yet accurate data linkage between DT and PT (Sect. 15.3.5).

Beyond automation of DT production and operation, we consider the wider engineering life cycle, and the need to keep an assured grasp on the large range of artefacts generated and maintained through a DT's existence. There is a need for research and innovation promoting dependable reuse and for methods that allow combined Development & Operations (DevOps) for DTs that are in service (Sect. 15.3.6), gaining the most value from the presence of multiple models in a DT (Sect. 15.3.7) and offering the right levels of assurance for DT applications (Sect. 15.3.8).

15.3.1 Automating Digital Twin Production

There are good arguments for bringing experience from Model-Driven Engineering (MDE) to bear on challenges in DT engineering. These include the need for management of heterogeneous models in DTs, synchronisation of the DT with the PT, and support for multi-stakeholder and multi-discipline engineering [9]. Any model-based framework to support DT engineering should be defined in terms of a basic architecture and set of language concepts and services, for instance, for integrating existing heterogeneous models that can be adapted and extended for specific DT developments. Among other things, the framework must enable the connection of the DT with system data on concrete platforms. The current frameworks (see Sect. 11.2) are mainly derived from corresponding Internet of Things (IoT) frameworks, and as a consequence the majority of the current automation is related to the communication of data from the PT to the DT. There is a need to at least introduce the same amount of automation for both simulations of models and checking inside monitors.

Such a framework requires identifying the information needed to deploy and configure key algorithms such as state estimation and monitoring automatically from a given set of models and metadata. Recent work suggests that the Functional Mockup Interface (FMI) – which provides an interface to interact with models regardless of the tool that built them – has potential here [26]. We need to evaluate the suitability of the FMI standard as the preferred way to interact with both the models and the monitors. Starting points could be implementing particle filter state estimators, which work with a wide range of continuous time and discrete event models and understanding their theoretical limitations [67].

15.3.2 Modelling Languages for Digital Twins

As outlined above, the MDE community has produced theories and techniques to combine and manage such heterogeneous languages but in the context of software systems. Moyne et al. identify DT requirements on DT frameworks in the medium and long term, including the capacity to use languages for DT definition, supporting reuse and portability, interoperability and interchangeability of DTs, and having the capacity to support validation and verification [47]. One of the languages that may be useful as a starting point for this strand of research is the Digital Twin Definition Language (DTDL)[1] targeted at describing Azure DTs. It is open-source, and may be a candidate for exploring where the limitations are for automation in DT construction [59]. DTDL is being extended with features such as DT configuration data. We imagine that these and other extensions to DT services can benefit the DT platforms using this.

[1] https://azure.github.io/opendigitaltwins-dtdl/

15.3.3 Incorporation of Ontologies and Knowledge Graphs

To engineer DTs effectively, we need techniques and tools to capture the *domain knowledge* required to reason correctly about and control these systems. One approach is to use *ontologies* to formalise these concepts and establish "a shared understanding of a domain" [62] in the form of a rich, semantic, and machine-readable expression of the domain concepts and relationships [33]. We foresee the use of ontologies in future DT engineering to provide the "semantic backbone" for aligning, integrating, and describing the components of the DT itself[2]. That is, ontologies can provide rich and traceable connections between the models, data, functions, and services of the DT, as well as the methodologies used to construct the DT itself. This backbone may aid integration of new information into DT platforms (Chap. 11), and can be used for (semi-)automatically producing high-quality experience reports [52].

Once ontologies have been constructed for a domain to provide the high-level concepts, the DT can also interact with a *Knowledge Graph* (KG) which aligns with these ontologies. The DT accesses these KGs as data repositories to provide *semantic* information, knowledge, and wisdom about the meaning of the data [53]. For example, the KG may contain temperature values which are not just *float* values, but can also be explicitly typed as *degree Celsius* values provided by a particular *sensor* component. The increased traceability enhances the DT's quality, as the KG "acquires and integrates information into an ontology and applies a reasoner to derive new knowledge" [22]. This reasoning can ensure the consistency and quality of the DT and its data [51], and can reduce the need to perform experimentation.

While the mathematical underpinnings of ontologies provide formal reasoning, they also bring practical challenges. These include scalability concerns, especially as KGs could contain millions of elements. Another factor is the expertise and tooling required to construct a domain ontology. The tool Protégé is widely used, even though it has been noted to "require significant experience in knowledge representation and semantics" [33]. This may change for future ontology construction, as the openCAESAR ontology-based modelling language framework [64] reduces the accidental complexity of ontological construction to aid the development process.

Another avenue for integrating domain knowledge is through the use of Large Language Models (LLMs). Once trained on a suitably-large corpus, LLMs could be used in the DT construction process by recommending DT components [70]. LLMs could also be employed as a natural language interface to the DT itself, where the user would provide questions as prompts. The LLM could then act as an agent to coordinate the KG search, the running of experiments, or executing simulations to obtain the answer. New research questions for LLMs come up, e.g., can we train LLMs to provide the level of robustness needed when composing unreliable or untrusted DT interfaces.

---

[2] Ontologies may work best in eco-systems in certain domains, such as space `https://mb4se.esa.int/`.

15.3.4 Distributed Simulation and Workload Distribution

Operating and executing DTs at high fidelity requires many resources and an appropriate workload distribution for simulating the DT models and monitors, for example in a High-Performance Computing (HPC) context. One possible research strategy is splitting the DT into components, distributing it to several devices and co-simulating the components. However, in such co-simulations it is also challenging to couple models with different levels of fidelity in a manner which is faithful in relation with the PT. Dynamic resource allocation will ensure optimal performance between edge, fog and cloud (see Sect. 11.8). Correct execution requires time-sensitive delivery of information and time-triggered networking and interaction. Improved workload distribution may require changing the fidelity of models (see Sect. 15.3.7). Relevant work includes serverless computing [55] and the cognitive compute continuum [28].

In particular, when the DT operator wants to carry out some of the more advanced DT services, such as what-if simulations (see Sect. 10.2), it will be necessary to use a distributed underlying hardware infrastructure. Research is needed to determine the optimal ways to use a distributed architecture based on the needs for the services requested from the DTs. For example, what-if analysis is only relevant if it can be conducted faster than real-time.

15.3.5 Bi-directional synchronisation with the actual system

The architecture of a DT and its PT requires automatic, bi-directional data linkage between the virtual and physical worlds [50]. Synchronisation is crucial for the DT to follow the changes in the PT's state. However, control, anomaly detection and behavioural optimisation require data to pass from the DT to the PT. The main challenge is in providing scalable solutions to support this bi-directional communication and synchronisation throughout the life cycle of the DT-enabled system. In particular, if errors of some kind start appearing at a PT, this will likely increase traffic to the DT, so here one can research alternative approaches for transferring/compressing the data. The possible time delay from PT data arriving at the DT can be so significant that it would only make sense to operate as a Digital Shadow as discussed in Sect. 1.2.2 and principles for ensuring this need to be researched.

In general, for configuring the data to be communicated and the capacity of the communication media used to a DT innovation in how to automate the set-up for potential compression/decompression of data if necessary (depending upon where the bottlenecks are present). It will also be interesting to research the extent to which it could be possible to change this dynamically and autonomously, since it is by no means obvious how this will best be accomplished.

15.3.6 Full Life Cycle Management

Establishing a DT-enabled system involves a multiplicity of artefacts, including DT assets and the associated documentation and assurance information. Innovation is necessary to ensure that all artefacts from the entire lifecycle are managed appropriately through version, and traceability control. The governance of models, data, and the relationships between them will be paramount. What is essential in this support is that whenever it makes sense, artefacts are reused in different phases (see Chap. 4).

The requirement to support collaborative development for DT-enabled systems applies throughout the life cycle. Thus, platforms must be able to support collaboration efficiently. When the DT-enabled system is deployed and operational, it is also possible to update services while the DT-enabled system is in use. In connection with DevOps it is essential that the PT that updates are pushed to is in a state where it is safe to provide the updates.

15.3.7 Leveraging Multiple Levels of Abstraction

A DT may use models that operate at different levels of abstraction, may have been intended for different purposes (see Sect. 1.4.2), and may offer different levels of computational performance. For example, a kinematic and a dynamic model of a mechanical system may provide similar predictions under some conditions and very different predictions under others. How can we take advantage of this wealth of models, tailoring them to the needs of each individual situation? To autonomously change between such alternative models, it would be necessary to capture the circumstances under which one wishes to swap between them. Ejersbo et al. [24] have proposed an approach allowing dynamic change of models at runtime. This includes replacing models as well as changing the structure of the entire DT (i.e., how individual models interact) while providing safety guarantees during the process.

Jones [37] poses the following research question: what level of fidelity is appropriate and realistic for a given use case in a DT setting? The answer must maximise benefit while minimising expense and the technical difficulty of implementation. The abstract level of fidelity raises some significant challenges. Can we twin before finalising the physical and virtual design? Can we create design prototype twins while the physical entity varies substantially in fidelity? Can we even twin a design concept or idea itself? The creation of abstract and early-stage DTs may have substantial benefits that are worth further exploration.

Models within the DT may operate at different levels of fidelity depending on the underlying requirements. Usually, the developer defines the degree of fidelity, fixing it for the model's lifetime. In this research area, we also want to investigate different fidelity levels to adjust models' fidelity at runtime. There will naturally be a tradeoff between simulation speed and precision so that if the deadline for performing a simulation is tight, a model with less precision may be advantageous. In general, model swapping poses significant challenges due to the dynamics of physical systems, so it is important to be able to identify the situations in which it is possible to swap between models of different fidelity. This ability may prove valuable in workload distribution (Sect. 15.3.4).

15.3.8 Certification of Digital Twins

How much assurance can we offer for a DT-enabled system? Much of the value in a DT comes from enabling services that include monitoring, decision support and predicting maintenance needs. All these capabilities potentially increase dependability. Whether as an advisory system or a partially autonomous controller, a DT may need to be trusted. Validation, Verification and Accreditation (VV&A) processes and methodologies which focus on models and simulations are a starting point here [19]. However, these need to be implemented, extended and assessed at the DT level. Here the process to be followed for the modelling and simulation capability needs to be credible and that can be documented following a specific Software Process Improvement and Capability Determination (SPICE) [23].

Building on this, the methodology of validating and verifying the services inside a DT needs attention for cases where the DT-enabled system requires assurance of its dependability properties. This will likely also entail semantic traceability between different artefacts as discussed in Sect. 15.3 above [44]. Research in this area will necessarily involve certification authorities.

15.4 Increasing the Level of Autonomy for Digital Twins

Part of the vision shared by the participants in our opening round-table discussion (Chap. 2) was of DTs having a degree of autonomy inclosing the loop to the PT. Such autonomous real-time decision-making requires a high degree of trust in the predictions and actions taken in the DT-enabled system. This way of potentially reducing the human supervision of the PT performance also allows more rapid decision-making so long as this does not compromise safety. However, providing guarantees of safety of the DT-enabled system can be very challenging because of uncertainties at all levels, and ever-changing environmental conditions in some DT application domains. Even specifying trustworthiness is challenging [1].

We consider human involvement in the feedback loop to the PT (Sect. 15.4.1) and the integration of a cognitive layer into DT-enabled systems (Sect. 15.4.2). Then Sect. 15.4.3 describes research challenges related to closing the reality gap between the current state in a PT and the perception within its DT, leading to a need to derive predictions even in the presence of inconsistencies. Sect. 15.4.4 considers research challenges for the perception of the state of the PT and its environment. Finally, Sect. 15.4.5 considers the possibility of adding robustness to the predictions performed inside the DT based on the uncertainty quantification bounds.

15.4.1 Reducing Human Supervision

As indicated in Sect. 2.5, it may make sense to initially have a human in the loop when a DT makes predictions. Research is needed into how one can gradually change the balance between the human and DT supervision when the level of trust in predictions is sufficiently high compared to the consequences of making wrong predictions.

A potential PT adaptation layer inside the DT can be responsible for reconfiguring the PT in response to novel environments while ensuring that the PT's performance is kept at the same level or improved. This layer, however, needs to set some requirements for it to operate without human supervision. We must verify new configurations against everything known about the PT and its environment [66, 69]. If these new configurations result from a design space exploration process (which is common), then one challenge is how to push the validation into the exploration process itself rather than just doing it at the end. Another challenge is representing and learning the safety properties we use to verify the new configuration. Signal temporal logic is an obvious candidate, but learning such properties is still in its infancy [36]). Furthermore, we need automation to create a runtime monitor to check whether the assumptions used to generate the new configuration remain valid as a PT operates [4]. If this is no longer true, we should roll back the system to a previously known safe configuration, as is commonly done in micro-service architectures [13]. Unfortunately it is often not feasible to roll back for many PTs, such the circumstances under which this is possible needs to be identified.

15.4.2 The Cognitive Digital Twin

A "Cognitive" DT has some autonomous capabilities, reasoning, learning and autonomously triggering reconfigurations as adaptations. Such a DT will typically apply deductive and inductive principles, the former based on formal rules, the latter on data science [71, 61]. To make such a vision a reality, a cognitive DT is defined to dynamically recognise complex and previously unpredicted behaviours with optimisation strategies. The expectation is that semantic technologies (such as the ontologies mentioned in Sect. 15.3.3), artificial intelligence, Industrial IoT and ubiquitous sensing technologies will make it possible to realise cognition capabilities at a certain level. Research on perception and prediction is needed to make decisions autonomously and react to the PT adaptively if we have sufficient trust in the perception/predictions being correct.

15.4.3 Awareness of the Reality Gap

Understanding uncertainties and considering their quantification allows users to adjust DTs and PTs accordingly (see also Sect. 15.2.2). Consequently, there is a need to investigate the capability of the autonomous system to utilise this information and obtain an awareness of the "reality gap" between the world as the DT perceives it, and reality. With this awareness, the system can potentially respond to operational shortcomings and explore options to mitigate its effects. Kounev et al. [38] has provided foundational work and an initial architecture for computational self-awareness while Bellman et al. [7] investigated self-aware properties specifically for CPSs. More generally, lessons learned from the autonomy research community will be an excellent starting point (see Chap. 14). In general, the DT will need to operate in the presence of inconsistent data, managing these inconsistencies in its assessment of the current situation. There is as yet only limited experience in this form of "situational awareness" in the DT setting (see also Sect. 1.3.3).

15.4.4 Capture and Representation of Causal Relations

The primary responsibility of a potential perception layer (above the data layer) is to gather information about the different modes of operation of the PT and environment. There is thus a need for work on how to represent the different system operation modes, and what the conditions are for the system to transition between those modes. Hybrid Automata [31] and Hidden Markov Models [5] and the theoretical framework thereof, can be used as starting points.

To enable scalable reasoning at the higher layers, it is crucial that the different operating modes can be clustered into classes of equivalence. For example, we must evaluate a new system configuration in all possible operating modes learned from the historical data (Sect. 15.4.1). A research objective is to understand what modifications we need to carry out in the representations so that we can merge similar operating modes without losing the ability to verify the properties.

15.4.5 Increased Robustness

As discussed in Sect. 15.2.5, a goal for future work is to have uncertainty bounds available for outputs from models in the DT. Decisions based on those models can also produce uncertainty bounds. Consequently, PT reconfigurations will include a level of certainty that can help in autonomous decision-making. Robustness is also required for controlling how much resource is dedicated to solving a particular task. The attempt to find an optimal configuration will involve considering many potential states and parameter combinations. There is therefore a trade-off between the search for an optimal solution and the benefits that the optimal solution may be expected to deliver.

Kulkarni et al. [41] present an Enterprise Simulation Language to develop DTs to support the decision-making process in complex systems. Oquendo [54]. Thelen et al. [60] give a good overview over recent methods and approaches to deal with the quantification of uncertainty in both areas, aleatoric and epistemic uncertainty. The discussed methods can be categorised in *parameter calibration*, *bias correction*, and the *KOH framework*, constructing a Gaussian process regression model to simultaneously quantify several sources of uncertainty. While uncertainty is present in any CPS, uncertainty quantification enables systems to identify them and with appropriate mitigation strategies, the robustness of the system can be improved.

15.5 Supporting Composition of Digital Twins

Part of our vision of the future of DT engineering (outlined in Sect. 1.4.2), is the composition of DT-enabled systems that may have been produced by different manufacturers. This gives rise to several challenges, typically because the organisations involved may be to varying degrees both collaborators and competitors. Can we enable an ecosystem in which organisations are not so scared of sharing data that they forego the potential benefits of collaborating DTs?

Broadly speaking, we anticipate that the more high-fidelity the models used inside a DT-enabled system, the closer the model is to its owner organisation's core IP. Consequently, the need for infrastructure that enables the establishment of DTs across different organisations *while securing such IP* becomes paramount. Here the recent developments for the smart home ecosystem with what is called thread[3] and matter[4] is an interesting and promising development. This is a particularly interesting initiative since it is led by a number of the "tech giants", and so it may be able to move more rapidly than traditional standardisation processes.

---

[3] `https://threadgroup.org/`

[4] `https://csa-iot.org/all-solutions/matter/`

In Sect. 15.5.1 we focus on the need to enable standardised interfaces for DTs as step towards an ecosystem of DT-based systems. In Sect. 15.5.2 we discuss how collaboration between DTs can be established, and in Sect. 15.5.3 consider the need for transparency between organisations in an SoS setting.

15.5.1 Need for Standardised DT Interfaces

Complete automation of interoperability between DTs requires consensual standardisation of DT interfaces, but there is a lack of standards, frameworks and regulations for DTs. Practical standards often emerge from well-established communities of developers and users. Such communities are starting to form: for example, the Digital Twin Consortium[5] is a part of the Object Management Group (OMG). This needs to be continued with different stacks of technologies so that the competition between the companies can be more on the services inside their DT-enabled systems rather than depending on the specific technology stack and alliances. In this context it is worth mentioning the efforts conducted for Version 2 of the SysML standard[6] since this is a significant progress for Model-Based Systems Engineering (MBSE).

Other standardisation efforts are underway (see Sect. 1.3), but such initiatives can be relatively slow, requiring considerable alignment. of baseline concepts. It will take a long time for new research results to be incorporated into such standards and in particular, there is still less focus on the interoperability between DTs. Additional research is needed at a cognitive layer to determine how collaboration between DTs can be enabled there.

15.5.2 Collaborative Digital Twins

CPSs usually do not operate in isolation but are embedded in a dynamic environment and the context of other computational systems. Diverse DT-enabled systems must collaborate and interact, causing other behaviours to emerge. Studying these behaviours might not require high-fidelity models. We may be able to replicate them with simpler models representing the collective. The first step is to identify collective behaviour and automatically create DTs of these collectives. A second step will investigate approaches enabling the reverse operation, where we extract individuals from the collective. Finally, runtime analysis will permit a smooth transition from individual to collective systems. Casadei et al. [10] propose an initial version of DT for collectives.

[5] https://www.digitaltwinconsortium.org/

[6] https://github.com/Systems-Modeling/SysML-v2-Release

DTs can support predictive maintenance, runtime monitoring and what-if scenarios in the environment. We can use them to explore and define the activities of autonomous CPSs. However, at this point, DTs are rarely used to describe collaborative behaviour, particularly between DT-enabled systems provided by different organisations. We could extend to collectives if individual systems had a DT for each other. More research is needed into opportunities to exchange and use DTs of other systems collaboratively. This includes using these DTs to perform what-if scenarios, enabling autonomous systems to conduct collaborative research without being explicitly designed for them. Prior work outlining initial ideas will be continued and intensified in this research area [25].

We expect collaborative behaviour and interaction among CPSs to speed up research. But it will also make the underlying systems prone to exploitation by others (e.g., obtaining valuable IP through repeated simulation requests). Our collaborative behaviour system aims to avoid exploitation among the individual systems. A starting point for this activity is provided in the work by Bellman et al. [6] and our preliminary work in enabling distributed co-simulations [57].

15.5.3 Openness of DTs

DTs play a crucial role in sustainment operations, such as enabling the prediction of the remaining useful life of a system and the prediction of what-if scenarios to evaluate particular operational conditions and rare events. In this setting, typically, reusing design artefacts (e.g., models) to build DTs is in general not applicable as such artefacts are not intended to perform predictions. Moreover, exploiting purely data-driven approaches is typically limited by the data availability and quality (i.e., representativeness) as well, and using purely physics-based models is generally limited by a lack of knowledge regarding physics processes, the complexity of the dynamics of the system or the fact that when models can be defined, the result is too complex to scale up to industrial applications. An emerging modelling paradigm in the literature (see, e.g., [43]) is to combine data-driven approaches (e.g., ML-based) with physics-based approaches, typically resulting in surrogate models. This research direction promises to tackle the challenges mentioned above if different stakeholders are willing to share data. This would also enable a higher degree of data fusion, which will be necessary for composing DTs.

In general, there is a challenge in managing data ownership in an SoS context while still being able to share data with other organisations. Openness and sustainability are necessary to compose DT-enabled systems correctly across different organisations. Research is needed to construct user communities for different application domains where there is potential for mutual gain from collaboration through DTs. Consequently, we would expect an increase in the adoption and commercialisation of composable DT-enabled systems.

15.6 Novel Applications of Digital Twins

The concept of the DT is being taken up much more widely than in the highly engineered systems that form the subject of this book. We conclude by considering two of them: DTs for humans, and the metaverse.

The application of DTs of humans can potentially transform patient care through personalised data-driven medicine [15, 11, 32]. The DT of a human can be subjected to experimental treatments conducted virtually. Data gathered from a multiplicity of (accurate) human DTs brings the potential to help assess treatments over populations. Combined with DTs of delivery mechanisms such as hospitals, this may lead to improved resource allocation and management of treatments. It is interesting to think about mapping the concepts that we have discussed at the level of engineered DTs in this book (such as fidelity, accuracy, robustness, uncertainty) to the biological context.

The *metaverse* is perhaps a rather slippery concept. Although widely used, the term appears to have no settled definition, but, broadly speaking, has been used to refer to networks of virtual worlds offering online experiences that include a sense of immersion, real-time interactivity and user agency [20]. The idea of an *industrial metaverse* is perhaps more obviously and closely related to the types of system considered in this book. Seen as a continuation of the digitalisation trend, it is a convergence of technologies including DTs, virtual and augmented reality, IoT and AI. Such a metaverse supports interactions with industrial processes and systems including prototyping, design, simulation, monitoring and process optimisation. DTs are a pivotal technology in this [45]. Key factors in realising the technology are seen as interoperability, data and integration, automation, workforce skills, costs, regulation and security [48] reputation and identity, data security, currency and payments, law and jurisdiction, ownership and property, community and IT infrastructure. Challenges include reputation and identity, data security, currency and payments, law and jurisdiction, ownership and property, community and network and time and space [42].

15.7 Concluding Remarks

In concluding, it makes sense to reflect on the vision outlined at the outset of our book in Sect. 1.4.2 and the views of DT technologies from the roundtable in Chap. 2. The chapters following them demonstrated the breadth of engineering processes and techniques that come together to create an effective DT of a CPS. There are many trade-offs to consider, for example between models, data and communications mechanisms. Indeed, the costs associated with establishing, operating and maintaining a DT needs to be carefully balanced with the value it brings to its users.

Governance of data and models is essential to success. In some application domains it may never happen, whereas in others regulatory pressures may force the ecosystems to be established (e.g., in order to save energy). DT platforms enabling support for this are essential to make such a vision a reality. We believe that initially most DTs will be realised as DSs and add value by means of the digital services they supply, with trusted automation in reconfiguring the PT still some way off at scale.

In situations where the PT is a part of a larger ecosystem there are more challenges to be overcome as a result of the need to balance IP with the benefits of interaction. However, in situations where the supplier of a PT is able to obtain the necessary data from the PT itself (via sensors) we expect that we will see a significant increase in new DT-enabled systems.

DT engineering, as it emerges, will be a remarkable discipline in that it brings together a great diversity of technologies, competencies and challenges. At first sight this presents a complex, almost intimidating, picture, but our experience is that taking a disciplined systems-based approach may help to realise the vision of the dependable DT-enabled system. The development of DT engineering will be an exciting journey to follow!

References

1. Abeywickrama, D.B., Bennaceur, A., Chance, G., Demiris, Y., Kordoni, A., Levine, M., Moffat, L., Moreau, L., Mousavi, M.R., Nuseibeh, B., Ramamoorthy, S., Ringert, J.O., Wilson, J., Windsor, S., Eder, K.: On specifying for trustworthiness. Commun. ACM **67**(1), 98–109 (2023). DOI: 10.1145/3624699
2. Acker, B.V., Meulenaere, P.D., Vangheluwe, H., Denil, J.: Validity frame–enabled model-based engineering processes. Simulation (2023). DOI: 10.1177/00375497231205035
3. Alcaraz, C., Lopez, J.: Digital twin: A comprehensive survey of security threats. IEEE Communications Surveys & Tutorials **24**(3), 1475–1503 (2022). DOI: 10.1109/COMST.2022.3171465
4. Bartocci, E., Deshmukh, J., Donzé, A., Fainekos, G., Maler, O., Ničković, D., Sankaranarayanan, S.: Specification-Based Monitoring of Cyber-Physical Systems: A Survey on Theory, Tools and Applications. In: E. Bartocci, Y. Falcone (eds.) Lectures on Runtime Verification, *Lecture Notes in Computer Science*, vol. 10457, pp. 135–175. Springer International Publishing, Cham (2018). DOI: 10.1007/978-3-319-75632-5_5
5. Baum, L.E., Petrie, T.: Statistical inference for probabilistic functions of finite state Markov chains. The annals of mathematical statistics **37**(6), 1554–1563 (1966)
6. Bellman, K., Botev, J., Diaconescu, A., Esterle, L., Gruhl, C., Landauer, C., Lewis, P.R., Nelson, P.R., Pournaras, E., Stein, A., et al.: Self-improving system integration: Mastering continuous change. Future Generation Computer Systems **117**, 29–46 (2021)
7. Bellman, K., Landauer, C., Dutt, N., Esterle, L., Herkersdorf, A., Jantsch, A., Taherinejad, N., Lewis, P.R., Platzner, M., Tammemäe, K.: Self-aware cyber-physical systems. ACM Transactions on Cyber-Physical Systems **4**(4), 1–26 (2020)
8. Bertot, Y., Casteran, P.: Interactive Theorem Proving and Program Development. Springer Verlag (2004)
9. Bordeleau, F., Combemale, B., Eramo, R., van den Brand, M., Wimmer, M.: Towards model-driven digital twin engineering: Current opportunities and future challenges. In: Ö. Babur,

J. Denil, B. Vogel-Heuser (eds.) Systems Modelling and Management, pp. 43–54. Springer International Publishing, Cham (2020)

10. Casadei, R., Pianini, D., Viroli, M., Weyns, D.: Digital Twins, Virtual Devices, and Augmentations for Self-Organising Cyber-Physical Collectives. Applied Sciences **12**(1) (2022). DOI: 10.3390/app12010349
11. Cellina, M., Cè, M., Alì, M., Irmici, G., Ibba, S., Caloro, E., Fazzini, D., Oliva, G., Papa, S.: Digital twins: The new frontier for personalized medicine? Applied Sciences **13**(13) (2023). DOI: 10.3390/app13137940
12. Chandola, V., Banerjee, A., Kumar, V.: Anomaly detection: A survey. ACM Comput. Surv. **41**(3) (2009). DOI: 10.1145/1541880.1541882
13. Chen, L.: Microservices: Architecting for Continuous Delivery and DevOps. 2018 IEEE International Conference on Software Architecture (ICSA) pp. 39–397 (2018). URL https://api.semanticscholar.org/CorpusID:50768519
14. Chou, Y., Sankaranarayanan, S.: Bayesian parameter estimation for nonlinear dynamics using sensitivity analysis. In: Proceedings of the Twenty-Eighth International Joint Conference on Artificial Intelligence, IJCAI-19, pp. 5708–5714. International Joint Conferences on Artificial Intelligence Organization (2019). DOI: 10.24963/ijcai.2019/791
15. Coveney, P., Highfield, R.: Virtual You: How Building Your Digital Twin Will Revolutionize Medicine and Change Your Life. Princeton University Press (2023)
16. Deantoni, J., Muñoz, P., Gomes, C., Verbrugge, C., Mittal, R., Heinrich, R., Bellis, S., Vallecillo, A.: Quantifying and combining uncertainty for improving the behavior of digital twin systems (2024)
17. Dehnert, C., Junges, S., Katoen, J.P., Volk, M.: A Storm is Coming: A Modern Probabilistic Model Checker. In: R. Majumdar, V. Kunčak (eds.) Computer Aided Verification, pp. 592–600. Springer International Publishing, Cham (2017)
18. Denil, J., Klikovits, S., Mosterman, P.J., Vallecillo, A., Vangheluwe, H.: The experiment model and validity frame in M&S. In: Proceedings of the Symposium on Theory of Modeling & Simulation, TMS/DEVS '17, pp. 1–12. Society for Computer Simulation International
19. Diem, J., Henninger, A., Youngblood, S.: VV&A in the M&S in the DoD: Moving from Art to Science. ITEA Journal (31), 324–327 (2010)
20. Digital, M.: Digital twins: The foundation of the enterprise metaverse. www.mckinsey.com/capabilities/mckinsey-digital/our-insights/digital-twins-the-foundation-of-the-enterprise-metaverse (2023). Accessed: 2023-11-12
21. Donzé, A.: Breach, a toolbox for verification and parameter synthesis of hybrid systems. In: T. Touili, B. Cook, P. Jackson (eds.) Computer Aided Verification, pp. 167–170. Springer Berlin Heidelberg, Berlin, Heidelberg (2010)
22. Ehrlinger, L., Wöß, W.: Towards a definition of knowledge graphs. SEMANTiCS (Posters, Demos, SuCCESS) **48**(1-4), 2 (2016)
23. Eichenseer, F., Heinkel, H.M., Benedikt, M., Ahmann, M., Holzner, M., Stadler, C.: Modeling & Simulation SPICE: Assessing the Capability of Credible Simulation Processes. INCOSE International Symposium **33**(1), 399–415 (2023). DOI: https://doi.org/10.1002/iis2.13029
24. Ejersbo, H., Lausdahl, K., Frasheri, M., Esterle, L.: Dynamic Runtime Integration of New Models in Digital Twins. In: 2023 IEEE/ACM 18th Symposium on Software Engineering for Adaptive and Self-Managing Systems (SEAMS), pp. 44–55 (2023). DOI: 10.1109/SEAMS59076.2023.00016
25. Esterle, L., Gomes, C., Frasheri, M., Ejersbo, H., Tomforde, S., Larsen, P.G.: Digital twins for collaboration and self-integration. In: 2021 IEEE International Conference on Autonomic Computing and Self-Organizing Systems Companion (ACSOS-C). IEEE (2021). DOI: 10.1109/ACSOS-C52956.2021.00040

26. Feng, H., Gomes, C., Sandberg, M., Macedo, H.D., Larsen, P.G.: Under what conditions does a digital shadow track a periodic linear physical system? In: A. Cerone, M. Autili, A. Bucaioni, C. Gomes, P. Graziani, M. Palmieri, M. Temperini, G. Venture (eds.) Software Engineering and Formal Methods. SEFM 2021 Collocated Workshops, pp. 143–155. Springer International Publishing, Cham (2022). DOI: https://doi.org/10.1007/978-3-031-12429-7_11
27. Feng, H., Gomes, C., Thule, C., Lausdahl, K., Iosifidis, A., Larsen, P.G.: Introduction to Digital Twin Engineering. In: Proceedings of the 2021 ANNUAL MODELING AND SIMULATION CONFERENCE. IEEE, Virtual Conference (2021). DOI: 10.23919/ANNSIM52504.2021.9552135
28. Ferrer, A.M.J., Becker, S., Schmidt, F., Thamsen, L., Kao, O.: Towards a cognitive compute continuum: An architecture for ad-hoc self-managed swarms. 2021 IEEE/ACM 21st International Symposium on Cluster, Cloud and Internet Computing (CCGrid) pp. 634–641 (2021). URL https://api.semanticscholar.org/CorpusID:232170583
29. Fidler, M.: Survey of deterministic and stochastic service curve models in the network calculus. IEEE Communications Surveys & Tutorials **12**(1), 59–86 (2010). DOI: 10.1109/SURV.2010.020110.00019
30. Fisher, R.A.: The Arrangement of Field Experiments, pp. 82–91. Springer New York, New York, NY (1992). DOI: 10.1007/978-1-4612-4380-9_8
31. Frehse, G.: An Introduction to Hybrid Automata, Numerical Simulation and Reachability Analysis, pp. 50–81. Springer Fachmedien Wiesbaden, Wiesbaden (2015). DOI: 10.1007/978-3-658-09994-7_3
32. Geddes, L.: How digital twins may enable personalised health treatment. The Guardian URL www.theguardian.com/science/2023/nov/12/digital-twin-personalised-medical-treatment?CMP=Share_iOSApp_Other
33. Hildebrandt, C., Köcher, A., Küstner, C., López-Enríquez, C.M., Müller, A.W., Caesar, B., Gundlach, C.S., Fay, A.: Ontology building for cyber–physical systems: Application in the manufacturing domain. IEEE Transactions on Automation Science and Engineering **17**(3), 1266–1282 (2020). DOI: 10.1109/TASE.2020.2991777
34. INCOSE: Systems Engineering Vision 2035: Engineering Solutions for a better World. https://www.incose.org/2023_redesign/publications/se-vision-2035 (2021)
35. Isermann, R., Münchhof, M.: Identification of dynamic systems: An introduction with applications (2010). URL https://api.semanticscholar.org/CorpusID:45986314
36. Jha, S., Tiwari, A., Seshia, S.A., Sahai, T., Shankar, N.: Telex: learning signal temporal logic from positive examples using tightness metric. Formal Methods in System Design **54**, 364–387 (2019)
37. Jones, D., Snider, C., Nassehi, A., Yon, J., Hicks, B.: Characterising the digital twin: A systematic literature review. CIRP Journal of Manufacturing Science and Technology **29**, 36–52 (2020). DOI: 10.1016/j.cirpj.2020.02.002
38. Kounev, S., Lewis, P., Bellman, K.L., Bencomo, N., Camara, J., Diaconescu, A., Esterle, L., Geihs, K., Giese, H., Götz, S., et al.: The notion of self-aware computing. In: Self-Aware Computing Systems, pp. 3–16. Springer (2017)
39. Kulik, T., Dongol, B., Larsen, P.G., Macedo, H.D., Schneider, S., Tran-Jørgensen, P.W.V., Woodcock, J.: A Survey of Practical Formal Methods for Security. Formal Aspects of Computing (2022). DOI: 10.1145/3522582
40. Kulik, T., Gomes, C., Macedo, H.D., Hallerstede, S., Larsen, P.G.: Towards secure digital twins. In: T. Margaria, B. Steffen (eds.) ISoLA 2022, *Lecture Notes in Computer Science*, vol. 13704. Springer (2022). DOI: 10.1007/978-3-031-19762-8_11
41. Kulkarni, V., Barat, S., Clark, T., Barn, B.S.: Digital twin as an aid for decision-making in the face of uncertainty. In: 2022 Winter Simulation Conference (WSC), pp. 1371–1385 (2022). DOI: 10.1109/WSC57314.2022.10015528
42. Labs, L.R.: Top seven metaverse challenges, issues, and solutions. lucidrealitylabs.com/blog/7-challenges-of-the-metaverse#

1-reputation-amp-identity-metaverse-challenges (2023). Accessed: 2023-11-12
43. Legaard, C., Schranz, T., Schweiger, G., Drgoňa, J., Falay, B., Gomes, C., Iosifidis, A., Abkar, M., Larsen, P.: Constructing neural network based models for simulating dynamical systems. ACM Comput. Surv. **55**(11) (2023). DOI: 10.1145/3567591
44. Liu, J., Cao, X., Zhou, H., Li, L., Liu, X., Zhao, P., Dong, J.: A digital twin-driven approach towards traceability and dynamic control for processing quality. Advanced Engineering Informatics **50**, 101395 (2021). DOI: https://doi.org/10.1016/j.aei.2021.101395
45. Lyu, Z., Fridenfalk, M.: Digital twins for building industrial metaverse. Journal of Advanced Research (2023). DOI: https://doi.org/10.1016/j.jare.2023.11.019
46. Mazor, E., Averbuch, A., Bar-Shalom, Y., Dayan, J.: Interacting multiple model methods in target tracking: A survey **34**(1), 103–123. DOI: 10.1109/7.640267
47. Moyne, J., Qamsane, Y., Balta, E.C., Kovalenko, I., Faris, J., Barton, K., Tilbury, D.M.: A requirements driven digital twin framework: Specification and opportunities. IEEE Access **8**, 107781–107801 (2020). DOI: 10.1109/ACCESS.2020.3000437
48. Müller, M., Bohné, T., Jamboula, T.: Identifying key factors for the development of the industrial metaverse. In: 2023 IEEE Smart World Congress (SWC), pp. 1–6 (2023). DOI: 10.1109/SWC57546.2023.10448572
49. Mustafa Bayram, T.P., Buyukoz, G.O.: Numerical methods for simulation of stochastic differential equations. Advances in Difference Equations (17) (2018)
50. Nwogu, C., Lugaresi, G., Anagnostou, A., Matta, A., Taylor, S.J.: Towards a Requirement-driven Digital Twin Architecture. Procedia CIRP **107**, 758–763 (2022). DOI: https://doi.org/10.1016/j.procir.2022.05.058. Leading manufacturing systems transformation – Proceedings of the 55th CIRP Conference on Manufacturing Systems 2022
51. Oakes, B., Gomes, C., Larsen, P.G., Denil, J., DeAntoni, J., Cambeiro, J., Fitzgerald, J.: Examining Model Qualities and Their Impact on Digital Twins. In: Annual Modelling and Simulation Conference, pp. 220–232
52. Oakes, B.J., Parsai, A., Meyers, B., David, I., Van Mierlo, S., Demeyer, S., Denil, J., De Meulenaere, P., Vangheluwe, H.: A Digital Twin Description Framework and its Mapping to Asset Administration Shell. In: Model-Driven Engineering and Software Development, vol. 1708, pp. 1–24. Springer Nature Switzerland. DOI: 10.1007/978-3-031-38821-7_1
53. Oakes, B.J., et al.: Structuring and accessing knowledge for historical and streaming digital twins. In: 1st Workshop on Ontology-Driven Conceptual Modeling of Digital Twins, pp. 1–13
54. Oquendo, F.: Dealing with Uncertainty in Software Architecture on the Internet-of-Things with Digital Twins. In: 9th International Conference on Computational Science and Its Applications (ICCSA 2019), LNCS 11619, Springer, 9th International Conference on Computational Science and Its Applications (ICCSA 2019), pp. 770–786. Springer, Saint Petersburg, Russia (2019). DOI: 10.1007/978-3-030-24289-3_57
55. Patros, P., Spillner, J., Papadopoulos, A.V., Varghese, B., Rana, O., Dustdar, S.: Toward sustainable serverless computing. IEEE Internet Computing **25**(6), 42–50 (2021). DOI: 10.1109/MIC.2021.3093105
56. Paulson, L.C.: Isabelle: A Generic Theorem Prover. Springer-Verlag LNCS 828 (1994)
57. Pedersen, N., Lausdahl, K., Sanchez, E.V., Larsen, P.G., Madsen, J.: Distributed Co-Simulation of Embedded Control Software with Exhaust Gas Recirculation Water Handling System using INTO-CPS. In: Proceedings of the 7th International Conference on Simulation and Modeling Methodologies, Technologies and Applications (SIMULTECH 2017), pp. 73–82. Madrid, Spain (2017). DOI: https://doi.org/10.1145/3179993. ISBN: 978-989-758-265-3
58. Platzer, A.: Logical Analysis of Hybrid Systems - Proving Theorems for Complex Dynamics. Springer (2010)
59. Schmidt, C., Volz, F., Stojanovic, L., Sutschet, G.: Increasing Interoperability between Digital Twin Standards and Specifications: Transformation of DTDL to AAS. Sensors **23**(18) (2023). DOI: 10.3390/s23187742

60. Thelen, A., Zhang, X., Fink, O., Lu, Y., Ghosh, S., Youn, B.D., Todd, M.D., Mahadevan, S., Hu, C., Hu, Z.: A comprehensive review of digital twin – part 2: Roles of uncertainty quantification and optimization, a battery digital twin, and perspectives (2022)
61. Unal, P., Albayrak, Ö., Jomâa, M., Berre, A.J.: Data-Driven Artificial Intelligence and Predictive Analytics for the Maintenance of Industrial Machinery with Hybrid and Cognitive Digital Twins, pp. 299–319. Springer International Publishing, Cham (2022). DOI: 10.1007/978-3-030-78307-5_14
62. Uschold, M.: Knowledge level modelling: concepts and terminology. The knowledge engineering review **13**(1), 5–29 (1998)
63. Van Acker, B., Oakes, B.J., Moradi, M., Demeulenaere, P., Denil, J.: Validity Frame Concept as Effort-Cutting Technique within the Verification and Validation of Complex Cyber-Physical Systems. In: Proceedings of the 23rd ACM/IEEE International Conference on Model Driven Engineering Languages and Systems: Companion Proceedings, pp. 1–10. ACM. DOI: 10.1145/3417990.3419226
64. Wagner, D.A., Chodas, M., Elaasar, M., Jenkins, J.S., Rouquette, N.: Ontological Metamodeling and Analysis Using openCAESAR, pp. 925–954. Springer International Publishing, Cham (2023). DOI: 10.1007/978-3-030-93582-5_78
65. Wang, Y., Su, Z., Guo, S., Dai, M., Luan, T.H., Liu, Y.: A survey on digital twins: Architecture, enabling technologies, security and privacy, and future prospects. IEEE Internet of Things Journal pp. 1–1 (2023). DOI: 10.1109/JIOT.2023.3263909
66. Weyns, D., Iftikhar, M.U., de la Iglesia, D.G., Ahmad, T.: A Survey of Formal Methods in Self-Adaptive Systems. In: Proceedings of the Fifth International C* Conference on Computer Science and Software Engineering, C3S2E '12, p. 67–79. Association for Computing Machinery, New York, NY, USA (2012). DOI: 10.1145/2347583.2347592
67. Wikle, C.K., Berliner, L.M.: A Bayesian tutorial for data assimilation. Physica D: Nonlinear Phenomena **230**(1), 1–16 (2007). DOI: https://doi.org/10.1016/j.physd.2006.09.017. Data Assimilation
68. Woodcock, J., Gomes, C., Macedo, H.D., Larsen, P.G.: Uncertainty quantification and runtime monitoring using environment-aware digital twins. In: T. Margaria, B. Steffen (eds.) Leveraging Applications of Formal Methods, Verification and Validation: Tools and Trends, pp. 72–87. Springer International Publishing, Cham (2021)
69. Wright, T., Gomes, C., Woodcock, J.: Formally Verified Self-Adaptation Of an Incubator Digital Twin. In: Leveraging Applications of Formal Methods, Verification and Validation. Practice: 11th International Symposium, ISoLA 2022, Rhodes, Greece, October 22–30, 2022, Proceedings, Part IV, p. 89–109. Springer-Verlag, Berlin, Heidelberg (2022). DOI: 10.1007/978-3-031-19762-8_7
70. Wu, J., Gan, W., Chen, Z., Wan, S., Yu, P.S.: Multimodal large language models: A survey (2023)
71. Zheng, X., Lu, J., Kiritsis, D.: The emergence of cognitive digital twin: vision, challenges and opportunities. International Journal of Production Research **60**(24), 7610–7632 (2022). DOI: 10.1080/00207543.2021.2014591

GPSR Compliance
The European Union's (EU) General Product Safety Regulation (GPSR) is a set of rules that requires consumer products to be safe and our obligations to ensure this.

If you have any concerns about our products, you can contact us on

ProductSafety@springernature.com

In case Publisher is established outside the EU, the EU authorized representative is:

Springer Nature Customer Service Center GmbH
Europaplatz 3
69115 Heidelberg, Germany

www.ingramcontent.com/pod-product-compliance
Lightning Source LLC
Chambersburg PA
CBHW061119270726
48828CB00005B/588

* 9 7 8 3 0 3 1 6 6 7 1 8 3 *